Life Cycle Engineering and Management of Products

José Augusto de Oliveira ·
Diogo Aparecido Lopes Silva ·
Fabio Neves Puglieri ·
Yovana María Barrera Saavedra
Editors

Life Cycle Engineering and Management of Products

Theory and Practice

Springer

Editors
José Augusto de Oliveira
Sao Paulo State University (UNESP)
São João da Boa Vista, São Paulo, Brazil

Fabio Neves Puglieri
Federal University of Technology - Parana
Ponta Grossa, Paraná, Brazil

Diogo Aparecido Lopes Silva
Federal University of Sao Carlos
Sorocaba, São Paulo, Brazil

Yovana María Barrera Saavedra
Federal University of Sao Carlos
Buri, São Paulo, Brazil

ISBN 978-3-030-78043-2   ISBN 978-3-030-78044-9 (eBook)
https://doi.org/10.1007/978-3-030-78044-9

This Springer imprint is published by the registered company Springer Nature Switzerland AG
The registered company address is: Gewerbestrasse 11, 6330 Cham, Switzerland

Foreword

The main concept of this book is the "life cycle" of a product. In product engineering and management, it is essential to consider Life Cycle Thinking to design, develop and produce products with less impact, i.e., with better environmental, economic and social performance. It is important to consider the entire life cycle of a product because many impacts can occur during its phases, by minimizing impacts from one phase, we may indirectly cause more impacts or even shift them from that phase to another; or simply not consider the phase in which the most significant impacts occur, generating significant losses for sustainability management and engineering in companies.

Both Life Cycle Engineering and Life Cycle Management are topics currently internationally recognized as integrated approaches, with a set of tools and methods that, based on the Life Cycle Thinking, are essential to implement the environmental, economic and social sustainability of products (services, organizations, systems, ...). The life cycle perspective is the basis for a holistic and systemic assessment of impacts, from the extraction of resources, through the production, distribution and use of a product, until its end-of-life management strategies selection through reuse, recycling, energy recovering or its final disposal. In a general perspective, products and their components can be produced in very distant locations, using different technologies, raw materials and energy sources, so the sustainability of a product will depend on its "life cycle" phases, which include the materials, processes and the various forms of energy used (in production, transport, use, ...).

The book *Life Cycle Engineering and Management of Products: Theory and Practice* presents and explains the main concepts, techniques and applications of LCEM of products (goods and services) to promote sustainability. This book was organized based on the most well-known and scientifically recognized approaches to perform LCEM, with emphasis on environmental Life Cycle Assessment (LCA), which popularized the life cycle concept, up to the most recent and emerging approaches, such as social LCA, product-service system (PSS), ecodesign and other key approaches for the implementation of circular economy systems, which are currently of great interest in the business sector and pose new challenges for Life Cycle Engineering and Management. In this context, new function-oriented systems (services provided) are being developed focused on adding more value (environmental, economic and social) to the entire product systems (value chains).

LCA has been used as a basis for developing approaches that include the social and economic aspects based on the triple bottom line model for the sustainability practical implementation. With the recent developments, LCA has considerably extended its environmental scope to a more comprehensive and multidimensional assessment of the sustainability.

This book was developed for undergraduate and graduate students to be used at the classroom level as basic academic material. It is also useful for researchers, engineers and managers of companies who seek to expand their knowledge in the area. This book shows a multidisciplinary nature in the context of sustainability, while the concept of the life cycle of products works as a guiding thread. Each book chapter connects with the others along with the text, but each chapter can also be read separately from the previous ones. After a first introductory chapter, each subsequent chapter of this book covers a specific LCEM topic. Chapter 2 is about cleaner production (CP). Then, the next four chapters (3, 4, 5, and 6) cover the Life Cycle Assessment (LCA) technique, each one covering the four LCA stages. Chapter 7 presents the emerging topic of social LCA. The following chapters are about ecodesign (Chap. 8), product-service systems (Chap. 9), corporate sustainability (Chap. 10), environmental management systems (Chap. 11) and green supply chain management (Chap. 12). The last chapter is about communication and environmental labeling (Chap. 13). Throughout its 13 chapters, this book shows several case studies developed on real and hypothetical situations, which will support and sustain the learning process of students. At the end of each chapter, a series of questions and practical exercises were proposed, which are certainly very useful for the study process.

This book assumes an important role in actively serving as a guide to the theory and practice of Life Cycle Engineering and Management. It is a complete and robust book, coordinated by four editors, and with the collaboration of 39 authors, recognized experts and researchers in the various LCEM topics.

Fausto Freire
Centre for Industrial Ecology (CIE)
University of Coimbra
Coimbra, Portugal

Preface

Since the 1960s, environmental management has been implemented and improved in different governmental, non-governmental and business instances. In the public sphere, it is worth mentioning the Environmental Impact Assessment (EIA), Environmental Licensing and Strategic Environmental Assessment (SEA) approaches, among others, which present the legal instruments for Public Environmental Management of public and private enterprises and services.

Corporate Environmental Management, on the other hand, started with a more reactive or end-of-pipe stance, moving to a stance with preventive practices, management systems, certifiable or not, to more proactive and strategic approaches that generate value and competitive advantages to business.

In recent years, there has been a convergence toward the Life Cycle Thinking (LCT), both in the academic and scientific areas, as well as in the business area as a whole. Thus, the object of analysis becomes the life cycle of products and services, which comprises all its phases and stages, from its conception to its recovery and return to the production cycle. Thus, two important and complementary areas are born, which are the central theme of this work: Life Cycle Engineering and Management (LCEM).

LCEM is composed of practices (concepts, techniques, tools, programs and strategies), all of which are applied in different phases of the product's life cycle, but which complement each other, mainly seeking to improve the environmental performance of this good or service and, more recently, applications have also been developed to improve the social performance of these productive systems.

Still recent, the bibliographic productions of these areas focus on articles, generally intended for the strictly scientific community and on books that present only some of the topics of LCEM and often in an isolated and superficial way. Thus, the idealization of this book by its organizers arose even when they were doctoral researchers, and they noticed the need to have a continuous, unique and complete work by EGCV and that it should be a textbook to be used in the undergraduate and postgraduate classroom as reference material. For this huge challenge, 30 more renowned authors were invited to contribute to each of the topics covered, in order to produce a deep and updated work, without, however, offering the reader a fragmented book. For this reason, the planning and execution of this book took more than four years, going through several individual reviews of each chapter and also of the work as a whole, seeking to make the book a homogeneous material for studies and research.

We hope that this work in the medium and long terms will contribute significantly to the continued training of new professionals with solid training in these themes that are so necessary in the face of the current environmental crisis and the macro trend of the circular economy.

Therefore, the organizers, on behalf of the more than 30 authors of this book, wish everyone an excellent reading, and most importantly, that everyone can learn and put into practice the themes developed here. It is a dream come true to be able to deliver this work to you!

São João da Boa Vista, Brazil — Prof. Dr. José Augusto de Oliveira
Sorocaba, Brazil — Prof. Dr. Diogo Aparecido Lopes Silva
Ponta Grossa, Brazil — Prof. Dr. Fabio Neves Puglieri
Buri, Brazil — Prof. Dr. Yovana María Barrera Saavedra

Acknowledgements

We are grateful for the support of the Brazilian Institute of Information in Science and Technology (IBICT) for the support to carry out this work. We also thank Eduardo Miele Dal Secco for technical review of the text.

The editors thank so much for the valuable author's contributions to this project.

Diogo Aparecido Lopes Silva thanks the support provided by the Conselho Nacional de Desenvolvimento Científico e Tecnológico (CNPq) under grant number 302722/2019–0.

Contents

About the Editors

José Augusto de Oliveira is a professor at Sao Paulo State University (UNESP), São João da Boa Vista/SP campus and a leader of the research group Center for Advanced and Sustainable Technologies (CAST).

Specialist in Life Cycle Assessment, ecodesign, cleaner production and end-of-life strategies for technology products.

Diogo Aparecido Lopes Silva is a professor at Federal University of São Carlos (UFSCar), Sorocaba/SP campus, in production engineering course.

Specialist in Life Cycle Assessment, sustainable manufacturing, eco-efficiency and ecodesign.

Fabio Neves Puglieri is a professor at Federal University of Technology—Paraná (UTFPR), Ponta Grossa/PR campus, in production engineering course.

Specialist in Life Cycle Management, strategic environmental planning and corporate strategies for sustainability.

Yovana María Barrera Saavedra is a professor at Federal University of São Carlos (UFSCar), Lagoa do Sino/SP campus, in administration, environmental engineering, agronomic engineering and food engineering courses.

Specialist in Life Cycle Management, ecodesign, end-of-life strategies, environmental management and waste management.

Chapter 1
Introduction to Life Cycle Engineering and Management (LCEM)

José Augusto de Oliveira, Diogo Aparecido Lopes Silva, Fabio Neves Puglieri, and Yovana María Barrera Saavedra

Do you know what a product Life Cycle is and how it can help sustainability? Well, before we talk about Life Cycle, let's define some important concepts about sustainability and how Life Cycle Thinking (LCT) fits into the 2030 Development Agenda.

The 2030 Development Agenda was released in 2015 with its 17 Sustainable Development Goals (SDGs) and 169 associated goals that are integrated and indivisible (Fig. 1.1). In these 17 SDGs, poverty eradication and Sustainable Development (SD) were used as the Agenda's core, recognizing that without a SD it will be difficult for the human being to continue progressing and developing in society.

In addition, we must also remember that in that same year the Paris Agreement was signed where a global agreement on climate change was signed and with the aim at keeping the increase in average global temperature below 2 °C.

But how will we be able to get to the 17 SDGs and keep the increase in global average temperature below 2 °C? Well, this is where LCT comes in to support the identification of priorities throughout the life cycle of an organization's products.

J. A. de Oliveira (✉)
Center for Advanced And Sustainable Technologies (CAST), Sao Paulo State University (UNESP), Campus of Sao Joao da Boa Vista, Av. Professora Isette Corrêa Fontão, 505, ZIP Code: 13876-750, São João da Boa Vista, SP, Brazil
e-mail: jose.a.oliveira@unesp.br

D. A. L. Silva
Department of Production Engineering, Federal University of São Carlos—UFSCar, Sorocaba Campus, Highway João Leme Dos Santos, Km 116, ZIP Code: 18052-780, Sorocaba, SP, Brazil

F. N. Puglieri
Department of Production Engineering, Federal University of Technology - Paraná—UTFPR , Rua Dr. Washington Subtil Chueire, 330, Jardim Carvalho, ZIP Code: 84017-220, Ponta Grossa, PR, Brazil

Y. M. B. Saavedra
Center for Natural Sciences-CCN, Federal University of São Carlos, Campus Buri-Lagoa do Sino, Lauri Simões de Barros Highway, km 12, SP-189, Bairro Aracaçú, Buri, São Paulo, Brazil

J. A. de Oliveira et al. (eds.), *Life Cycle Engineering and Management of Products*,
https://doi.org/10.1007/978-3-030-78044-9_1

Fig. 1.1 Sustainable development goals. 2020 *Source* Adapted from NU-Brazil ()

We will see later in this book a technique called Life Cycle Assessment (LCA). As it allows identifying, calculating, and measuring the environmental performance of products, services, and processes, it can be used to identify the highest points of concentration of emissions and analyze which actions can be taken to achieve their reduction more efficiently, with less impact and with better management of the possible trade-offs that can be created by these scenarios and among the 17 SDGs.

Thus, we begin to notice how Life Cycle Engineering and Management (LCEM), supported by the LCA, can help countries to comply with the SDGs. This agreement had the main goal of strengthening the global response to the climate change threat and reinforcing the countries' capacity to deal with the impacts resulted from these changes.

So, we may say that with the LCT the main critical points for improving the efficiency of resources can be identified with benefits in the three pillars of sustainability. Most of the 17 SDGs offer opportunities to integrate the life cycle approach, however, the SDG11 stands out on sustainable cities and communities and the SDG12 on sustainable consumption and production. With the use of LCEM, it is possible to promote business models with more sustainable technologies and where tools such as communication and environmental labeling can be used to have more transparent communication with consumers. In addition, the integration of ecodesign in the development of these products can help with environmental improvement and closing their life cycle (UNEP 2019).

Well, we know that we have not yet learned enough to understand in a deep and full way how LCEM can contribute to sustainability, but with this brief contextualization we hope to have stimulated your curiosity so that you can fully enjoy this book and, at the end of your studies, answer for yourself how LCEM can contribute to the fulfillment of the SDGs and then promote sustainability in the productive and business environment.

So, let's go back to the question of what the life cycle of a product is. Consider this book which you are reading and taking as a product. Now, we are going to invite you to some reflections based on the following questions:

- Do you know the story behind the production of this book?
- Do you know what materials have been used in its production?
- Do you know how its final disposal is carried out after the end of its lifespan?
- Do you know if this paper will be recycled or not, or even if recycling itself is the best technological alternative available to process paper waste?
- If it will be recycled, does this process of material recovery at the end of the life cycle present a better environmental performance than the final destination of this waste (landfill disposal, for example)?

All of these questions come out precisely when we start thinking or developing a thought about the life cycle of the products we buy and consume. This book has been designed to present, explain and detail the main concepts, techniques, and applications of LCEM to contribute to the stakeholders—business and economic sectors, governments, non-governmental organizations, and civil society.

Returning to the example of the book, Fig. 1.2 represents all the steps that the book goes through, and it is important to highlight that, regardless of the product, some sort of impact (negative or positive) will always take place throughout its entire cycle, that is, from the extraction of the resources necessary for its conception and production to the distribution, use, and disposal in post-consumption. Thus, it is possible to have a broader and more realistic view of its interaction with the environmental, economic, and social pillars, that is, with sustainability. But why is life cycle important for

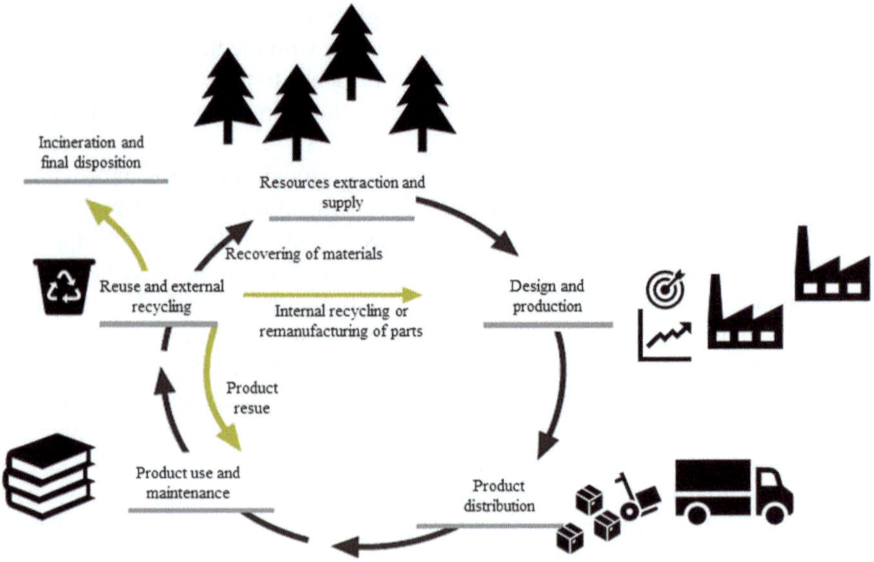

Fig. 1.2 Product life cycle—example in case of a paper book. *Source* Adapted from UNEP (2007)

Sustainability? Think of the current context, think of how the negative impacts are increasing more and more due to the exaggerated consumption of products and the short life cycle, that is, the increase in generating waste. This scenario makes it urgent to implement LCEM in practice.

However, before continuing it is important to define the main concept of our book which is life cycle, also known worldwide as LCT (Life Cycle Thinking). According to Life Cycle Initiative, the definition of the LCT concept is:

> Thinking about the life cycle is going beyond the traditional focus on the production place and on manufacturing processes to include environmental, social and economic impacts of a product throughout its life cycle (UNEP 2019).

UNEP (2016) adds that LCT aims at guiding the transition from an isolated analysis of the stages of a product to a systemic analysis of the product with all its stages and all its activities and, with that, ensuring improvements and reducing the use of resources throughout the product life cycle. With LCT it is possible to have a better understanding of the modern world where everything is interrelated and where it is necessary to identify the environmental, economic, and social impacts throughout the entire life cycle of a product, process, and even an organization. In addition, implementing LCT is thinking about the entire supply chain and engaging stakeholders which may vary from policy makers to environmental managers or product designers and engineers.

Although it seems obvious, the LCT became more prominent only after the development of the Life Cycle Assessment (LCA) technique, which allows a quantitative assessment of all the potential environmental impacts that occur throughout a product life cycle. Thus, returning to the example of the paper book life cycle, through an LCA, it would be possible to verify where its main environmental impacts occur and what would be its scale of magnitude for each life cycle stage. LCA allows you to generate answers to relevant questions for this case, such as: which is better, doing the final disposal of the paper book after the end of its lifespan, or recycling it? Where do you generate more impact, in cultivating trees that will supply the raw material necessary to produce the books, or in the process of manufacturing the product? What is the relevance of logistical processes of raw material transportation and product distribution for the paper book life cycle impacts?

However, LCA comes from a broader concept defined as Life Cycle Engineering (LCE). According to Alting and Legarth (1995), LCE consists of "designing product life cycles through choices about concepts, structures, materials, and product manufacturing processes". We understand that the choices mentioned refer to the design, development and production of products using concepts, structures, and materials with less potential for impact, always in search of continuous improvement in decision-making processes. In a more recent concept, Jeswiet (2014) defines LCE as:

> Engineering activities that include the application of technological and scientific principles for manufacturing products with the aim at protecting the environment, conserving resources, encouraging economic progress, keeping in mind social concerns and the need for sustainability, optimizing the product life cycle and minimizing pollution and waste (Jeswiet, 2014).

The LCE covers five basic lines, which are: product design, product manufacturing, enterprise profitability, environmental impact, and social impact of activities (Jeswiet, 2003). These five lines must be considered when thinking about sustainability.

Tripple Botton Line, or Tripod of Sustainability, aims at the balance between economic growth, environmental responsibility, and social equity. In this sense, we must think of how LCE should be treated for the promotion of this tripod, aiming at its acceptance by the business world and its implementation in the real scenario, so that it is not just a theory without applicability. For this aim, we come across the LCE Triad, which is illustrated in Fig. 1.3.

In short, this triad is represented by the integrated analysis of the three expected performances of a product or process. If on the one hand, we think about designing, developing, and producing products with less potential for environmental impacts, that is, with greater and better[1] environmental performances, we also need to analyze what will be the impacts coming from these improvements in economic performance and in technical/technological performance when they are about products and operational performance when they are about processes. In addition, evaluate the social aspects of products and processes to promote sustainability.[2]

[1] A performance may be better, from a qualitative point of view, when there is improvement, or it may be better, from a quantitative point of view, when the indicators rise.

[2] Social performance has not been included in Fig. 1.2 as it is difficult to measure in LCE and comparing to the other performances observed in the same figure.

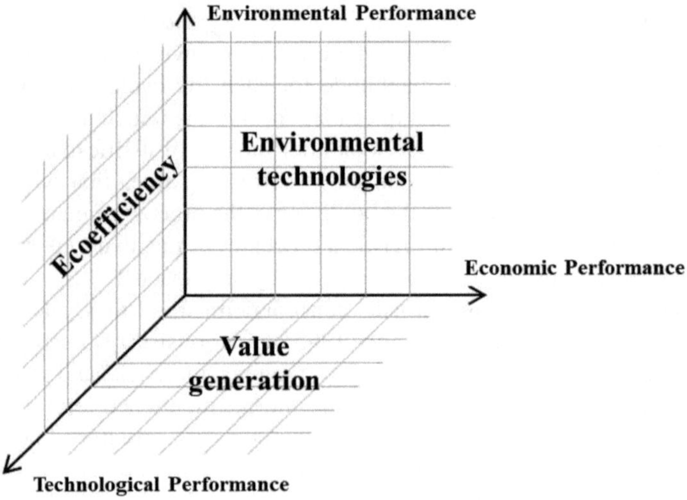

Fig. 1.3 Life cycle engineering triad. *Source* Authors

But why so? Well, if we want to develop an applied science, we need to think of the target audiences for this application, that is, the market. For obvious reasons, companies expect to improve the environmental performance of products but do not want the practices developed to have a negative impact on the technical/technological performance of their products, on the operational performance of their processes, much less on the economic performance. Of course, we will often have some kind of negative impact from these performances to the detriment of improvements in environmental or social performance, but even so, these need to be studied in order to be able to minimize and manage them from a practical point of view.

But why adopt LCE? From an environmental point of view, we know that the Support Capacity and Resilience Capacity of the Biosphere is limited and they regenerate in cycles, just as it occurs with biogeochemical cycles, for example, the Carbon (C) cycles, Nitrogen (N), Water (H_2O), etc. However, because of our current linear and high model of resource extraction, the increase in production and consumption, and consequently the disposal of products, these cycles are not taking the time to regenerate and if we continue with this model and do not adopt a cyclical view in order to reduce the extraction of resources and the disposal of waste, whether solid, liquid or gaseous, without options for recovering them, we know that the Biosphere will no longer support human activities over time.

And from an economic point of view, how may LCE help? Well, we know that the economy we currently live in basically consists of the economic system of Capitalism, which is based on production (Y), which in turn occurs as a function of four factors of production, as shown in Eq. 1.1:

$$Y = f(\text{NR}, C, L, T) \tag{1.1}$$

where

NR = Natural Resources,

C = Capital,

L = Labor,

T = Technology.

Well, but what does Eq. (1.1) have to do with LCE? We know that the capitalist economy mostly predicts an increase in C, right? We also know that demographic growth naturally increases the L variable exponentially. These first two variables require a constant search for technological advancement, that is, an increase in T. On the other hand, we know that the supply of NR through the Biosphere behaves like a relatively decreasing[3] function. Thus, we verified a tradeoff, which consists of general terms, in an inverse relation which may also be inversely proportional. That is, while C, L and T rates increase, NR rate decreases.

But the relation between Economy and LCE is not clear yet, right? If we have a tendency to reduce NR due to its finitude, we must increase the efficiency in its use, that is, we must use each NR more often to generate more value for products and services, according to Eq. 1.2.

$$\text{Efficiency in the use of NR} = \frac{\text{Generated products or services}}{\text{Used NR}} \tag{1.2}$$

Thus, if we use and reuse NR, we will keep their rates constant and if we increase the generated products and services, we will have an increase in efficiency rates in the use of NR.

Well, now it starts to make sense. But, what about C and T? Clearly, if the value of the generated products and services increases, a greater return on capital invested in the form of assets is obtained, thus, increasing the values of C. T is still missing, right? Well, technology plays a fundamental role in LCE as it may behave as a multiplicative factor for other production factors, as suggested by Eq. 1.3.

$$Y = T.f(\text{NR}, C, L) \tag{1.3}$$

Technology, if applied for sustainability, may also increase the production function exponentially or even multiply, generating a very positive impact on the efficiency in the use of NR, C and L.

Thus, it is observed that technology plays a fundamental role for LCE and for Economy, divided into two main approaches: Sustainable Technologies; and Technologies for Sustainability. But what is the difference between them? Technologies that are said to be sustainable are those that have less potential for environmental impacts and may be replaced over time in order to increase the sustainability performance of products and processes that are carried out by these technologies. As examples, we can mention the changes in a production process, the reuse of some

[3] Natural Resources have a decreasing function in relation to the other variables that increase with greater intensity. Thus, Natural Resources are finite at the expense of their demand by human actions.

waste from process X in a process Y, the conversion of thermal energy lost in the form of heat into electrical energy to be used by the company itself, etc. On the other hand, Technologies for Sustainability are those which have the main focus on preventing, reducing, or mitigating the environmental impacts of products and processes. As examples for this approach, we can mention Biotechnologies such as the development of biodegradable polymers and/or produced from renewable sources, the use of microorganisms for the treatment of waste and effluents, along with others we will see throughout this book.

To summarize, with the inefficient use of production factors the economy as a whole will decrease and, thus, its negative effects will be observed over time not only from an environmental perspective but from an economic and social point of view. So, practicing LCE is not only a strictly environmental issue, although this focus by itself has great relevance, but LCE is extremely necessary to ensure economic growth and social equity. Thus, it is considered that the decision-making processes on product concepts, product designs, manufacturing projects, as well as their materials and structures, must be systematized and must consider the LCE Triad in an integrated manner.

On the other hand, Life Cycle Management (LCM), which correlates with LCE, is defined as a framework aimed at analyzing and managing the sustainable performance of products and organizations throughout their entire life cycle (UNEP 2009). In other words, it is an approach for managing environmental and socioeconomic problems of products and services, and the organization itself, considering its entire value chain (UNEP 2007).

It is important to realize that LCM is not a tool or method. In fact, it is a collection of management systems and dissemination of information related to products and processes, based on programs, strategies, practices, and concepts (UNEP 2007). In short, LCM is a way of implementing management practices with a focus on operationalizing the LCT, as illustrated in Fig. 1.4.

Among other goals, LCM aims at integrating product-related policies to help the organization to improve the sustainability of products, processes, and businesses, besides improving communication and relationship with stakeholders, both internally and externally. LCM can also help in processes of innovating the search for socioenvironmental improvements to an organization's products.

According to Sonnemann and Margni (2015), LCM demands a holistic view and a complete understanding of the interdependence of businesses in order to support relevant decisions and actions to improve sustainability performance that considers environmental and social benefits and, at the same time, offers opportunities of generating business value. It is understood here that LCM covers LCE techniques, tools, and practices. Or rather, LCM plays the fundamental role of implementing and managing the application of LCE throughout the product life cycle. In other words, LCE is applied at the microlevel of products and processes, whereas LCM manages these practices at the macro interorganizational and intraorganizational level, according to the scheme in Fig. 1.5.

LCM can be implemented through a continuous improvement cycle, that is, based on the "Plan-Do-Check-Act" cycle, or PDCA cycle, as shown in Fig. 1.6.

Fig. 1.4 Life cycle management. *Source* Adapted from Ometto and Saavedra (2012)

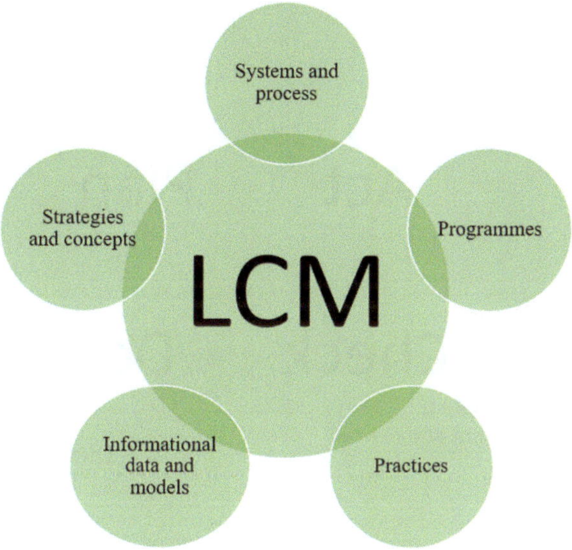

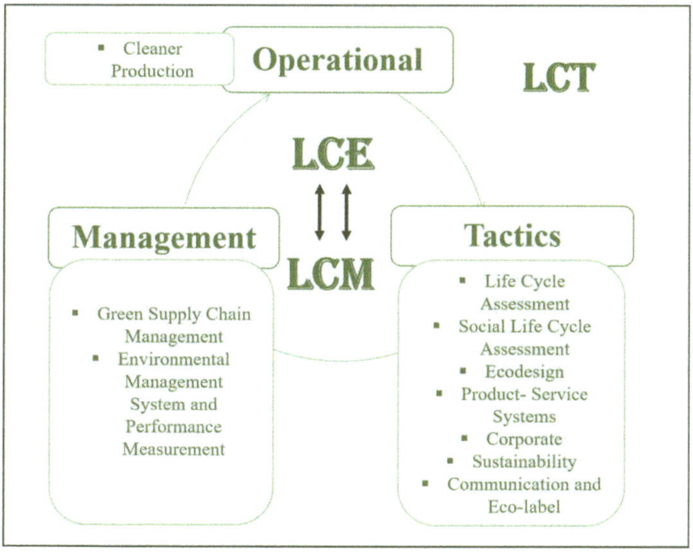

Fig. 1.5 Relation between LCE and LCM. *Source* The authors

The entire planning part (Plan), that is, the definition of policies oriented to a life cycle approach, awareness of people in the organization, determination of competitive business strategies, in addition to the definition of goals and action plans, may be seen in Chaps. 10 and in 11 on this book.

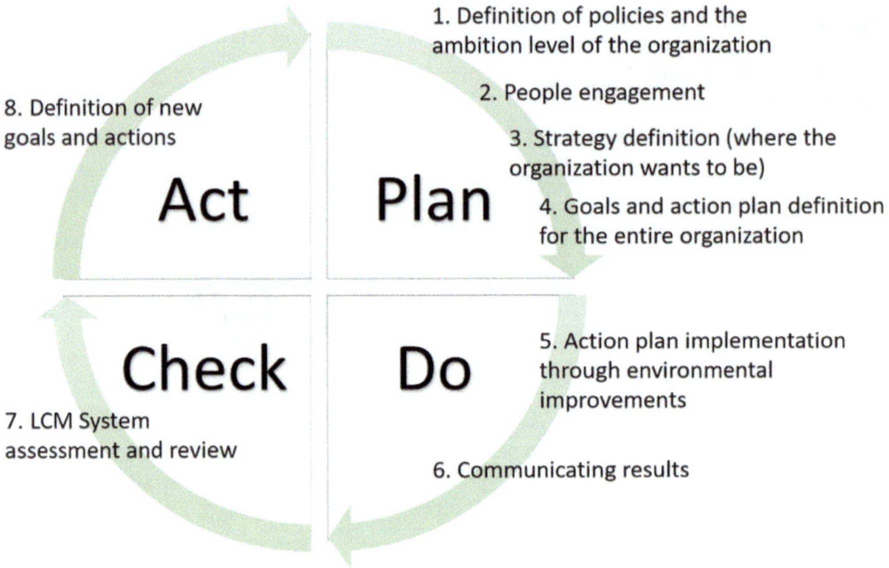

Fig. 1.6 LCM implementation steps. *Source* Adapted from UNEP (2007)

The first will mainly deal with the attributions of the company's top management (main stockholders, presidency, and direction) from a top-down approach, of how a strategic plan oriented to improvements in the life cycle and the proposal of an action plan can be elaborated. In other words, corporate sustainability acts primarily in the planning (Plan) of long-term business actions (from 5 to 10 years), so that it can become more sustainable.

The strategic plan, created from the strategic planning for sustainability, involves determining the organization's commitment to socioenvironmental improvement in life cycle, identifying the company's current situation, going through the intended future situation, and defining strategic and competitive goals based on the Circular Economy concept. This strategic plan is translated into an action plan, that is, the determination of sustainable business models that are priority for the business.

In its turn, in chapter, "Environmental Management Systems and Performance Measurement" all the stages of this continuous improvement cycle are built up, as so concepts related mainly to ISO 14001 and 14031 Standards are worked on, in which both are structured from a PDCA cycle. Thus, in this chapter you will be able to find the whole process that involves the establishment of an environmental policy, the identification of opportunities for improvement in the organizations' products and processes, the definition of an action plan and goals, the putting of the action plan into practice and system assessment using indicators. However, unlike Corporate Sustainability, the direction of actions here is more at the tactical and operational level.

It is worth mentioning that since the last update of the ISO 14001 Standard, in 2015, it requires the inclusion of the LCT, thus making up a management instrument strongly related to the LCM.

The part of our PDCA cycle that refers to the communication of results you may follow in Chap. 13. In this chapter, through environmental labels, ways of communicating the environmental performance of products to customers, final consumers, and the entire public of the organization will be discussed. One way to carry out this type of communication is through the Environmental Product Declaration (EPD), in which results about the environmental performance of products throughout their life cycle are communicated to consumers. These results are generated from an LCA study and can be disseminated on products' packages in the same way that you currently find nutritional information when you pick up a food product at the supermarket, for example.

Finally, many of these strategic, tactical, and operational opportunities highlighted in the PDCA cycle for implementing LCM can be quite limited if there is no commitment to environmental issues of all links in the organization's value and supply chain. Then, in Chap. 12 we will see what refers to the actions that can be taken with suppliers, product recovery, including in this case reverse logistics, and the selection of end-of-life product strategies.

And how can LCT and LCEM contribute to different economic sectors? Facilitating access to LCT and LCEM knowledge for small and medium-sized companies allows them to identify their main critical points and may improve and create more sustainable businesses by improving their credibility, sustainability, and competitiveness.

In its turn, broader approaches such as organizational LCA can provide information for the elaboration of reports to be able to guide priorities such as, for example, public and private purchasing and as a support to mitigate climate change by low carbon purchasing (LCI, 2020).

The books addressing these themes deal with Environmental Management or Corporate Environmental Management in general, as a conceptual "black box", which do not deepen into techniques, tools, and strategies which, in turn, hinders their understandings and practical implementations both in the scope of research and in business scenario. This book aims at advancing into this theoretical and practical gap, bringing definitions of concepts but also going deeper into the LCEM techniques, tools, and strategies.

Thus, in this book, besides covering in detail the methodology and state-of-the-art involving the use of LCA from Chap. 3 to Chap. 6. Other LCE techniques and tools will also be presented, such as Cleaner Production (CP) in Chap. 2, Ecodesign in Chap. 8, the development of Product-Service Systems in Chap. 9, and Social LCA in Chap. 7.

In addition, LCM techniques and tools will also be covered in this book such as, for example Corporate Sustainability explained in Chap. 10; the development of Environmental Management Systems (EMS) and the organizational Performance Measurement Systems (PMS) according to Chap. 11; the Green Supply Chain Management which will be presented in Chap. 12, and the communication and environmental labeling which will be addressed in Chap. 13.

It is important to highlight that both LCE and LCM complement each other, generating a synergistic effect for the organizations which adopt them. While LCM is seen as a management system, LCE is defined as the art of designing more sustainable processes and products.

In this first chapter, we hope we have elucidated the concepts of LCEM as well as its techniques, tools, and applications in products and services in order to promote sustainability in its broadest sense. We invite you to continue your studies and read our next chapters. After this introduction, we will present and detail LCE. Next, we will study LCM.

For this book, we bring together renowned national and international researchers with a vast research history and deep expertise in each theme that makes LCEM up. In this proposal, we intend to bring the state-of-the-art of each of these themes, suggesting practical applications and exercises for the assimilation and understanding of the LCEM concepts, tools, and techniques.

We wish you all good studies and may this book contribute to your more sustainable professional performance!

References

Alting, L., Legarth, J. B.: Life cycle engineering and design. CIRP Ann. **44**(2), 569–580 (1995). https://doi.org/10.1016/S0007-8506(07)60504-6

Jeswiet, J.: A definition for life cycle engineering. In: International Seminar on Manufacturing Systems, 3 June 2003, Saarbrucken. Proceedings. Saarbrucken: 2003. Plenary Speech, pp. 17–20 (2003)

Jeswiet, J.: Life cycle engineering. In: Laperrière, L., Reinhart, G. (eds.) The International Academy for Production Engineering. CIRP Encyclopedia of Production Engineering. Springer, Berlin, Heidelberg (2014)

Nações Unidas Brasil: Transformando Nosso Mundo: A Agenda 2030 para o Desenvolvimento Sustentável. Disponível em: https://nacoesunidas.org/pos2015/agenda2030/ (2020). Acesso 03 Mar 2020

Ometto, A.R., Saavedra, Y.M.B.: Gestão e Engenharia do Ciclo de Vida. In: Ometto, A.R., Peres, R.B., Saavedra, Y.M.B. (org.) Ecoinovação para a melhoria ambiental de produtos e serviços: experiências espanholas e brasileiras nos setores industrial, urbano e agrícola. Diagrama Editorial, São Carlos, pp. 25–38 (2012)

Sonnemann, G., Margni, M.: Life Cycle Management. Springer, Dordrecht (2015)

UNEP, SETAC: Opportunities for national life cycle network creation and expansion around the world (2016)

UNEP: Life Cycle Management—A Business Guide to Sustainability. UNEP/SETAC Life Cycle Initiative, Paris (2007)

UNEP: LCI—Life Cycle Initiative: The Business Case for Life Cycle Thinking (2019)

UNEP; SETAC: Life Cycle Management: how business uses it to decrease footprint, create opportunities and make value chains more sustainable (2009)

Chapter 2
Cleaner Production (CP)

José Augusto de Oliveira, Antonio José Gonçalves da Cruz, Andreza Aparecida Longati, and Letícia Barbosa Fidanza

Cleaner production (CP) is "the continuous application of an environmental strategy integrated with processes, products, and services to increase efficiency and reduce risks to humans and the environment" (UNEP, 2002).

Therefore, it is a proactive strategy focused on the root causes of environmental aspects, thus aiming at avoiding and preventively minimizing environmental impacts in production processes.

2.1 Emergence and Importance of Cleaner Production

There are many reports in the literature about the origin of proactive approaches and strategies. There is still a lack of definition about the themes and concepts which define the preventive strategies for environmental impacts on production, which depend a lot on the different schools, founding organizations, and geographic

J. A. de Oliveira (✉) · L. B. Fidanza
Center for Advanced And Sustainable Technologies (CAST), Sao Paulo State University (UNESP), Campus of Sao Joao da Boa Vista, Av. Professora Isette Corrêa Fontão, 505, CEP: 13876-750, São João da Boa Vista, SP, Brazil
e-mail: jose.a.oliveira@unesp.br

A. J. G. da Cruz
Department of Chemical Engineering, Federal University of São Carlos, Rodovia Washington Luís, km 235, CEP 13565-905, São Carlos, SP, Brazil

A. A. Longati
Department of Materials and Bioprocess Engineering, School of Chemical Engineering, University of Campinas, Av. Albert Einstein, 500, CEP 13083-852, Campinas, SP, Brazil

J. A. de Oliveira et al. (eds.), *Life Cycle Engineering and Management of Products*,
https://doi.org/10.1007/978-3-030-78044-9_2

locations. The origin of the term CP is associated with the origins of environmental management itself and, therefore, permeates environmentalism and its history.

In this historical context, it is worth noting that, during the 1970s, a concern with the environment was intensely debated stimulating business performance in the face of its negative environmental impacts. This trend favored the emergence, in 1975, of the 3M pollution prevention pays (3P) initiative, one of the first proactive approaches to environmental management, a program based on the action of 3M company employees, for the implementation of innovative projects aimed at eliminating pollution using product design and communication practices applied to environmental management.

It is necessary to contextualize that, before this initiative, industrial environmental issues were first ignored, then diluted (1970s), and then dealt with at the end of the processes or *end of pipe* in the 1980s. When 3M created its innovative design, industries were a mix of these three approaches until then.

In 1989, the United Nations Environment Programme UNEP started its CP initiative, trying to conceptualize its real term instead of focusing only on "clean technology." UNEP's Technology, Industry and Environment Division highlighted the importance of effective management and organization in addition to the need for constant improvement in CP performance. UNEP was already striving to provide leadership and encourage partnerships to promote this concept on a worldwide scale. Specifically, at that time, they involved the provision and exchange of information as well as means for training and technical assistance and promotion of CP strategies (UNIDO 2002).

Already in 1990, the term CP was officially defined by its founding body, UNEP. This definition has been used as the working definition of all programs related to the promotion of a CP, and it remains a valid definition until the present day.

According to UNIDO/UNEP (2004), some important points about CP must be considered:

I. It must be a continuous process and not a specific and isolated activity;
II. It is not limited to specific industries, sizes, or business sectors;
III. It directs production toward a balance between availability and consumption of materials (including water) and energy. It does not deny growth but insists that it is environmentally sustainable;
IV. It refers to an approach for goods and services with a minimum of environmental impacts within the technological and economic limits;
V. It is not limited to minimizing waste, but rather uses a broader context and uses the term "environmental impacts" in the life cycle;
VI. In addition to the environmental aspects of the life cycle, CP also addresses health and safety issues still emphasizing the risk of reduction;
VII. It is efficient (in terms of increasing outputs immediately) and effective (in terms of positive results in the long run); and
VIII. It is a "win–win–win" strategy that protects the environment, communities (i.e., health and safety of workers, consumers, and the neighborhood) and business (i.e., their profitability and image). Therefore, CP addresses social,

economic, environmental issues and should not be considered just as an environmental strategy.

CP's priority differs from *end of pipe* practices which only try to mitigate the environmental damage caused by an unsustainable production system, remedying its effects without combating the causes that produced them, that is, proposals for reactive and selective solutions, generally introduced to meet emissions or environmental quality standards set by government regulations.

It may be inferred that CP aims at eco-efficiency in the applying of its practices. Eco-efficiency is an evolving concept that allows companies to adapt to the current market's changing dynamics. Organizations are increasingly focused on meeting their stakeholders' demands. In this way, those who implement eco-efficient practices will be able to respond more positively to competitive pressures and anticipate customers' needs while, at the same time, protecting environment, health, and safety of employees.

In simple terms, the vision of eco-efficiency is on "producing more with less." Reducing waste and pollution, using less energy, and using less raw materials are obviously good for the environment. It is also good for business because it reduces company costs and eventually avoids potential environmental liabilities. It is, therefore, a prerequisite for the business's long-term sustainability.

The concept of eco-efficiency is better understood by analyzing the concept of traditional efficiency, that is, producing more with less use of resources. However, while productive efficiency focuses on a strictly quantitative analysis of the relation between outputs and inputs, that is, the relation between added value and resources used. Eco-efficiency focuses on the qualitative and quantitative consideration of the relationship between industrial services and the environmental impacts of these activities. Thus, an eco-efficient process is one that produces more products using fewer resources and generating less environmental impacts. Equation 2.1 presents the basic formula for calculating eco-efficiency.

$$\text{Eco-efficiency} = \frac{\sum \text{Product}}{\sum \text{resources} + \sum \text{environmental impacts}} \tag{2.1}$$

It should be noted that Eq. 2.1 is an illustrative way of measuring eco-efficiency, and its application is complex due to the lack of heterogeneity in the units of each set of variables.

2.2 Steps for the Leading and Practice of Cleaner Production

The application of CP must occur following a decreasing order of the established priority levels, as shown in Fig. 2.1.

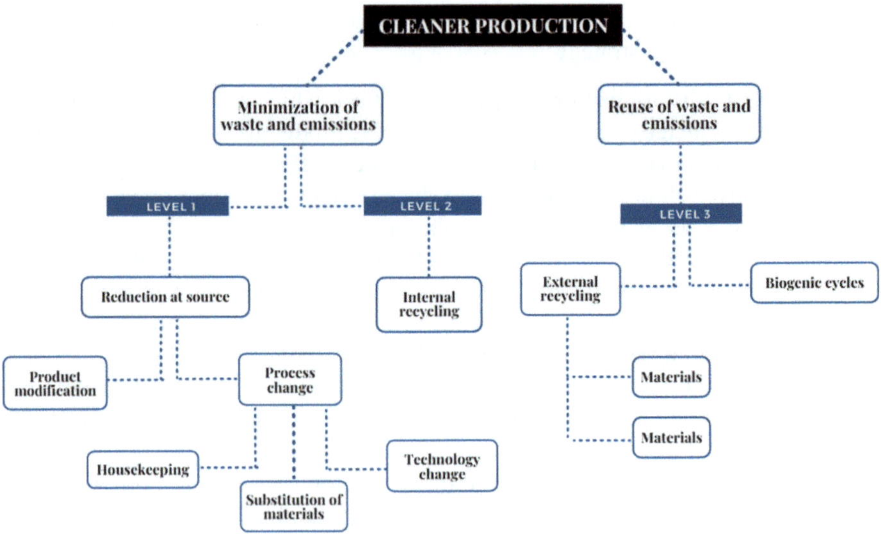

Fig. 2.1 Application levels of cleaner production. *Source* Adapted from UNIDO/UNEP (2004)

The order of application levels that companies must follow for the applying of CP must prioritize the options that present the greatest environmental and economic performance within their technological limits, that is, prioritizing from level 1 to 3. Environmental performance and economic performance of CP options are reduced from level 1 to level 3; that is, eco-efficiency tends to be minimized.

The CP implementation program can be divided into five steps (UNEP, 2007):

1. Planning and organization—provide and establish activities and resources necessary for the program;
2. Diagnosis—aims at knowing the company's current level in relation to environmental management practices;
3. Assessment—gathers up-to-date concrete data, analyzing and determining what are the existing CP options for adjusting the previously prioritized aspects;
4. Feasibility study—assesses the opportunities identified in the previous step and selects the most viable ones for implementation, in view of the greatest possible eco-efficiency potential;
5. Implementation and monitoring—put into practice the options selected in the previous step, compare the results obtained on the planning, and intervene to make the necessary changes, aiming at increasing eco-efficiency.

Figure 2.2 summarizes these steps in a logical flow for implementing CP.

These five steps for the implementation of CP are in accordance with UNEP/UNIDO methodology and will be described in Sects. 2.2.1–2.2.5.

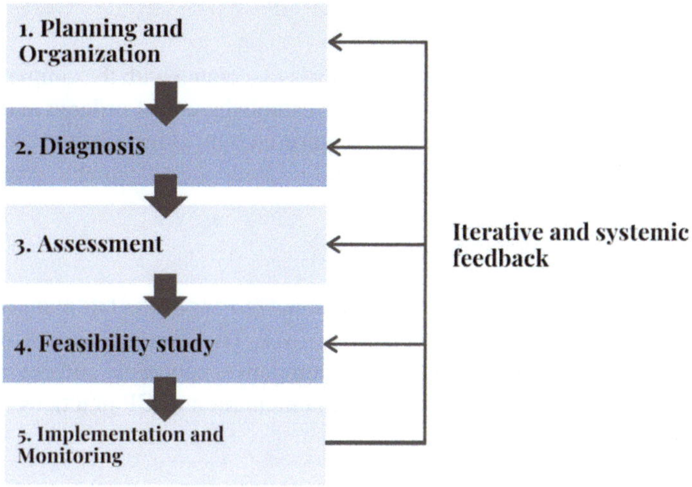

Fig. 2.2 Implementation of cleaner production. *Source* Adapted from UNIDO/UNEP (2004)

2.2.1 Planning and Organization

The aim of this step (Step 1, Fig. 2.2) is at obtaining commitment to the project, allocate resources and detail the work to be carried out. This step is very close to some requirements of NBR ISO 14001: 2015. For more details on environmental planning in the business context; see Chap. 11. The Planning and Organization step comprises the following activities:

I. Definition of legal requirements

The inadequate disposal and the large amount of waste generated have led many countries to adopt legal measures for waste management. Germany was the pioneer in terms of solid waste management. Thus, inspired by German legislation, other developed countries sought ways to incorporate laws and actions to meet the theme. In Brazil, for example, the National Solid Waste Policy (in Portuguese *Política Nacional de Resíduos Sólidos*—PNRS), Law No. 12305 establishes guidelines for waste management. In the planning phase, it is of great importance to consider the requirements imposed by the current legislation in relation to the shared responsibility of institutions and the destination and treatment of their solid waste. It is worth noting that non-compliance with legal requirements may generate major impacts not only in financial terms but also in environmental terms and in relation to the institution's public image. In addition to these, CP practices need to be focused on meeting legal requirements established by legislation on: Nationals Environmental Policies; Water use; Particularities of the undertaking place; Requirements established in the Environmental License; Resolutions of the National Environment Council, Resolutions of the Environmental Companies; NBR 10004; etc.

II. Definition of the environmental policy

The determination of an environmental policy is associated with the company's focus on the implementation of CP. It is about the establishment of ordered and practical actions that must be taken to guarantee the enterprise's most sustainable development. It is about establishing the practices adopted to guide the sustainable growth of the organization.

III. Top management commitment

The involvement phase and the consequent commitment of the top management are of great importance in the implementation of CP. The identification of the importance of practices for the image and for the company's economic and environmental profitability promotes continuity for the implementation of CP practices.

IV. Involvement of employees

It is a fundamental activity for the implementation of CP as it promotes a change of organizational and cultural character in employees. From the moment that the employee recognizes himself as a necessary part, his attitude becomes proactive regarding gains for the company and for the employee's own routine.

XXII. Formation of the cleaner production team

This team is commonly called ecoteam by companies and may even be formed by the team responsible for the environmental management system (EMS). For the team's formation, there is a need for variability in relation to the performance areas of the CP team members in order to form a multidisciplinary group. Sectors such as finance, maintenance, engineering, quality, and research and development are essential to identify points for improvement in the system. There are many ways to carry out steps I to V.

VI. Presentation of methodology

It includes incorporating a series of technical meetings in which the CP team sets goals and means to achieve them. It is about specifying activities and structuring the actions that must be taken in addition to sectors that are expected to be reached in the organization. Here, in general, trainings are given on the use of certain techniques and tools that will assist in the process of practical implementation of CP. Examples of some typical tools that may be adopted by the CP team during training are available in Silva et al. (2013).

VII. Opportunity identification

In adopting CP practices, companies may face different types of opportunities in different categories. These barriers may be external to the company and internal, such as organizational, systemic, attitude, economic, technical, among others. Thus, it is of great importance to clarify the practices implemented and their benefits for the company and employees.

2.2.2 *Diagnosis*

The CP assessment (Step 2, Fig. 2.2) begins with the "diagnosis" of process, a step in which deficiencies and their causes are identified as well as options on how to improve this process is found. The diagnosis step consists of pre-assessing the processes aiming at identifying opportunities for CP. This step is subdivided into:

I. Elaboration of the process flowchart:
   - First, qualitative;
   - Second, quantitative.

II. Input and output assessment:
   - Pointing of inputs, outputs, and losses by means of mass and energy balance calculations.

III. Pre-determination of CP focuses:
   - Indication of possible performance focuses of CP practices.

The phases of step 2 will be detailed below.

Pre-assessment

The aim of the pre-assessment step is at obtaining an overview of the company's productive and environmental aspects. Production processes are better represented by drawing up a flowchart showing the inputs, outputs, and losses. Then, the action focuses of the CP practices may then be determined.

I. **Elaboration of the process flowchart**

The first pre-assessment phase is the elaboration of the process flowchart, which consists of a graphical representation of all the steps of a process and how they relate to each other. The elaboration of the flowchart is a fundamental step in the assessment, which is made first of a qualitative flowchart and then a quantitative flowchart.

The qualitative flowchart presents the various activities related to the execution of a specific task or series of actions. It also records the progress of processes across one or several sections or departments of the company. In this flowchart, it is possible to identify the inputs and raw materials used, the products and by-products produced, the energy and water consumption, and also the waste and emissions generated. In other words, through this flowchart all flows of material and product under elaboration are viewed. The flow of information and the path taken in the various processes are also known. To prepare this flowchart, existing operational data such as production reports, audit reports, and site plans are used.

Then, the quantitative process flowchart is prepared. The aim of this step is at collecting quantitative data on production and environmental aspects using all available and existing sources related to processes, such as estimates of the purchasing sector, sales sector data, and production planning history. The process flowchart is intended to provide an overview and therefore must be accompanied by quantitative input (raw materials, water, energy, and other inputs) and output (products, by-products, emissions, effluents, and waste) data for each operation or process sector considered in the analysis, in addition to data on the company's environmental situation. It is also important to include in this flowchart some activities that are often overlooked, such as cleaning, storage and material handling, auxiliary operations (refrigeration, steam, and compressed air), maintenance and repair of equipment, some material streams (catalysts, lubricants, etc.), and by-products released into the environment.

This flowchart will be the basis for the mass and energy balances that will be carried out in the subsequent step, the assessment step. Some examples of production flowcharts used in LCEM studies can be found in this book in Chap. 3. These flowcharts are essential as they represent the mapping of operations of interest either to carry out CP or to conduct other improvement initiatives toward more sustainable production.

II. **Assessment of inputs and outputs**

Much of the information needed to complete the quantitative process flowcharts, previously described, can be obtained during a step-by-step inspection in the company. This inspection must follow the process from the beginning to the end, focusing on areas where products, waste, and emissions are generated. During the analysis, all the information obtained must be listed and, if there are obvious solutions to the existing problems, they must be observed. It is important that special attention is paid to free and/or low-cost solutions. These solutions must be implemented immediately without waiting for a detailed feasibility analysis.

III. **Pre-determination of cleaner production focuses**

In this step, which is the last step of the pre-assessment step, the problems that will be dealt with in a later step are known and it aims at establishing the focus for future CP work. In an ideal case, all unit processes and operations should be assessed. However, some restrictions such as time and financial resources may make it necessary to select the most important aspects or process areas. Thus, the CP study focuses are defined by several factors, and some of the criteria to be used are: level of danger to the environment, cost of raw materials, submission to current and future regulation and taxation, cost of waste and emission management (treatment and disposal), amount of waste, hazards to the safety of employees and neighboring areas, in addition to the budget available for assessment. Thus, the team will define the steps, processes, products, and/or equipment that will be prioritized for the effective assessment.

Another criterion for the selection of CP focus may be given by calculating the current eco-efficiency of the mapped processes. Those processes with less eco-efficiency must then be prioritized for further assessment and suggestions for improvements on the production system.

2.2.3 Assessment

The purpose of the assessment step (Step 3, Fig. 2.2) is to collect (or survey) data and assess the company's environmental performance and production efficiency. In this assessment, measurements and balances of mass and/or energy will be performed. The data collected on management activities may be used to monitor and control the overall efficiency of the process, set goals, and calculate monthly or annual indicators. The data collected on operational activities may be used to assess the performance of a specific process.

Assessment is comprised of the phases:

I. Mass (material) and energy balances;
II. Assessment of causes;
III. Identification of CP opportunities;
IV. Selection of CP opportunities.

These phases will be detailed below.

I. **Mass (material) and energy balance**

Mass balances (also called material balances) are based on the mass conservation law proposed by Antoine Laurent Lavoisier (1743–1794): "in nature nothing is lost, nothing is created, everything is transformed." Thus, nature imposes restrictions on the physical and chemical transformations of matter which need to be taken into account when assessing new processes or analyzing an existing one. Energy balances are analyzed based on the energy conservation law or the first law of thermodynamics.

When an accountant makes a company's financial statement it counts the money that comes in, the money that goes out, the money that the company gains or losses in investments, and the money that the company accumulates in cash. The mass and energy balances are similar; however, instead of accounting for the cash flow (money) of a particular firm, they account for the mass and energy in a given control volume.

The balances involve a **system or control volume (CV)**; that is, a portion of the space on which the analysis will be performed. The professional needs to limit his study to the relevant part of its process or even to the process as a whole. The control volume is separated from the rest of the universe by a **boundary or control surface (CS)**.

When there is an exchange of matter between the control volume and its surroundings it is said the **system is open**. An example of an open system is a lake, there is always water leaving through evaporation or entering through rivers or rain. If there is continuous transfer of matter across the boundary of a system, as in the case of the river, it is called streams. When there is no transfer of matter across the boundaries, the **system is called closed**. A safe is an example of a closed system. When there is no transfer of matter and energy across the boundaries, the **system is isolated** (Fig. 2.3).

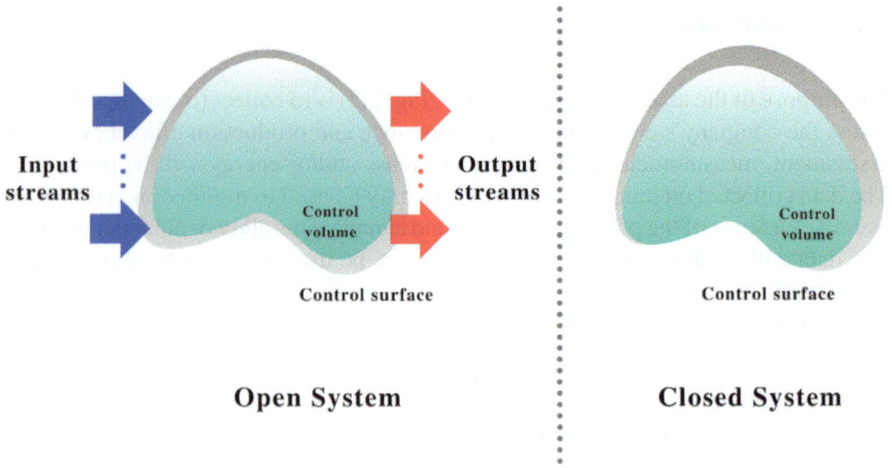

Fig. 2.3 Open and closed systems. *Source* Authors

It is possible to classify a chemical, physical, or biochemical process in a similar way to how a system is classified depending on how exchange of matter occurs between the control volume where the process occurs and the medium.

A process in which matter passes through the system boundaries by incoming and outgoing streams is called **continuous process**. Most industrial-scale process units in engineering are designed to operate continuously. Examples of units which operate continuously include heat exchangers, separation units, and most reactors. A more everyday example is an electric shower, the mains water enters continuously through the plumbing and exits continuously through the shower nozzle. Continuous processes are classified as **open systems**.

A process in which there is no continuous passage of matter across the process boundaries and feeding and unloading take place at once at the beginning and at the end of the process is called **discontinuous or batch**. This type of process is uncommon in the industry and is most commonly used in laboratory experiments, for example, a chemical reaction taking place in an Erlenmeyer flask. The preparation of a portion of cookies in an oven is also an example of a batch process. On an industrial scale, batch processes are open systems at the beginning (filling of the system) and at the end (discharging of the system) of the process, but they are closed systems for the rest of the time.

A process in which there is only input of matter but not output, or only output but not input, is called **semicontinuous**. An important example is the production of ethanol. First, the fermentation yeasts are introduced into a fermentation vat. Then, a stream containing the sugar (sucrose) which will be fermented is fed continuously to the tank over a few hours, until it is filled. This process is classified as semicontinuous. Then, the feeding is interrupted and the process becomes discontinuous for a few more hours in order to provide the time necessary for the yeasts to consume all sugar. At the end, the tank is quickly unfilled. Another more familiar example is a glass of soda. While you drink, there is only one output, there is no input of soda into the glass. Semicontinuous processes are open systems as long as there is input or output streams and can switch from open to closed systems over time. Some important observations:

- batch and semicontinuous processes occur in a non-steady state (transient regime). This is because in both cases there is a change in the values of the process variables over time, such as mass, volume, and concentration, for example.
- continuous processes are designed to occur in a steady state (permanent regime). However, in some situations such as when starting the process (start-up) or when there are changes in operational conditions (setup), continuous processes occur in a transient regime.
- a process may occur in a permanent regime in relation to a given variable and be in a transient regime in relation to other(s). Thus, the process must be analyzed from the perspective of the variable of interest.

To carry out a mass or energy balance in a system, the inputs and outputs of matter and/or energy (in the form of heat and/or work) must be taken into account. It is also possible to identify the values of the streams which may be unknown or difficult to be known without this technique. The conservation law is based on the fact that matter cannot disappear or be spontaneously created. Although the total mass of a system cannot be created or destroyed (unless by nuclear reactions), chemicals may convert from one species to another, so a mass balance made component by component may reveal substances being generated and consumed (Fig. 2.4). Thus, in general, the balance for any substance may be written according to Eq. 2.2:

$$\text{accumulate} = \text{in} - \text{out} + \text{appear} - \text{disappear} \tag{2.2}$$

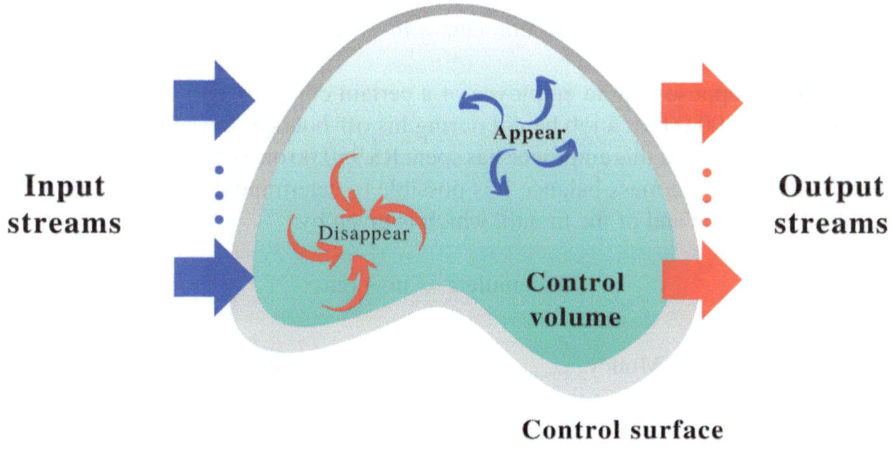

Fig. 2.4 Mass balance for a substance in an open system. *Source* Authors

Equation 2.2 is the **general balance equation**. In this equation, the terms "+ *appear*" and "− *disappear*" may be written as "+ *generated*" and "− *consumed,*" respectively. Equation 2.2 may be simplified to represent some special systems:

Closed system, by Eq. 2.3:

$$\text{accumulate} = +\text{appear} - \text{disappear} \tag{2.3}$$

Continuous operation, by Eq. 2.4:

$$\text{accumulate} = \text{in} - \text{out} + \text{appear} - \text{disappear} \tag{2.4}$$

System without reaction, by Eq. 2.5:

$$\text{accumulate} = \text{in} - \text{out} \tag{2.5}$$

Material balances: The quantity of matter may be counted in mass or molar terms. In addition, it is possible to take stock of the quantity of matter totally or by components. The choice between the mass basis and the molar basis depends on the process studied. It is generally preferable to use molar basis in processes that involve chemical reactions or phase equilibrium and mass basis in others. Either way, it is possible to convert between mass basis and molar basis through the molar mass of the substance or the average molar mass of the mixture.

When applying the general balance equation for the total mass of a system, the terms of appearance and disappearance may be discarded. Except for nuclear reactions and quantum or relativistic phenomena, which are not the focus of industrial processes, mass creation or destruction is not observed. Thus, the **total mass balance** may be written, by Eq. 2.6, as:

$$\text{accumulate} = \text{in} - \text{out} \tag{2.6}$$

Example 1 Suppose that an employee of a certain company earned a R$ 1000.00 salary and R$ 400.00 for a job he did during his off-hours at the end of a month. In addition, consider that this employee has spent R$ 950.00 on expenses throughout the month. Applying the mass balance it is possible to determine the amount of money accumulated at the end of the month, which is given by:

$$\text{accumulate} = \text{in} - \text{out}$$

$$\text{Money}_{\text{accumulated}} = \text{Money}_{\text{in}} - \text{Money}_{\text{out}}$$

$$\text{Money}_{\text{accumulated}} = R\$\ 1000.00 + R\$400.00 - R\$950.00$$

$$\text{Money}_{\text{accumulated}} = R\$450.00$$

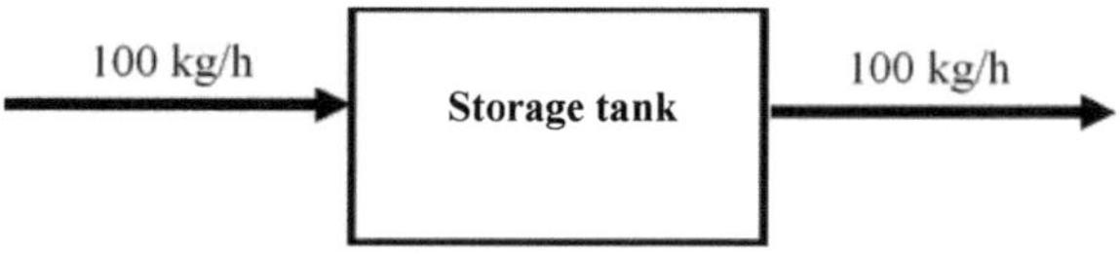

Fig. 2.5 Scheme that represents the first part of Example 2. *Source* Authors

Therefore, at the end of the month, the employee accumulated four hundred and fifty reais. Note that, in this case, a cash balance has been made. In the case of a material balance, matter accounting is carried out. Let us now see the example in which the same concept of global balance is applied to an industrial water storage tank.

Example 2 Let the industrial tank be represented in Fig. 2.5. What would be the water accumulation for the case in which input and output streams equal to 100 kg/h are fed.

Applying mass balance:

$$\text{accumulate} = \text{in} - \text{out}$$

$$\text{mass}_{\text{water, accumulated}} = \text{mass}_{\text{water. in}} - \text{mass}_{\text{water, out}}$$

$$\text{mass}_{\text{water, accumulated}} = (100 - 100)\ \text{kg/h}$$

$$\text{mass}_{\text{water, accumulated}} = 0\,\text{kg/h}$$

As can be seen, there is no water accumulation in this tank since all the water that enters the tank comes out in the same amount, that is, mass flow. This system is classified as continuous (there are material streams crossing the boundaries while the process takes place) and in a permanent regime or steady state (there is no change in the mass value inside the tank over time).

For the case where that same tank has input (100 kg/h) and output (90 kg/h) mass flow rates, we have (Fig. 2.6):

Applying mass balance:

$$\text{accumulate} = \text{in} - \text{out}$$

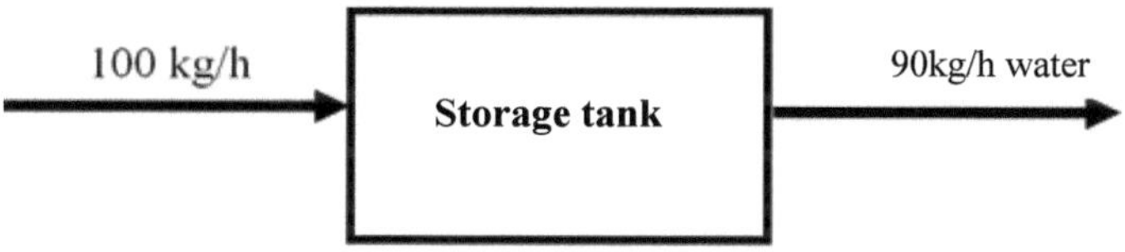

Fig. 2.6 Diagram showing the second part of Example 2. *Source* Authors

$$\text{mass}_{\text{water, accumulated}} = \text{mass}_{\text{water. in}} - \text{mass}_{\text{water, out}}$$

$$\text{mass}_{\text{water, accumulated}} = (100 - 90)\ \text{kg/h}$$

$$\text{mass}_{\text{water, accumulated}} = 10\ \text{kg/h}$$

In this case, the water output mass flow is less than the input and, thus, the water accumulation occurs in the storage unit. The system may be classified as continuous and in transient regime (there is a change in the water mass in the system over time).

As with total mass, the general balance equation may also be applied to a particular substance in a mixture. This type of balance is called **component mass balance** and is represented in Eq. 2.7:

$$\text{accumulate} = \text{in} - \text{out} \pm \text{react} \tag{2.7}$$

Example 3 Let us have a solution of water and sucrose. This solution is fed into a continuous reactor filled with a certain amount of invertase enzyme inside (the invertase enzyme catalyzes the reaction of transforming sucrose into glucose and fructose). Thus, each sucrose molecule is "broken" into a glucose and a fructose molecule. However, the balance for sucrose will include a disappearance term. If we call sucrose by index A, the mass balance for sucrose is represented by Eq. 2.8:

$$\frac{\mathrm{d}m_A}{\mathrm{d}t} = \dot{m}_{A,\text{in}} - \dot{m}_{A,\text{out}} - r_A \cdot V \tag{2.8}$$

where $\dot{m}_{A,\text{in}}$ is the mass (or molar) flow of sucrose that enters the CV and $\dot{m}_{A,\text{out}}$ is the mass (or molar) flow of sucrose that exits the CV, these terms have mass/time or mol/time units. The term that presents the derivative $\left(\frac{\mathrm{d}m_A}{\mathrm{d}t}\right)$ is the amount of sucrose mass m_A accumulated at time t inside CV. The term r_A is the sucrose hydrolysis rate (reaction rate, with unit of concentration by time, mass/(volume · time) or mol/(volume · time)), and V is the volume.

Unlike total mass balance, in which there is no creation or destruction of mass, in molar balance there may be the creation or disappearance of chemical species and/or changes in the number of moles in the system even in a closed system. In balances that involve reactions, it is more practical to work on mols. It is observed, in this case, that stoichiometry imposes restrictions on the process. Suppose the following reaction stoichiometry:

$$1\text{A} + 2\text{B} \rightarrow 3\text{C} + 2\text{D}$$

In this case, we have by Eq. 2.9:

$$\frac{r_A}{1} = \frac{r_B}{2} = \frac{r_C}{3} = \frac{r_D}{2} \tag{2.9}$$

General guidelines for calculating material balances:

Material balance problems involve the determination of quantities and properties of output streams from quantities and properties of input streams and vice versa, in a given CV. Equating these problems is often very simple. However, they may involve a large number of variables and equations, which can cause some complications. A procedure that may facilitate the solution of this type of problem is presented below in six steps (Badino and Cruz 2013):

1. Draw a block diagram of the process.
2. Number or name all streams and chemical species involved.
3. Write the values and units of all known variables on your block diagram or on a table.
4. Indicate unknown variables in the block diagram or in a table.
5. If there is a mixture of variables on mass and molar basis, convert them to the same appropriate dimension. If there are volumetric flows, convert them to mass or molar basis.
6. Write the total and component mass balance and equations of the mixture.

Example 4 Consider a combustion chamber (Fig. 2.7) where 10 kmol of methane, CH_4 (hydrocarbon) per hour, and an air current are fed (consider air as a mixture containing 21% O_2 and 79% N_2 in molar basis) at 500 °C and 1 atm at a flow of 300 m^3/h. In this chamber, a current of combustion products (CO_2 and H_2O) is generated. Assuming that the burning is complete (or total) and knowing that the N_2 fed with air does not participate in the combustion reaction (N_2 is an inert), apply the procedures for calculating the material balance to represent the problem. The first step is to draw the flowchart (Fig. 2.7):

The second step is to number or name all streams and chemical species involved. Numbering the streams, we have Fig. 2.8.

Assigning symbols to species:

Methane (M), oxygen (O_2), nitrogen (N_2), carbon dioxide (CO_2), and water (H_2O).

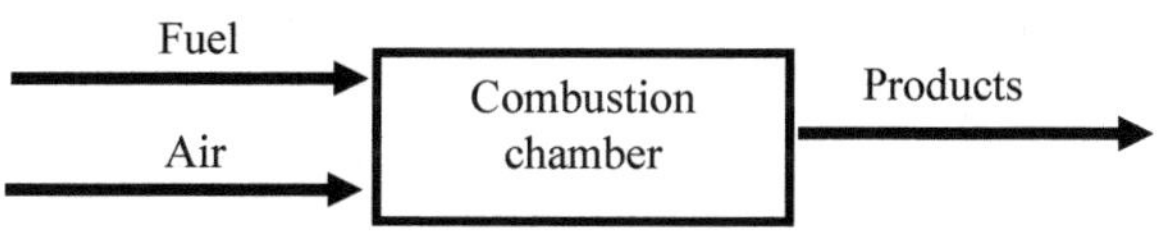

Fig. 2.7 Combustion chamber. *Source* Authors

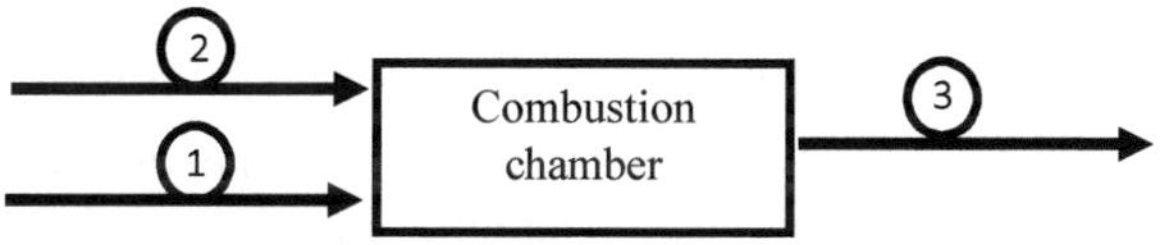

Fig. 2.8 Combustion chamber with numbering. *Source* Authors

Table 2.1 Information on the input (1 and 2) and output (3) streams

Stream		
1 (air)	2 (fuel)	3 (products)
$T_1 = 500\ °C$	$n_2 = 10$ kmol/h	n_3 (kmol/h)
$P_1 = 1$ atm	$y_M = 1.0$ kmol C/kmol	$y_{O_2,3}$(kmol O_2/kmol)
$V_1 = 300\ m^3/h$		$y_{N_2,3}$ (kmol N_2/kmol)
$y_{O_2,1} = 0.21$ kmol O_2/kmol (21% O_2)		$y_{M,3}$ (kmol C/kmol)
$y_{N_2,1} = 0.79$ kmol N_2/kmol (79% N_2)		$y_{CO_2,3}$ (kmol CO_2/kmol)
[a] $n_1 = 4.73$ kmol/h		$y_{H_2O,3}$ (kmol H_2O/kmol)

[a] Mass flow calculated from the volumetric flow using $n_1 = \frac{P_1 \cdot V_1}{R \cdot T_1}$ where R is the ideal gases constant 0.08206 atm m^3/(kmol K)

The third, fourth, and fifth steps are described in Table 2.1. It contains the values and units of all known and organized variables (third step), and it also indicates unknown variables (fourth step) and also presents all variables for the same basis (fifth step).

Finally, the last step consists of writing the mass balance equations for the mixture components having the complete combustion of methane ($CH_4 + 2O_2 \rightarrow CO_2 + 2H_2O$) reaction as stoichiometry. It is important to highlight that, for mass or molar fractions, it must be used the implicit relation that the sum of the fractions is equal to 1.

Energy balance: Energy E has many forms. In industrial process engineering, the most common forms are mechanical energy (kinetic and potential) and internal energy. The kinetic energy (K) is the mechanical energy associated with the movement of the center of mass of the macroscopic system. The potential energy (Φ) is associated with the position of the system center of mass in relation to a gravitational field. The internal energy (U) is defined as the total energy of the microscopic components of a macroscopic system. More specifically, internal energy is the energy due to all molecular, atomic, and subatomic motions and interactions. In most applications, variations in internal energy are much higher than variations in mechanical energy, which may be despised ($E = K + \Phi + U \approx U$) (Badino and Cruz 2013).

The system's internal energy may be altered by energy transfers in the form of heat (Q) and work (W). Thus, the energy balance in closed systems is given by Eq. 2.10:

$$\Delta K + \Delta\Phi + \Delta U = Q + W \tag{2.10}$$

Some important observations according to Badino and Cruz (2013):

(a) A system's internal energy (U) depends on the chemical composition, the state of aggregation, and the temperature of the components that form it. U is independent of the pressure for ideal gases and practically independent of the pressure for liquids and solids. Therefore, if there are no changes in temperature of phase and chemical composition in the process, or if the process materials are all solids, liquids, or ideal gases, in these cases, $\Delta U = 0$.

(b) If a system and its surroundings are at the same temperature or if the system is perfectly isolated, we have an adiabatic process; therefore, $Q = 0$.

(c) Work done on or by the closed system is accompanied by a movement of the boundary against a resistance force or by a generation of current or electrical radiation beyond the boundaries of the system. If there are no moving parts or current generation for a closed system, we have $W = 0$.

(d) If variations in potential energy that are not due to the difference in height (movement against an electrical resistance force or an electric or magnetic field) occur, the terms to account for it must be included in the term Φ of the energy balance.

By definition, in open systems there is matter crossing the boundaries during the process. We will adopt, by convention, that in open systems the rates of heat transfer ($\dot{Q}$) and work ($\dot{W}$) have a positive sign when they enter the control volume and a negative one when they leave. The work may be divided into three components: the axis work applied on the system by agitators, compressors, among others ($\dot{W}_s$), the frontier work performed by the expansion or contraction of the control volume ($\dot{W}_b = -P\frac{\mathrm{d}V_R}{\mathrm{d}t}$), and the work of the input and output streams with different pressures and flow rates ($\dot{W}_f = \dot{V}_0 P_0 - \dot{V} P$). This last component may be combined with internal energy, resulting in the enthalpy variable ($\underline{H} = \underline{U} + P\underline{V}$). Thus, the general energy balance for an open system is given by Eq. 2.11.

$$\frac{\mathrm{d}E}{\mathrm{d}t} = \dot{n}_o E_0 - \dot{n} E + \dot{Q} + \left(\dot{W}_s + \dot{W}_b + \dot{W}_F\right) \tag{2.11}$$

where $\dot{n}_o$ is the molar flow that enters and $\dot{n}$ is the molar flow that leaves.

As mentioned earlier, in most industrial applications variations in internal energy are much greater than variations in mechanical energy (kinetic and potential) and, therefore, may be despised. Thus, we may approximate the total energy of the system (E) by the internal energy (U) according to Eq. 2.12:

$$\frac{\mathrm{d}}{\mathrm{d}t}(K + \Phi + U) \approx \frac{\mathrm{d}U}{\mathrm{d}t} \tag{2.12}$$

Replacing Eq. 2.12 in Eq. 2.11, we have:

$$\frac{\mathrm{d}U}{\mathrm{d}t} = \dot{n}_o U_0 - \dot{n} U + \dot{Q} + \left(\dot{W}_s + \dot{W}_b + \dot{W}_F\right) \tag{2.13}$$

Recalling that $\dot{W}_f = \dot{V}_0 P_0 - \dot{V} P$ and replacing in Eq. 2.13, we obtain:

$$\frac{dU}{dt} = \dot{n}_o U_0 - \dot{n}U + \dot{Q} + \left(\dot{W}_s + \dot{W}_b + \dot{V}_0 P_0 - \dot{V}P\right) \tag{2.14}$$

Rearranging Eq. 2.14:

$$\frac{dU}{dt} = \dot{n}_o\left(\underline{U}_0 + P_0\underline{V}_0\right) - \dot{n}\left(\underline{U} + P\underline{V}\right) + \dot{Q} + \left(\dot{W}_s + \dot{W}_b\right) \tag{2.15}$$

Knowing that enthalpy is given by $\underline{H} = \underline{U} + P\underline{V}$, we finally have:

$$\frac{dU}{dt} = \dot{n}_o\underline{H}_0 - \dot{n}\underline{H} + \dot{Q} + \left(\dot{W}_s + \dot{W}_b\right) \tag{2.16}$$

Some important simplifications (Badino and Cruz 2013):

(a) If there are no moving parts in the system, $\dot{W}_s = 0$;
(b) If the system and its neighborhood are at the same temperature, $\dot{Q} = 0$;
(c) If the linear rates of the streams are the same, $\Delta K = 0$;
(d) If all streams enter and leave the process at the same geometric position, $\Delta\Phi = 0$.

Summing up, the total mass and/or energy at the input is composed of the sum of all materials and/or energy streams that enter the system: inputs and raw materials, auxiliary materials, energy (work and/or heat) and water, among others. The total mass and/or energy at the outlet comprises the sum of all the masses and/or energy that leave the system: products, by-products, waste, effluents, and emissions. The accumulated mass and/or energy refers to aspects coming from accumulated productive steps. Thus, through the balance (of matter and/or energy) it is possible to identify and quantify the consumption of matter and energy in the process as well as the losses, waste, and emissions resulting from the process, providing an indication of its sources and causes (Badino and Cruz 2013). Making it possible, thus, to point out areas of inefficient use of resources and with poor management of waste generated in production and, consequently, to identify the main inflows/outflows that make the polluting process-based. Thus, future decisions to be made during the cause assessment step (next topic) may be made based on quantitative data.

Cause assessment

Based on the data collection and the application of mass/energy balance in the prioritized steps and/or sectors, a deep analysis must be carried out in order to ascertain/determine the real source and causes of generation of waste, emissions and contaminants, losses (of matter and/or energy), and water streams. At this step, one must know why, how, when, and where each of these streams was generated.

It is important to mention that the generation of CP options may be limited since efforts may be concentrated on only one equipment that emits the current under assessment (it may be a current of waste, contaminants, emissions, losses, and water). However, this equipment may be responsible for producing only a part of this current.

Thus, depending on the results of the initial assessment, the team may see the need to know more specific details on the generation of these streams, which may be done by conducting a detailed audit.

As CP is a proactive strategy focused on generating sources, one must investigate the problems and opportunities previously listed to identify the causes of these problems and find possible opportunities. For that, some tools commonly used in quality management may be used, especially the cause–effect diagram, also known as Ishikawa diagram, or the tree diagram, or even fishbone diagram, which is shown by Fig. 2.9.

The cause–effect diagram, shown in Fig. 2.5, makes it possible to know the causes of the respective effects (problems and/or opportunities) that are related to the method, the machine, the measure, the surrounding environment, the material, or the labor. These six groups are related to the process for the implementation of CP. The causes may occur in one or more groups. This diagram helps a lot in the identification of causes that lead to the process effect that causes aspects and later environmental impacts, reducing project time and costs as well as enabling greater effectiveness and efficiency of CP practices.

However, there is still a need to identify root causes, that is, what is the fundamental cause that in turn generates the effect. For this, the tree diagram must be used as shown in Fig. 2.10.

To use the tree diagram shown in Fig. 2.10, start from the effect and open the possible causes in n levels in the vertical direction of primary causes and also deepening in the horizontal direction in N levels until one or some root causes of the problem studied are found. It is precisely this group or this fundamental cause that must be tackled so that the environmental aspect(s) is (are) eliminated and its (their) environmental impact(s) is (are) prevented. Such stratification of root causes must be carried out by the CP team.

II. **Identification of cleaner production opportunities**

Knowing the characteristics of the process, the sources and causes of waste and contaminating emissions, CP enters the creative phase. Based on the process mapping

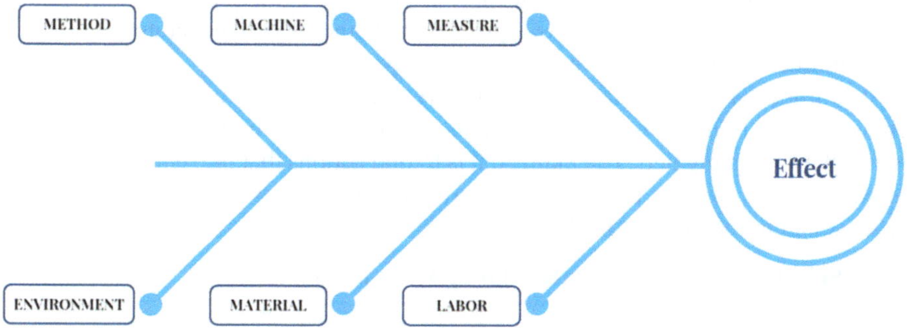

Fig. 2.9 Ishikawa diagram. *Source* Authors

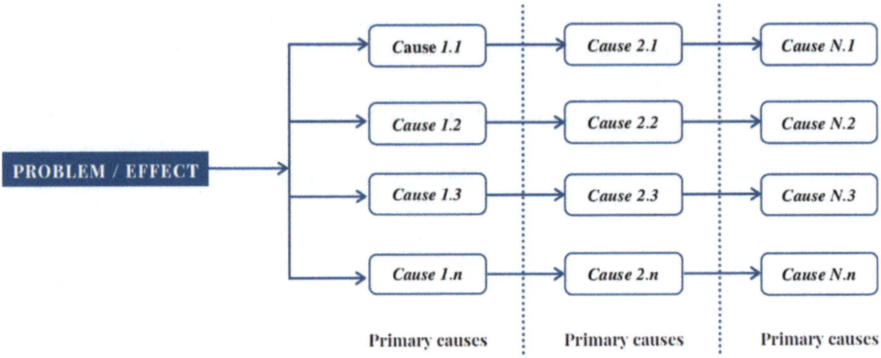

Fig. 2.10 Tree diagram. *Source* Authors

and the mass/energy balance, the operating unit, material, waste or emissions can be chosen; that is, where the CP implementation is to be started. Thus, this step consists of identifying opportunities to improve the environmental performance of this process and outline the ways of generating CP options. The generation of CP opportunities is made by tracking the five characteristics that influence the process and that may serve as focal points. These characteristics are: (1) input materials; (2) technology; (3) process execution; (4) products; and (5) waste and emissions. Based on these five characteristics, the corresponding action points may be used to improve the environmental performance of the process:

1. Changes in input material or replacement of inputs;
2. Technological changes;
3. Good maintenance and/or good operational practices;
4. Product changes;
5. Recycling.

A very efficient way to produce ideas for CP opportunities is to hold a brainstorming session. These brainstorming sessions are most effective when managers, engineers, process operators, and other employees, as well as some outside consultants, work together without hierarchical restrictions. Many CP solutions are obtained by carefully analyzing the cause of a problem. Another way to generate ideas for CP opportunities is to focus on preventive practices.

It is important to emphasize that priority must be given to actions that aim at eliminating or minimizing waste, effluents, and emissions in the production process where they are generated. The main goal is to find actions that prevent the generation of waste at the source. These actions may include changes both in the production process and in the product itself.

IV. **Selection of cleaner production opportunities**

Once several CP opportunities have been suggested, generated, and recorded, they must be classified between those which may be implemented directly and those which require further investigation. It is useful to follow the steps:

1. Organize options according to unit operations or process areas, or according to input/output categories (e.g., problems that cause high water consumption, problems that cause high energy consumption, problems that emit large amount of waste, among others).
2. Identify any options for mutual interference as the implementation of one option may affect the other.
3. Opportunities that are free or low cost, that do not require an extensive feasibility study or that are relatively easy to implement, must be immediately implemented.
4. Opportunities that are unfeasible or that cannot be implemented must be removed from the list of options for further study.

2.2.4 *Feasibility Analysis*

After CP opportunities are identified and selected, the next phase is the feasibility analysis (Step 4, Fig. 2.2) of each of the options identified. The purpose of this assessment phase and feasibility study is to assess the proposed CP opportunities and select those options that are appropriate for implementation. The opportunities selected during the assessment phase must be assessed according to their technical, economic, and environmental merits. Thus, the feasibility step is composed of the following phases:

I. Technical assessment;
II. Economic assessment;
III. Environmental assessment; and
IV. Selection of CP opportunities.

These four phases that make up the feasibility analysis are detailed below:

I. **Technical assessment**

In the technical assessment, the properties and requirements that the raw materials, in addition to other materials, must present for the product to be manufactured are considered, so that modifications may be suggested. In the technical assessment, it is important to consider the possible impacts of the proposed action on products, processes, productivity, safety, quality, etc. In addition, laboratory tests or trials may be necessary when the options significantly change the existing process. It is also interesting to consider the experiences of other companies with the option that is being studied (i.e., to perform competitive benchmarking). A technical assessment will also determine whether the opportunity requires personnel changes, training, or additional maintenance to the process. If it is technically possible to implement the option, an economic assessment is carried out.

II. **Economic assessment**

The aim of this step is at evaluating the cost–benefit ratio of CP opportunities. This step intends to quantify the gains and the return on investment in local currency through the financial quantification of the elements identified in the data collection. The economic assessment makes it possible to verify whether there have been economic gains for the company and to assess the return on investment. Thus, economic feasibility is often the key parameter that determines whether a CP opportunity will be implemented or not.

In the economic assessment, it is important to consider: the necessary investments; the operating costs and revenues of the existing process, as well as the operating costs and projected revenues of the actions to be implemented; and the company's economy with the reduction/elimination of fines. Some of the economic analysis techniques that may be applied include the payback period, the net present value (NPV), and the internal rate of return (IRR).

The payback can be simple or discounted. The simple method considers the current time for the cash flow (CF) of the CP project to have its investment value recovered, that is, when CF = 0. The discounted payback method also considers the depreciation of the invested capital due to financial losses, examples of which include the rate of inflation, the opportunity cost of not investing this amount in another project and/or investment, etc. Thus, we include in the calculation of the discounted payback method the discounted cash flows (DCF) due to these depreciations of the capital invested in the CP project. Equation 2.17 illustrates the calculation of the discounted payback.

$$\mathrm{DCF} = \frac{\mathrm{CF}}{(1+i)^n} \tag{2.17}$$

where i is the MARR or the discount rate and n is the duration of the project.

Other tools such as NPV and IRR may also be applied here at this phase, and it is suggested to consult Assaf Neto and Lima (2014) for further details on the subject.

III. **Environmental assessment**

The aim of the environmental assessment is at determining the positive and negative environmental impacts of the proposed CP option. In many cases, the environmental benefits are obvious: a reduction in toxicity and/or amount of waste or emissions. In other cases, it may be necessary to assess whether an increase in electricity consumption, for example, would overcome the environmental advantages of reducing material consumption.

In the environmental assessment, it is important to consider:

(a) The amount of waste, effluents, and emissions that will be reduced;
(b) The quality of the residues, effluents, and emissions that have been eliminated—check if they contain less toxic substances and reusable components and also check if there has been a change in their degradability;
(c) The reduction in the use of natural resources;

(d) The reduction in consumption of inputs, raw materials, water, and energy; and
(e) Changes in the degradability of waste or emissions.

In many cases, it will be impossible to collect all the data necessary for a good environmental assessment. In such cases, a qualified assessment will have to be made based on the existing information. Given the wide range of environmental issues, it will probably be necessary to prioritize the issues of greatest concern, since depending on the country's or company's environmental policy some issues may have a higher priority than others.

In addition, recent studies have deepened the application of life cycle assessment (LCA) to generate scenarios aimed at improving the production process. These scenarios can be generated on a simulation basis, and thus, it is possible to identify the environmental feasibility of the CP proposals before their implementation.

IV. **Selection of cleaner production opportunities**

The results found during the technical, economic, and environmental assessment make it possible to select the actions of viable CP opportunities according to the criteria that have been established. Thus, the most promising options must be selected in close collaboration with the company's administration.

2.2.5 Implementation and Monitoring

The implementation (Step 5, Fig. 2.2) of the CP practices must be systematic and continuous in order to integrate with the companies' management systems and strategies and tools. The isolated practice of CP projects without this integration generates inefficiency and unsatisfactory results for organizations. The implementation and monitoring of CP occur basically following the continuous improvement cycle, also known as the deming cycle or PDCA cycle composed of the plan, do, check, and act phases. Figure 2.11 shows a PDCA designed for CP becoming a continuous improvement program for companies.

Regarding the monitoring of improvement actions, it is worth mentioning the importance of monitoring the eco-efficiency a posteriori of the implementation of CP projects. Thus, it becomes possible to measure the projects effectiveness and conclude on the greatest added value to the least impact on the environment for the company. The use of performance indicators is essential for the fulfillment of this step. For more details, see Chap. 11.

2.3 Case Studies in Cleaner Production

The Environmental Company of the State of São Paulo (CETESB) is an agency of the state government responsible for monitoring and verifying pollution-generating

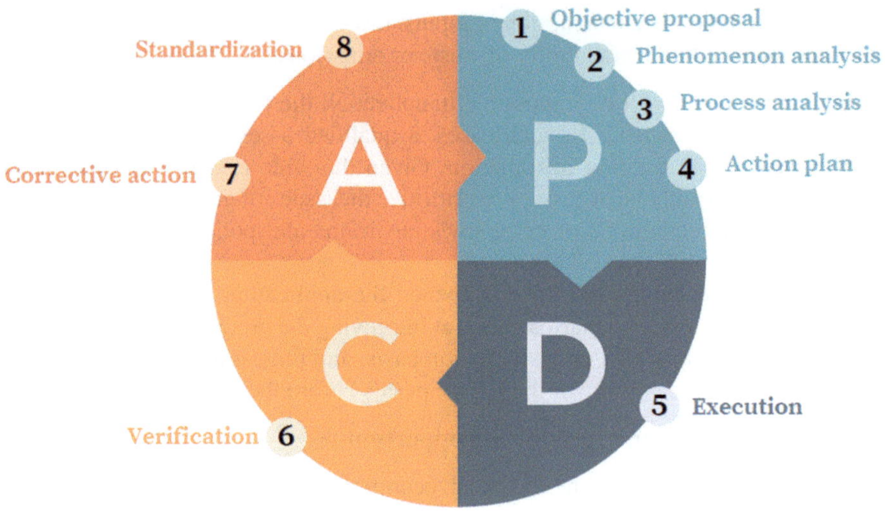

Fig. 2.11 PDCA application to CP. *Source* Authors

activities with the main motivation on preserving the quality of water, air, and soil. This agency is also responsible for promoting studies and publicizing successful cases of adopting measures and CP and of sustainable production and consumption in the State of São Paulo. In this way, we will discuss some successful cases of implementing CP actions in companies of different areas and of different sizes showing that the application of CP practices does not depend on such factors.

Case study 1: Reduced use of sodium cyanide in the degreasing bath.

Characterization of the opportunity: The company operates in the business of manufacturing costume jewelry for personal use; in this case, we will call it Company A. Company A, small, participated in a pollution prevention pilot project promoted by CETESB itself. In this way, the implementation and application of CP practices took place together with the team of the environmental agency, with identification and opportunities to reduce the environmental aspects covered by the company. The original process, carried out by the company, used a degreasing bath based on sodium cyanide presenting a high degree of toxicity which may cause serious environmental problems if released directly into the environment. Thus, a monthly volume of 270 L with a concentration of 75 g/L was sent to the company's effluent treatment system as a preventive action and to guarantee the product quality, it was sent even if it did not know whether it had run out or not its degreasing function, representing a high cost to the company.

Actions adopted by Company A: Evidencing the cost of treatment and the associated environmental impact, the company identified in the CETESB Project the opportunity to improve its processes. The disposal previously done monthly started to be studied for a continuous treatment. The alternative adopted was the inclusion of a filtering process for cleaning the effluent in addition to the use of parameter correction using necessary chemicals. The company started to adopt such procedures in cases where the solution no longer presents satisfactory characteristics of time and results in the degreasing of the treated parts. A reduction in the concentration of cyanide solution was also identified without any compromise in the final result of the process.

There was no investment by the company, only readjustments in its processes.

Results: Company A showed a reduction of 80% in the use of sodium cyanide since the solutions went from 75 to 15 g/l saving about R$ 1620/year in the purchase of cyanide salts. There was a reduction in the use of chemical reagents used by the company in addition to other non-quantified gains related to the reduction of costs in the management of effluents and waste.

As characteristics of continuous improvement, the company has sought constant optimization of its processes aiming at both environmental and economic aspects in its production practices (SOURCE: CETESB 2019).

Case study 2: Successful case of adoption of cleaner production actions in the use of recycled PET bottles in the manufacture of enamels and synthetic varnishes.

Characterization of the opportunity: The large company, here illustratively called Company B, which operates in the field of paints and varnishes manufactures resins which are applied in the formulation of enamels and varnishes in its real estate paint line. The constitution of such resins is alkyd by reaction between a polyacid, a polyalcohol, and a vegetable oil.

Since polyethylene terephthalate (PET) has similar properties (polyacid + polyalcohol), and considering the large supply of PET bottles in the environment where their incorrect disposal has a great environmental impact, the company identified the potential use of bottles for the formulation of resins with consequent cost reduction in the manufacture of enamels and varnishes.

Actions adopted by Company B: The company needed to adapt its processes to meet some operational problems such as difficulty in filtering resins in addition to continuous supply of raw materials. After structuring, the productivity measure is about 6 PET bottles of 2 L for each gallon of varnish/synthetic enamel with 3.6 L.

The investments made by Company B were adaptations of infrastructure and specific training for labor and operation. These are investments common to any type of change in a production line, so there were no significant financial investments in terms of costs for the company.

Results: As environmental contributions, the company started to reiterate in the life cycle of its products PET bottles that are difficult to use and have a great environmental impact. BASF, as the biggest contributor in the recycling process, consumes 3% of the total PET generated in Brazil. There are about 50–60 million bottles that are not inappropriately disposed in landfills or in rivers and sewers. There was also a reduction in the volume of the company's effluents (about 40%), water that was generated in the esterification reaction. The reduction in raw materials from non-renewable sources, such as oil naphtha, was approximately 3000 t/year.

Regarding economic results, the company earned about R$ 3 million/year by reducing production costs by replacing raw materials. The productivity increase was 13% regarding the reduction of filtration time and dispersion of pigments. Improvements in product quality were also identified in terms of resistance to ultraviolet (UV) rays, yellowing, and brightness.

It is also worth mentioning the positive social impact with the generation of approximately 550 new jobs in the PET recycling chain.

The prospects for future implementations by the company are aimed at further reducing water in the reactions as well as promoting its reuse in the process (SOURCE: CETESB 2019).

In these two cases, realize that it was not possible to measure or assess the environmental impacts avoided by CP practices. What was observed was the prevention of environmental aspects which in turn generated adverse environmental impacts. Calculations and the application of techniques are necessary for the identification and quantification of these avoided environmental impacts, as for example, through the use of LCA, as will be worked out in Chaps. 3–6.

Even so, there is a positive impact on the environmental performance of the production processes described in these CETESB success cases.

2.3.1 Proposed Exercises

Question 1. What is the importance of qualitative and quantitative flowcharts for CP practices?

Question 2. What are the steps for implementing CP?

Question 3. Why is it important for a team made up of professionals from different sectors of the company for implementing CP practices?

Question 4. Case study for the application of cleaner production practices.

The present case study was taken from the article "Economic assessment of the CP application in an effluent treatment plant: case study in a mechanical metal company," written by Leite and Neto (2018). After the description of the practical case, some exercises for development and analysis will be presented.

Many studies and analyzes point to the large amount of water used by the metallurgical industries in their production processes, which consequently generates various types of waste that may lead to environmental contamination. The description given below is of a company, located in the city of São Paulo, which we will illustratively call Company C. Company C may be characterized as medium size acting mainly in the secondary sector in the manufacture of clamping elements by the process of cold forming, using materials such as carbon steel and aluminum.

The manufacturing starts with the manufacturing of steel and aluminum pieces and mandrels phase, the pieces proceed sequentially for degreasing, washing, heat treatment, and polishing. Then, both the pieces and the pins go to the assembly in order to make the union using presses, where they are tested by sampling. In subsequent processes, the products are packaged and sent to stock and later dispatch.

Company C makes use of several chemical products in the described process, using, for example, degreasers, rinse aid, and oils, for example. It is evident that the main steps related to the generation of liquid waste are: manufacture of pieces and mandrels, heat treatment, polishing, washing, and degreasing, discarded in public area. The legislation provides that, in case of disposal in public area, the concentrations must meet minimum standards; therefore, the company must implement an effluent treatment station (ETS).

The liquid waste treatment process begins with the targeting of the water used in the production process, which is relocated to the insoluble oil and water separating grating box. The next step, with oil-free water, is to adjust the pH and transparency with air agitation in which chemical reagents and solutions (acid, base, and flocculants) are added. The next step is to pass through the mixed sand and activated carbon filter to eliminate possible solvents. This liquid is then pumped into the settling tank with a conical bottom where the precipitation of metals and suspended solids occurs, directing the sludge to disposal drums and the treated water to the public area, meeting the environmental disposal legislation according to CETESB (2017).

(a) The flowchart allows a graphical representation of all steps of a process and how they are interrelated. Sketch a flowchart of the processes described in the case study.
(b) For implementing CP in your production process, what are the opportunities for implementing CP in Company C? Justify.

Question 5. 50,000 kg/h in weight of a mixture (F) composed of 50% benzene and toluene are processed in a distillation column. The top product of the column, also called distillate (D), is composed of 92% benzene in weight. The bottom product (B) of the column is composed of 8% benzene in weight. Determine:

(a) The mass flow of the top stream (T) and the mass flow of the bottom stream (B) of the column.
(b) The percentage of benzene recovery in the distillate stream.

Question 6. Suppose blackberries contain 15% solids and 85% water in mass. To industrially manufacture this fruit's jam, the following process occurs: (1°) mix the crushed blackberries and sugar in the proportion 40:60% in mass (40 parts of blackberries and 60 parts of sugar); (2°) heat the mixture obtained in the first step to evaporate the water until the mixture reaches a concentration of solids equal to 70% in mass. Considering that sugar consists of 100% solids, we ask:

(a) Draw a process flowchart;
(b) Determine the blackberry mass needed to produce 1 kg of jam.

Note: Assume that evaporation takes place without dragging solids.

Feedback

Question 1. The flowchart allows the visualization of all material and product flows present in the assessed process. Through it, it is also possible to visualize the flow of information and the path taken in the various processes. To prepare this flowchart, existing operational data such as production reports, audit reports, and site plans are used. For implementing CP practices, it becomes essential for the team to visualize the possible points of implementation and optimization of practices.

Question 2. The implementation of CP practices can be divided into five stages, which are: 1. planning and organization; 2. diagnosis; 3. assessment; 4. feasibility; and 5. implementation and monitoring.

Question 3. The variability in relation to the team members is mainly due to their different views given their company's operating area. The aim is at forming a multidisciplinary group composed of representatives of the sectors such as finance, maintenance, engineering, and research and development, which are essential to identify points for system improvement.

Question 4.

(a) The processes described in the case study can be described in the form of the following flowchart:

(b) Reuse of its waste in the production process of Company C with the implementation of a treatment system to improve water quality using, for example, a more efficient filtration system. Other aspects of optimizations would also be the study of the concentrations of the solutions used in each stage of the process. Here, it is worth mentioning that the CP techniques go beyond those ensured by law, that is, the minimum requirements for minimum effluent treatments.

Question 5.

(a) The mass flow of the top stream (T) and the bottom stream (B).

Total mass balance:

$$F = T + B$$

$$50{,}000\,\text{kg/h} = T + B$$

$$B = 50{,}000\frac{kg}{h} - T \tag{2.18}$$

Mass balance for the components:

- Benzene:

$$0.5 \cdot F = 0.92 \cdot T + 0.08 \cdot B$$

$$0.5 \cdot 50{,}000\,\text{kg/h} = 0.92 \cdot T + 0.08 \cdot B \tag{2.19}$$

- Toluene

$$0.5 \cdot F = 0.08 \cdot T + 0.92 \cdot B$$

$$0.5 \cdot 50{,}000\,\text{kg/h} = 0.08 \cdot T + 0.92 \cdot B \tag{2.20}$$

Replacing (2.19) in (2.18), we have:

$$0.5 \cdot 50{,}000\,\text{kg/h} = 0.92 \cdot T + 0.08 \cdot (50{,}000\,\text{kg/h} - T)$$

$$25{,}000\frac{kg}{h} = 0.92T + 4000\frac{kg}{h} - 0.08T$$

$$T = 25{,}000\frac{kg}{h} \tag{2.21}$$

Therefore, replacing (2.21) in (2.18) we find the value of the mass flow of the bottom stream (B):

$$B = 25{,}000\frac{kg}{h} \tag{2.22}$$

(b) The percentage of benzene recovery in the distillate.

The mass balance for benzene is given by:

$$0.5 \cdot F = 0.92 \cdot T + 0.08 \cdot B$$

where a represents the mass of benzene present in the feed, b represents the mass of benzene in the top stream, and c represents the mass of benzene in the bottom stream. The percentage of benzene recovery in the distillate is given by:

$$\%\ \text{benzene recovery in the distillate} = \frac{\text{recovered benzene mass}}{\text{fed benzene mass}}$$

$$\%\ \text{benzene recovery in the distillate} = \frac{0.92 \cdot T}{0.5 \cdot F} = \frac{0.92 \cdot 25{,}000}{0.5 \cdot 50{,}000}$$

$$\%\ \text{benzene recovery in the distillate} = 92\%$$

In this way, 92% of all fed benzene is recovered in the top stream (T). The rest goes out in the bottom stream (B).

Question 6.

(a) Draw the process diagram;

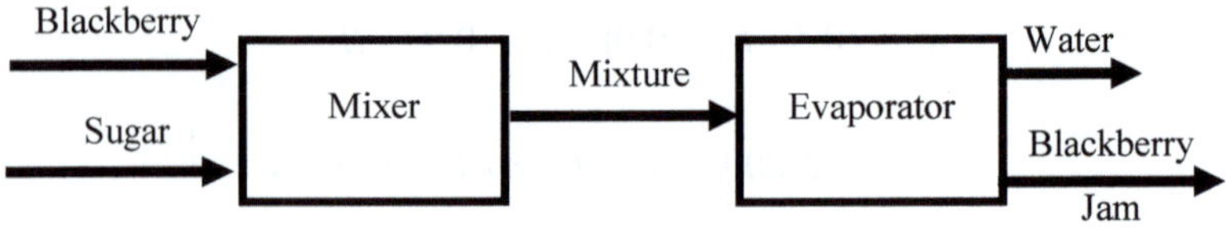

(b) Determine the blackberry mass needed to produce 1 kg of jam.

We will solve this example using the general guidelines for calculating material balances. The first and second steps which consist of drawing the block diagram and numbering or naming each stream were made in letter (a) of this exercise. The third step is to indicate the known and unknown variables in the block diagram (or in a table) followed by the step of writing the values and units of all known variables in your block diagram (or in a table).

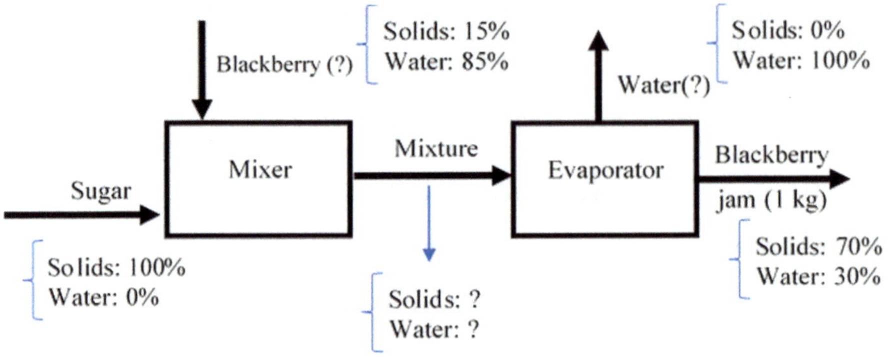

Where all the points that appear in the question mark (?) are the places with unknown amounts. It is important to highlight that the statement "assume that evaporation takes place without dragging solids" means that pure water comes out from the evaporator and all the solid remains in the jam.

The fifth step is to convert the variables to the same appropriate dimension. In this case, all variables are already in the same dimension. As a last step, the process streams total (and for the components) mass balance equations are then written. Considering as control volume (CV) only the mixer, we can find the solids and water composition of the material that leaves the mixer which is the same that enters the evaporator. Remembering that there is no accumulation, and the regime is stationary; therefore, input = output.

$$\text{sugar} + \text{blackberry} = \text{mixture}$$

Thus, the amount of solids in the mixture is given by:

$$\% \text{ solids}_{\text{sugar}} \cdot \text{proportion}_{\text{sugar}} + \%\text{solids}_{\text{blackberry}} \cdot \text{proportion}_{\text{blackberry}} \\ = \%\text{solids}_{\text{mixture}} \cdot \text{proportion}_{mixture}$$

$$100\% \cdot 0.6 + 15\% \cdot 0.4 = 66\% \cdot 1.0$$

As 66% of the mixture formed in the mixer is solids, 34% of the mixture is water. Considering now as CV only the evaporator, we have:

$$\text{mixture} = \text{water} + \text{jam}$$

$$\text{mixture} = \text{water} + 1\,\text{kg}$$

The component balance (solid component) provides:

$$\%\text{solids}_{\text{mixture}} \cdot \text{mass}_{\text{mixture}} = \%\text{solid}_{\text{water}} \cdot \text{mass}_{\text{water}} + \%\text{solid}_{\text{jam}} \cdot \text{mass}_{\text{jam}}$$

$$66\% \cdot \text{mass}_{\text{mixture}} = 0\% \cdot \text{mass}_{\text{water}} + 70\% \cdot \text{mass}_{\text{jam}}$$

$$66\% \cdot \text{mass}_{\text{mixture}} = 0 + 70\% \cdot 1\,\text{kg}$$

$$\text{mass}_{\text{mixture}} = \frac{70\% * 1\ \text{kg}}{66\%} = 1.06\,\text{kg}$$

Thus, we found that to produce 1 kg of jam the mixture (sugar and blackberry) must weigh 1.06 kg.

Knowing that:

$$\text{mixture} = \text{water} + 1\ \text{kg}$$

$$1.06\,\text{kg} = \text{water} + 1\,\text{kg}$$

We have, then, that the total evaporated water corresponds to 0.06 kg. The total global mass balance of the process as a whole is:

$$\text{blackberry} + \text{sugar} = \text{water} +$$

$$\text{blackberry} + \text{sugar} = 0.06\,\text{kg} + 1.0\ \text{kg}$$

$$\text{blackberry} + \text{sugar} = 1.06\,\text{kg}$$

In which, according to the statement, 40% is blackberry and 60% is sugar. Therefore, the amount of blackberry required to produce 1 kg of jam is 0.424 kg (1.06 kg * 0.4).

References

Assaf Neto, A.A., Lima, F.G.: Curso de Administração Financeira, 3ª ed. Atlas (2014)

CETESB—Companhia de Tecnologia de Saneamento Básico. Produção mais Limpa. Casos de Sucesso. Título: Projeto PET—Utilização de Garrafas de PET Recicladas na Fabricação de Esmaltes e Vernizes Sintéticos. Disponível em: https://cetesb.sp.gov.br/consumosustentavel/wpc ontent/uploads/sites/20/2015/01/caso34.pdf. Acesso: 15 Oct 2019

CETESB—Companhia de Tecnologia de Saneamento Básico. Produção mais Limpa. Casos de Sucesso. Título: Redução de químicos e eliminação de descarte de banhos de ativação. Disponível em: https://cetesb.sp.gov.br/consumosustentavel/wp-content/uploads/sites/20/2015/01/caso43.pdf. Acesso: 15 Oct 2019

Leite, R.R., Neto, G.C.d.O.: Avaliação Econômica da Aplicação da Produção mais Limpa em uma Estação de Tratamento de Efluentes: Estudo de caso em uma Empresa Metal Mecânico. Produção Online Revista Científica Eletrônica de Engenharia de Produção, Florianópolis **18**, 1445–1469 (2018)

Nilson, L., Persson, P.O., Rayden, L., Darozhka, S., Zaliausklene, A.: Cleaner Production—Technologies and Tools for Resource Efficient Production. Book 2 in a series on Environmental Management Book Series. The Baltic University Press (2007)

Peters, M., Timmerhaus, K.: Plant Design and Economics for Chemical Engineers, 4th edn. McGraw-Hill, New York (1991)

Senai, R.S.: Cinco fases da implantação de técnicas de produção mais limpa. Porto Alegre, UNIDO, UNEP, Centro Nacional de Tecnologias Limpas SENAI, 103p. il. (Série Manuais de Produção mais Limpa) (2003)

Badino Junior, A.C., Cruz, A.J.G.: Fundamentos de Balanços de Massa e Energia, 2ª ed. Edufscar (2013)

Chapter 3
Life Cycle Assessment (LCA)—Definition of Goals and Scope

Diogo Aparecido Lopes Silva

3.1 Introduction to Life Cycle Assessment (LCA)

Until the mid-1990s, little had been said about life cycle assessment (LCA). However, this has changed thanks to the rise of LCEM theme in recent years as discussed in Chap. 1: Introduction to Life Cycle Engineering and Management (LCEM). Boosted by issues such as the scarcity of resources (material and energy) and the impacts generated due to human activities, not only in the manufacturing process but where these impacts are relevant, currently LCA is considered the main LCEM technique for the quantification of the environmental aspects and impacts of the products' life cycle. More recently, an LCA has also been developed for the inclusion and measurement of social and economic aspects and impacts, thus allowing an integrated assessment with a focus on the sustainability of the products' life cycle.

Specifically about the development of the LCA methodology in the environmental field, its use has evolved over time starting in 1969 when a study called Resource and Environmental Profile Analysis (REPA) was carried out aiming at analyzing and comparing the effects on the environment from the use of different types of beverage package. This work was conducted in the USA for the Coca Cola company quantifying the consumption of resources and emissions in the package production. Later, according to Ferreira (2004), the REPA studies continued to be developed and, in 1974, another study was commissioned by the U.S. Environmental Protection Agency (USEPA) in order to compare different beer packages. This was considered the most ambitious REPA ever made since not only the package manufacturing process was taken into account but also the logistical distribution processes and

D. A. L. Silva (✉)
Department of Production Engineering, Federal University of Sao Carlos (UFSCar), Campus of Sorocaba, Highway João Leme dos Santos—km 116, CEP: 18052-780, Sorocaba, SP, Brazil
e-mail: diogo.apls@ufscar.br

J. A. de Oliveira et al. (eds.), *Life Cycle Engineering and Management of Products*,
https://doi.org/10.1007/978-3-030-78044-9_3

product use in addition to the inclusion of data on the production chain of several materials such as glass, aluminum and plastic.

According to Hunt and Franklin (1996), the acronym LCA, however, was used for the first time only in the early 1990s, in the USA, by the English term Life Cycle Assessment (LCA). According to Guinée et al. (2011), the 1990s was a period of great development for LCA worldwide and this is mainly due to the development of the standards of the *International Organization for Standardization* (ISO) at that time. In 1992, in order to normalize the LCA methodology as an environmental management technique, ISO created the Technical Committee TC207/SC 5 (Tibor and Feldman 1996) and this committee has been divided into five groups to create the following standards: group 1—ISO 14040:1997 (Environmental management—Life cycle assessment—Principles and framework); group 2 and group 3—ISO 14041:1998 (Environmental management—Life cycle assessment—Goal and scope definition and inventory analysis); group 4—ISO 14042:2000 (Environmental management—Life cycle assessment—Life cycle impact assessment); and group 5: ISO 14043:2000 (Environmental management—Life cycle assessment—Life cycle interpretation).

These standards were valid until 2005 and have been later replaced by ISO 14040:2006 (Environmental management—Life cycle assessment—Principles and framework) and ISO 14044:2006 (Environmental management—Life cycle assessment—Requirements and guidelines).

Thus, throughout these 50 years of LCA development, a company today can already use this technique in several applications. Today, it is already possible to calculate through LCA what a new product's environmental impact will be during its use or postuse phase, for example, regarding impacts on climate change or in relation to impacts on human health. Such estimation of life cycle impacts during the design of new products is currently developed in studies on product ecodesign as will be addressed in Chap. 8: Product Ecodesign. Other applications of LCA (see Fig. 3.1)

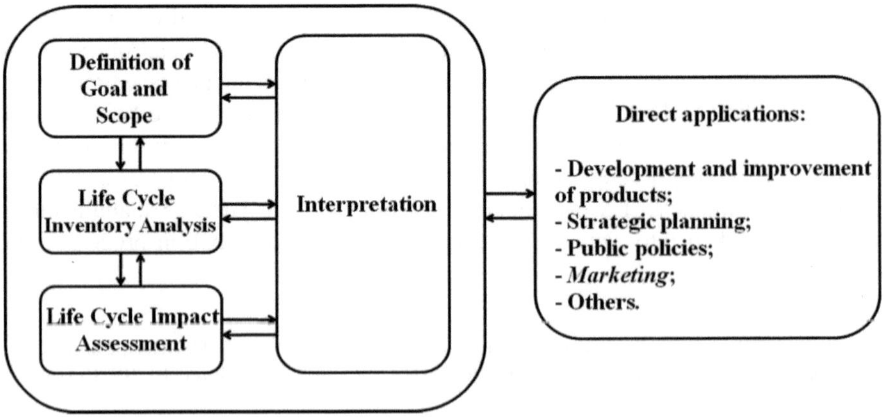

Fig. 3.1 Typical phases for conducting an LCA. *Source* ISO (2006)

also involve the planning of new sustainable business models, for example, aiming at ecoinnovation as will be discussed in Chap. 10: Corporate Sustainability—defining strategies and business models based on the Circular Economy, or the use of LCA in the development of environmental product statements for marketing issues, as detailed in Chap. 13: Environmental Communication and Labeling.

Finally, it is noteworthy that LCA follows the phases in Fig. 3.1 which each will be covered in detail throughout this book. If LCA is perceived to be iterative, that is, as it progresses in its phases, new information and data are obtained and this may lead to changes in the LCA modeling process. After all, as you progress through these phases, more knowledge will be gained about the product under study and, thus, the initial scope definitions usually need to be refined or revised.

According to ISO standards (ISO 2006a, b), LCA is defined as a technique for "*compiling and assessing the inputs, outputs and potential environmental impacts of a product system throughout its life cycle*". Inputs include the measurement of material and energy consumption, and outputs, product and byproduct flows and emissions to air, water and soil, among other flows, which will make up the life cycle inventory (LCI) of the product studied. Having consolidated the inventory, the environmental impacts are then calculated through the product life cycle impact assessment (LCIA), and the environmental hotspots (critical points) are identified aiming at the subsequent interpretation of results according to the LCA goals.

3.2 Definition of Goals and Scope

This chapter will cover the first phase of LCA, the Definition of Goals and Scope. The purpose here is to carry out a planning of the LCA study so that all essential elements are properly established and, thus, to avoid errors and rework in the LCA. Two elements are essential in this stage: (i) definition of goals and (ii) definition of scope.

3.2.1 Typology of LCA

The definition of goals aims at declaring the intended application, the reasons for carrying out the study and the target audience (ISO 2006a). For the Joint Research Center of the European Commission (EC-JRC 2010), in the manual International Reference Life Cycle Database System (ILCD), the definition of goals is essential as it guides the definition of all aspects that will compose the scope of the LCA. Therefore, the ILCD manual recommends to firstly define the descriptive character of the study and suggests three target situations:

- **Situation A**: It aims at providing information for **decision support at the microlevel**, that is, providing information at the process and product level covering applications such as: ecodesign, analysis of a product/process's weaknesses, selection of environmental indicators, development of environmental labeling criteria, product comparisons at the microlevel, etc. In this case, decisions from the LCA must generate no or only small-scale changes in other systems of the economy, which would not change its structure directly or indirectly. See examples of LCA goals of situation A in Chart 3.1.

- **Situation B**: The target situation of type B covers the **decision support at the meso-, macro- and strategic levels**, that is, it aims at selecting strategies with wide-ranging consequences that may affect not only the studied product system but also others external sectors. Situation B involves more global applications such as the development of public policies and, thus, with large-scale impacts on the economy. Chart 3.1 provides examples of LCA goals that fit the situation B archetype.
- **Situation C**: Lastly, situation C is seen as an "accounting" whose aim is at monitoring or quantifying the aspects and impacts of the products' life cycle without having an interest in addressing decision-making at the micro- or macro-level. Situation C involves applications of this type: monitoring environmental impacts of a nation, region, industrial sector, product or group of products, generation of documents on the consumption of resources and on the generation of waste and emissions, among other applications such as presented in Chart 3.1.

In Chart 3.1, it is important to mention that the target situations A and C, in general, when chosen the LCA must be of the attributional type. Target situation B, on the other hand, usually requires LCA to be of the consequential type. On the differences between attributional and consequential LCA, the ILCD manual (EC-JRC 2010) defines:

- **Consequential modeling**: Principle of modeling the product life cycle inventory that identifies and models all processes in the background system of the product system as a result of decisions made in the foreground system. The product system includes a foreground and a background system, being that the modeling uses marginal data for the life cycle inventory based on a generic supply chain in order to analyze the consequences that a decision in the analyzed product system may lead in other processes and systems of the economy. In this modeling, it is analyzed how direct and indirect changes in the supply and demand of the studied product may cause changes in the demand and supply of other products.
- **Attributional modeling**: It is the descriptive accounting of the inflows and outflows of all processes of a product system as they occur in order to assess the environmental performance of the system without analyzing decisions at the meso/macro or strategic level for the economy. The modeled product system is less dynamic than in consequential modeling, being that the modeling makes use of historical, measurable data based on real facts, therefore, the product system is modeled as it is or was (or as expected it will be). It is worth mentioning that, throughout this book, the most adopted approach for the proposed examples and exercises will involve only attributional LCA. Figure 3.2 illustrates the difference between the two (attributional and consequential) approaches in LCA.

Chart 3.1 Examples of goals in LCA studies

Typical goals	Practical examples	Target situation
Identifying opportunities for improving the products' environmental performance	• Replacing material X with an alternative material Y which is more environmentally friendly over the life cycle of a company's product	A
	• Establishing scenarios in the purchase decision and choose among the preselected models the one with the least environmental impact throughout the life cycle	A
Reducing costs by replacing and/or optimizing the use of materials	· Reducing costs in the consumption of material X by optimizing its consumption throughout the life cycle of a company's product	A
	• Reducing costs with transporting materials X and Y by changing modes and choosing alternative routes	A
Assisting the development of ecodesign	• Developing a new product by replacing material A with material B that is more environmentally friendly compared to its competitors in the market	A
	• Developing material A to replace material B and, thus, reducing environmental impacts during the use and postuse phases of the final product	A
Providing detailed level of information to decision makers in industry, governmental and non-governmental organizations, with a view to strategic planning, setting goals and priorities and adapting to environmental legislation	• Creating public policy that prohibits or encourages the disuse of materials that are environmentally harmful to the environment and human health	B
	• Creating tax incentive mechanisms for companies and final consumers to promote sustainable consumption of products	B
	• Identifying groups of products with the greatest potential for environmental improvement and/or for the prohibition of their production and consumption	B
Selecting relevant environmental performance indicators	• Monitoring impacts instead of environmental aspects. Example: monitoring the eutrophication potential of fresh water instead of just monitoring water consumption in the production process	C

(continued)

Chart 3.1 (continued)

Typical goals	Practical examples	Target situation
	• Monitoring the potential for ecotoxicity and human toxicity in the product's life cycle instead of just monitoring the generation of toxic waste in the process	C
	• Generating corporate or local environmental reports	C
Obtaining an environmental label or elaborate an environmental product declaration	• Creating a label that informs the consumer about the carbon footprint of the product purchased and consumed	A

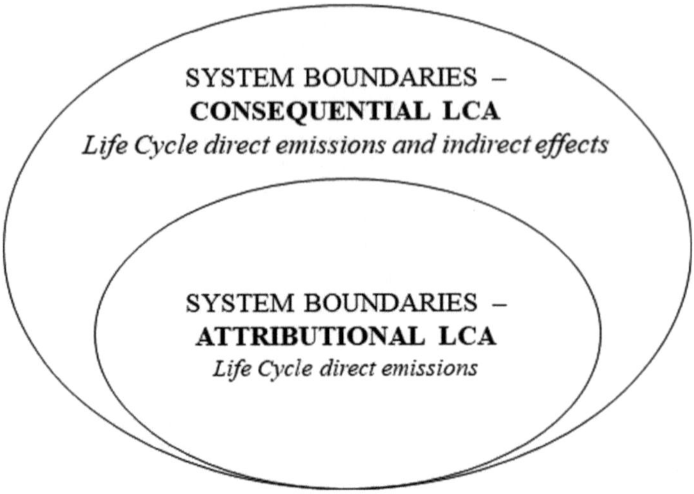

Fig. 3.2 Product systems boundaries under the attributional and consequential approach. *Source* Adapted from Brander et al (2009)

The big difference between the two approaches is in the object of the assessment. While attributional LCA applies to the product system, consequential LCA is focused on the changes that the product system causes. Brander et al. (2009) highlight that the consequential approach focuses on changes both inside and outside the life cycle of the product system under assessment. In a simplified way, it can be understood that the first assessed the system in a static way like a portrait; the second considers the product system as something dynamic influenced by factors and influencing factors such as the market (an important agent of changes in product life cycles). Due to these conditions, there are limitations in both approaches: For attributional LCA, product systems are subjective due to factors such as multifunctionality and geographical boundaries, whereas for the consequential there is much uncertainty and instability related to changes.

Based on the goal outlined for the LCA, the definition of scope will involve the detailing of several items, from the definition of the product system, the geopolitical, temporal and technological contexts, the quality of the data that will integrate the product life cycle inventory, the definition of the methods and categories of impact for the environmental impact assessment of the product, to the definition of criteria for the interpretation of results. Thus, these mandatory elements of the scope definition according to ISO (2006a, b) are:

- The product system to be studied;
- The functions of the product system, or, for comparative studies, of the systems;
- Functional unit;
- System boundary;
- Allocation procedures;
- LCIA methodology and types of impacts;
- Interpretation to be used;
- Data requirements;
- Assumptions;
- Choice of values and optional elements;
- Limitations;
- Data quality requirements;
- Type of critical review, if applicable; and
- Type and format of the report required for the study.

Although the normative guidelines of ISO 14040 and ISO 14044 provide the main requirements for scoping, they do not provide technically detailed recommendations in this regard. In EC-JRC (2010), this detail is provided through the publication of a technical manual on the use of LCA in practice—the ILCD manual published by EC-JRC (2010). In addition, this book provides examples for the proper formatting of the LCA definition of goals and scope with exercises and practical examples.

3.2.2 Definition of Function, Functional Unit and Reference Flow

In order to define the function of the product system, ISO (2006) mentions that the purpose for which the product is intended must be taken into account by surveying its performance characteristics during the use phase.

Thus, the functional unit must be defined based on the function that the product proposes to fulfill. It is established to provide a baseline in relation to the one on which the inventoried input and output data for each elementary process will be converted in a mathematical sense, and it is essential that it is measurable. In other

Chart 3.2 Typical examples of function, functional unit and reference flow in LCA studies

Product	Function	Functional unit	Reference flow	Authors
Alcohol fuel	Serving as fuel in transport vehicles	10,000 km course	1 ton of alcohol fuel	Ometto et al. (2009)
Nitrogen fertilizer—urea	Providing an amount of nitrogen for soil fertilization	1 ton of nitrogen nutrient	2.17 tons of urea	Ribeiro (2009)
Ceramic floor tiles	Provide the correct floor covering for a building	1 m^2 coverage with tiles over a period of 40 years	Tiles weighing 18 kg/m^2	Nicoletti et al. (2002)
Particleboard	Intermediate product, base for the production of wooden furniture	1 m^3 of panel	1 m^3 of panel	Silva et al. (2013)

words, **the functional unit quantifies and qualifies the function defined for the product system**, making it necessary to know the product's technical aspects.

It must also be established a reference flow that measures the amount of product needed to fulfill the function expressed by the functional unit. Chart 3.2 shows different examples of function, functional unit and reference flows in LCA.

Comparisons between products in LCA can only be made for product systems defined based on the same initial conditions, that is, the same function, quantified by the same functional unit. However, different reference flows can be adopted in comparative LCA studies. For example, Ribeiro (2009) assessed the production of nitrogen fertilizers in Brazil and, for the functional unit of 1 ton of nitrogen, the urea fertilizer reference flow was 2.17 tons, while the fertilizer reference flow ammonium sulfate was 4.72 tons.

In Chart 3.2, the last case presented is an LCA for an intermediate product and not a final product. The functional unit was equal to the reference flow itself. This is due to the fact that the wood panel's use stage was not included in the product system assessed by Silva et al. (2013); therefore, the product's function could not be analyzed. **The inclusion or not of the product's use stage within the product system generates changes in the functional unit definition**.

3.2.3 Defining Product System Boundaries

A product system is a set of elementary processes with all their input and output flows, which performs one or more defined functions and which shapes a product's life cycle (ISO 2006b). But, what is an elementary process? An elementary process is understood as the smallest portion of the product system (ISO 2006b), which facilitates the visualization and management of all flows.

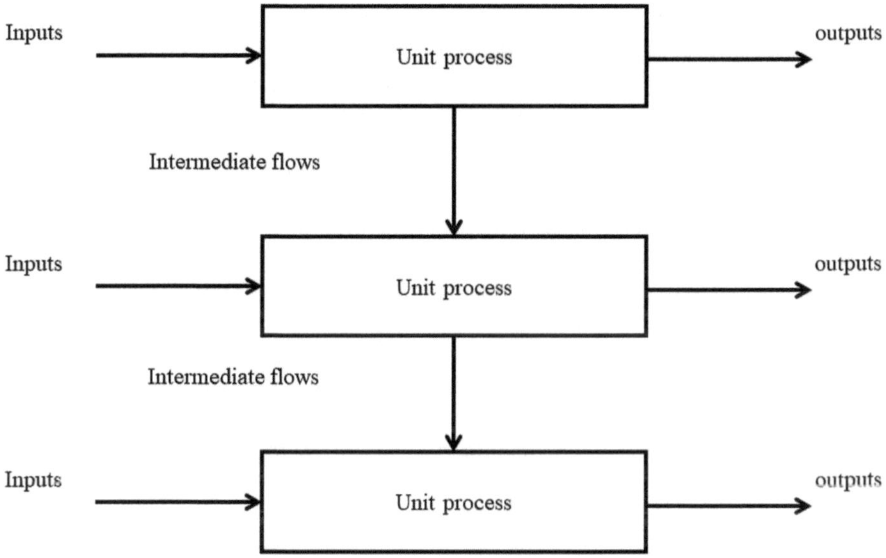

Fig. 3.3 Set of elementary processes that make up a product system. *Source* ISO (2006b)

Input, output and intermediate flows in Fig. 3.3 can comprise elementary flows, raw materials, energy, auxiliary inputs, products, intermediate products, energy losses and emissions to air, water and soil.

For ISO (2006b), an intermediate flow may be a byproduct or an intermediate product. An intermediate product can be understood as an output of an elementary process that constitutes an input to another elementary process and that requires further transformation within the product system, for example, the byproducts. Among the inflows and outflows are the elementary flows, which are materials or energy that enter or leave the product system without undergoing the previous transformation by human interference, such as the extraction of crude oil, solar radiation, emissions to the air, water and soil. Auxiliary inputs are the materials used in elementary processes, but which are not part of the product, for example, package of inputs and catalysts.

The construction of a product system in LCA must be based on a process flow diagram (or process flow chart). For that, it is first necessary to carry out a process mapping for the product under study. So, to map a process is necessary to design it in a logical way so that other people can observe and understand the process as well. The flowchart obtained will be the product system's backbone, in general, designated in the literature by the term *foreground system*. The foreground system will be those elementary processes that will present inventory data (inputs and outputs) fundamentally of primary origin, that is, obtained *in loco*. An example of a product system will be given in Fig. 4.1 of Chap. 4: LCA—Life Cycle Inventory Analysis and Databases.

For ISO (2006a), the boundary established must be consistent with the study's goal and the criteria adopted in its determination must be identified and explained. The exclusion of life cycle stages, processes, input and output flows must be clearly recorded and the reasons and implications of such limitations justified as will be highlighted in Sect. 3.2.6.

A product system may be of the type:

- ***Cradle-to-grave***: This approach indicates that the study will be complete in terms of its product's life cycle stages as the system will take into account elementary processes that will represent all the steps shown in Fig. 3.4. There is also the *cradle-to-cradle* (or *closed loop production*) variant, which assumes product recycling, reuse or remanufacturing as end-of-life strategies rather than final disposal;
- ***Cradle-to-gate***: It is equivalent to the term "from the cradle to the factory gate" as it considers elementary processes from the stages of extraction and processing of natural resources to the manufacture of intermediate products and/or the final product. The inclusion of the background system occurs only for the steps upstream of the modeling, excluding the steps downstream of the factory gate;
- ***Gate-to-gate***: It means "from the gate to the factory gate" as it considers only the stage of the product's manufacture in the LCA study;
- ***Gate-to-grave***: It takes the "from the factory gate to the grave" approach as it only assumes elementary processes that will represent the stages of distribution, use and the product's end of life (disposal in landfill, incineration, recycling, remanufacturing, etc.). In this case, the inclusion of the background system occurs only for the steps downstream of the factory gate, as schematically illustrated in Fig. 3.4.

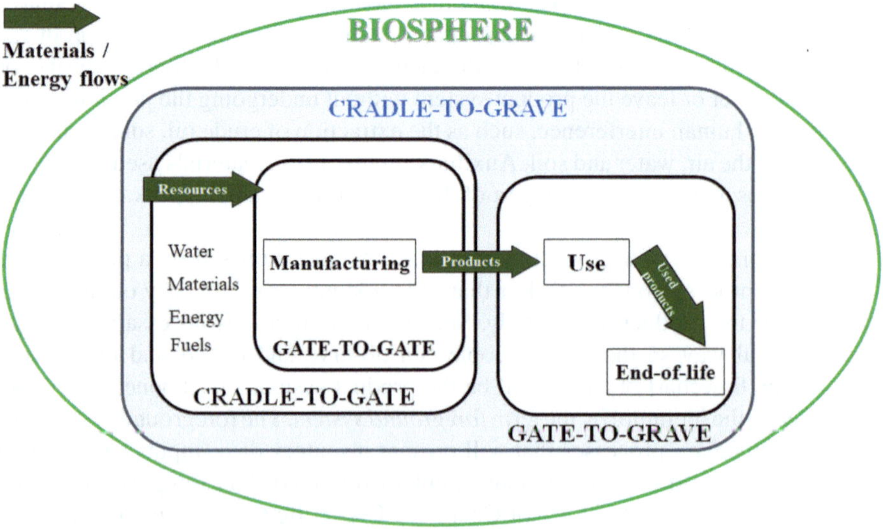

Fig. 3.4 Types of product systems in LCA. *Source* Author

Thorn et al. (2011) explain that this diversity at the product system's boundaries is due to the complexity and the large expenditure of resources (financial, time, labor) to carry out a complete LCA study of the *cradle-to-grave* type. In addition, they add that when studying intermediate products, in general, the studies tend to be of the *cradle-to-gate* or *gate-to-gate* type. Studies that are more concerned with product end-of-life strategies can focus only on the *gate-to-grave* perspective in Fig. 3.4. For example, a *gate-to-grave* LCA study could be the assessment of end-of-life strategies for particleboard waste, including from landfill disposal to material recycling for the manufacture of new wood panels.

For EC-JRC (2010), **the boundaries of a product system must be defined taking into account aspects such as the geographical, technological, temporal and the use of cut-off criteria**. Thus, in order to properly define the boundary of a product system, several dimensions must be taken into account. Curran (2017) praises the following dimensions:

- **Boundaries in relation to the natural system**: These are the limits that indicate the beginning and end of a product's life cycle, that is, it establishes whether the study is on an approach *cradle-to-grave*, *cradle-to-gate*, *gate-to-gate*, etc.;
- **Boundary in relation to other systems**: A product system is associated with a network of other systems, from the extraction of raw materials, energy matrix, generated byproducts, recycling processes, etc. For this reason, it must be delimited within this network which systems must or must not be added to the product system. In short, the product system's secondary systems must be established;
- **Boundary in relation to capital and personal goods**: This boundary is applied to determine the inclusion or not of the infrastructure related to the system under study, such as the buildings of the industrial units, roads, industrial machinery, etc. Personal goods include food, employee transportation, administrative consumer goods (paper, equipment, packages, etc.);
- **Geographical boundary**: The importance of geography in life cycle studies is relevant as physical realities such as climate, relief, etc. can affect an LCA results. Thus, it is important to register the geographical scope of the study, emphasizing whether it is local, regional, national, continental or worldwide. For example, in Silva's (2012) LCA study of particleboards, the geographical boundary was based on technical visits to three large companies in the sector located in the states of São Paulo and Minas Gerais in Brazil, corresponding to 57% of the product's national production at the time;
- **Temporal boundary**: It is also necessary to fit a LCA study within a time scope, as industrial practices, consumers' profile and aspects of regulations and legislation vary over time. The product's lifetime must be declared as well as the expiration date for the data that make up the inventories of each elementary process of the product system. For that, a deadline period must be determined for the results, conclusions and recommendations of the LCA and, for that, aspects related to the technological, economic development of new legislation are taken. For the case study involving the particleboards in Brazil, Silva (2012) highlights an deadline of 8 years for the LCA results in view of the technological changes predicted at

the time and which would affect the boundaries of the studied product system and, so, would also significantly alter the deadline of the conducted LCA results;
- **Technological boundary**: It is about characterizing the product system within the technological context in which it is inserted. It is necessary, here, to detail each elementary process modeled in the product system from the foreground to the background processes. For the foreground processes, a greater level of detail can be provided, since, in general, these are processes mapped with information collected *in loco* in companies. In turn, in technological boundary related to the processes of the background system, the bibliographic sources consulted must be highlighted, since, in general, the background processes are modeled from data collected in the literature and in LCA databases. About the LCA databases, in Chap. 4: LCA—Life Cycle Inventory Analysis and databases, several examples of the main existing databases will be presented.

Finally, the cut-off criterion can be used to assist in defining the system boundaries. ISO (2006a) mentions that the function of the cut-off criterion is to assist in the decision on which inputs will be included in the modeling of the product system, that is, the cut-off criterion helps in defining the boundary in relation to the inclusion of background systems. There are several types of criteria, the main ones being: the mass, energy and environmental relevance criteria. In order to apply the mass (or energy) criterion, in general, the Pareto Chart can be used as shown in Chart 3.3.

Chart 3.3 shows examples of elementary processes that integrate the foreground system with their respective inflows and quantities consumed per functional unit. Here, the data were obtained and adapted from Silva (2012) for the production of particleboards using 1 m^3 of panels as a functional unit.

The entries highlighted in bold in Chart 3.3 account for 97.29% of the total inflows. Therefore, for these highlighted flows it is necessary to include the production chain of these materials and energy for consolidating the product system boundaries, that is, for these flows it is necessary to include the background system. In this example, inflows with a relative share greater than 1% of the total inputs consumed were used as a reference.

Regarding the criterion of environmental significance, for its adoption in LCA studies it is necessary to know previously typical environmental problems associated with the use of certain input resources in each elementary process (e.g., toxic materials or energy or with a scarcity risk in the long-medium term), which when consumed by a productive system can cause serious damage to the environment regardless of the quantity consumed and its representativeness. In the case of Chart 3.3, the herbicidal inputs, diesel and urea and ammonium sulfate fertilizers presented low mass representativeness; however, it could be considered that they have high environmental significance and, therefore, be included in the product system boundaries.

Finally, it is important to know that because the product system is a model used to represent a product's life cycle, such modeling will be as representative the better the level of detail added to the system. On the other hand, as more details are put into the life cycle modeling, as the product system becomes more extensive, the level of

Chart 3.3 Example of definition of cut-off criteria based on the mass of the total resources that enter a product system—the case of particleboards

Elementary processes	Inflows	Inflows per functional unit (kg/m^3)	Percentage (%)	Cumulative percentage (%)
Seedling production	Diesel	6.69E−02 kg	0.06	–
	Herbicides	1.42E−04 kg	0.00	–
Soil preparation	Diesel	6.00E−01 kg	0.53	–
	Herbicides	2.17E−02 kg	0.02	–
	Limestone	**1.20E+01 kg**	**10.69**	**10.69**
	Potassium chloride	2.50E−01 kg	0.22	–
	Urea	1.00E−01 kg	0.09	–
	Single superphosphate	**2.75E+00 kg**	**2.45**	**13.14**
Seedling planting	Diesel	13.20E−02 kg	0.12	–
	Lubricants	1.00E−04 kg	0.00	–
Forest maintenance	Diesel	10.39E−01 kg	0.93	–
	Herbicides	1.33E−01 kg	0.12	–
	Limestone	**8.00E+00 kg**	**7.12**	**20.26**
	Potassium chloride	**2.50E+00 kg**	**2.23**	**22.49**
	Ammonium sulfate	4.00E−01 kg	0.36	–
	Single superphosphate	**3.02E+00 kg**	**2.69**	**25.18**
Harvest and transportation of wood	**Diesel**	**2.47E+00 kg**	**2.20**	**27.38**
	Lubricants	3.75E−04 kg	0.00	–
Industrial production	Diesel	2.85E−01 kg	0.25	–
	Urea formaldehyde resin	**71.65 kg**	**63.81**	**91.19**
	Ammonium sulfate	**1.38E+00 kg**	**1.23**	**92.42**
	Paraffin	**5.47E+00 kg**	**4.87**	**97.29**
	Lubricants	1.80E−02 kg	0.02	–
TOTAL		**1.12E+02 kg**	**100**	**97.29**

Source Adapted from Silva (2012)

complexity of the jobs also increases as does the time to complete the jobs and the need for resources to complete the system.

3.2.4 Multifunctionality and Allocation Procedures

According to ISO (2006b), the multifunctionality of product systems must be treated in order to avoid the use of allocation procedures. But, what does it mean in practice? Well, first, before understanding what allocation is it is necessary to understand what multifunctionality is based on the scheme in Fig. 3.5, extracted from EC-JRC (2010).

Figure 3.5 shows an elementary multifunctional process, that is, a process that provides more than one function—that is, it provides several products. When there is a multifunctional process, it is said that it has several byproducts such as Product A and Product B. EC-JRC (2010) cites several examples of multifunctional processes, such as:

- Incineration of waste which results in the production of electricity (Product A) and steam (Product B) as byproducts;
- Electrolysis of a sodium chloride solution which provides several byproducts such as the solution of sodium hydroxide, chlorine gas and hydrogen gas;
- Production of multifunctional products such as a smartphone—which can be used to make phone calls, listen to music, watch videos, etc. In this case, the difference between this and the two previous examples is that multifunctionality is not in the process, but in the product.

However, in general, LCA studies are only interested in one of the byproducts and not both (e.g., only the production of Product A or only the production of Product B). Thus, to deal with multifunctionality, ISO (2006b) suggests three steps:

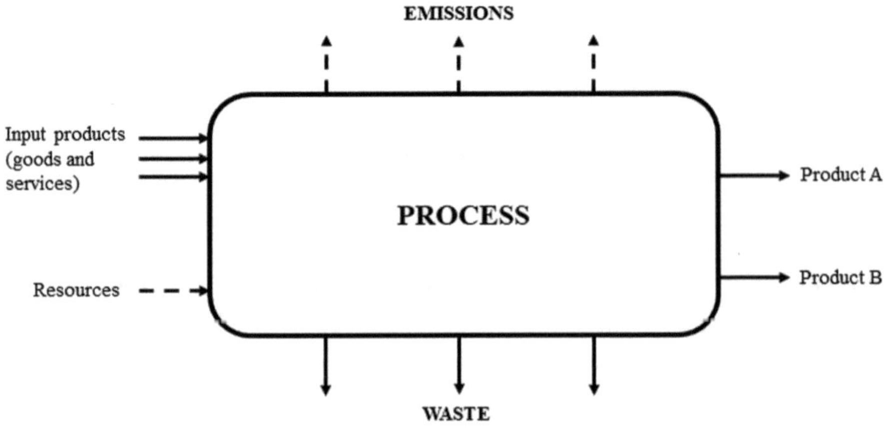

Fig. 3.5 Example of a multifunctional elementary process. *Source* EC-JRC (2010)

- **Step 1:** Allocation should be avoided whenever possible through:
  - Subdivision of multifunctional processes; or
  - System expansion (including replacement of multifunctions).
- **Step 2**: When allocation cannot be avoided, it is advisable that the inputs and outputs of the system are subdivided between their different products in order to reflect the underlying physical relationships between them (allocation based on mass/energy);
- **Step 3**: When a physical relationship alone cannot be established or used as a basis for allocation, it is advisable that the inputs are allocated between products in a way that reflects other relationships (economic value/market value).

Regarding step 1, the subdivision must be the first approach adopted to try to treat multifunctionality in the LCA study.

Subdivision means dividing the elementary processes to be allocated into two or more subprocesses (subdivision) and collecting inputs and outputs related to these subprocesses separately, according to the scheme in Fig. 3.6. In short, a subdivision aims at converting multifunctional processes into monofunctional processes.

In the upper part of Fig. 3.6, it can be seen that the process was initially of the black-box type, producing several byproducts. Thus, through subdivision it is possible

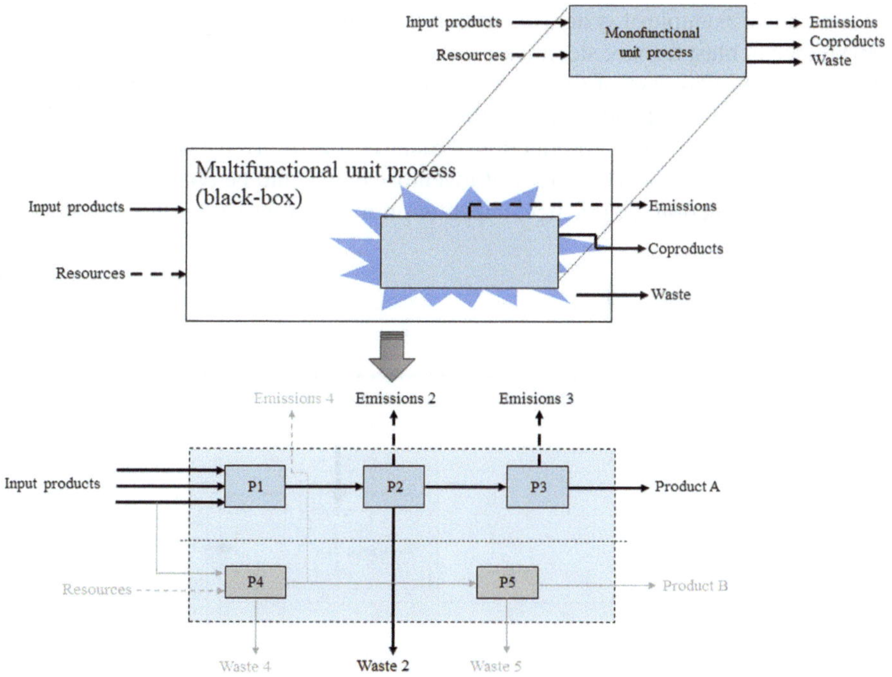

Fig. 3.6 Subdivision approach for the treatment of multifunctional processes. *Source* Adapted from EC-JRC (2010)

to separate the consumption of resources, products, and the generation of waste and emissions by each process now monofunctional and no longer multifunctional. Thus, at the bottom of Fig. 3.6, there are the monofunctional processes P1 to P5 with processes P1 to P3 referring to the production of Product A, while P4 and P5 perform the production of Product B. With the subdivision, the effectively necessary processes are released and the multifunctionality problem is solved since the subdivision allows the exclusive production of the process chain of the elementary processes "P1" to "P3", which result in the analyzed Product A (EC-JRC 2010).

However, the subdivision cannot be used when the elementary processes are of a single operation, that is, when they are not black-box. Finally, it is worth mentioning that the subdivision is seen as the best way to avoid allocation in attributional LCA for EC-JRC (2010) and Curran (2017).

Regarding system expansion, the second option to avoid the need for allocation, it means expanding the product system to include additional functions related to byproducts (alternative products). The expansion is done including the replacement of the multifunctional process's multifunctions, as illustrated by Fig. 3.7.

The system expansion includes another function not provided by the multifunctional process (Process for A and B), thus, expanding the system boundaries by the addition of the alternative *Process for* B. Then, the function that is not needed is subtracted from the expanded system by an alternative way of providing it, that is, the replacement is done by system expansion as schematically shown in Fig. 3.8.

A practical example of system expansion replacement using the solution in Fig. 3.8 can be seen in blast furnace steelmaking. According to EC-JRC (2010), blast furnace slag is a joint byproduct of steelmaking and is normally used in the production of cement (replacing Portland cement) or in paving roads. Thus, to treat the multifunctionality of steel and slag co-production, the slag co-function must be eliminated from the life cycle inventory. Thus, a set of inventory data could be obtained exclusively from steel production. In practice, substitution means subtracting the inventory of

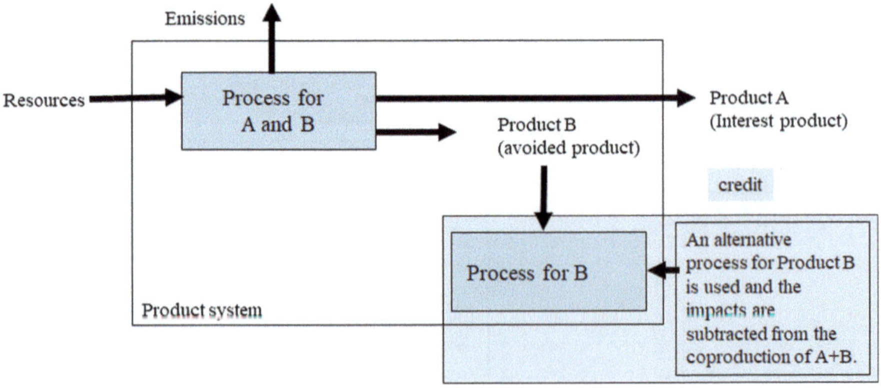

Fig. 3.7 System expansion approach including replacement for the treatment of multifunctional processes *Source* Adapted from EC-JRC (2010)

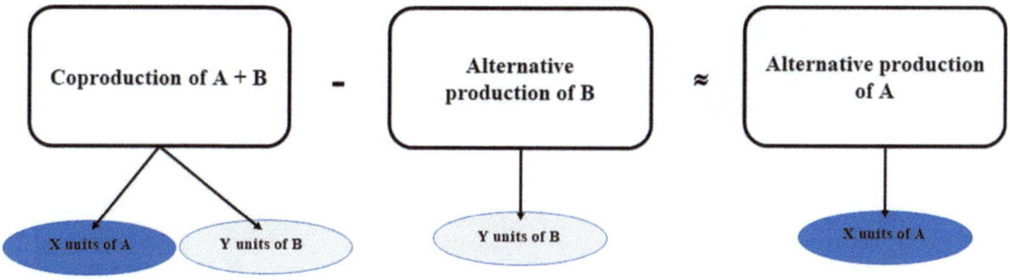

Fig. 3.8 Solution for the multifunctionality problem via replacement by system expansion. *Source* Adapted from EC-JRC (2010)

another system (production of Product B) from the analyzed system (co-production of A + B), and this procedure generally results in negative flows in inventory if the quantity of *Y units of B* is greater than *X units of A*.

However, it is not always possible to apply substitution by system expansion. For example, there may be no other existing form/technology for the alternative production of the unwanted co-function. Thus, as a last resort for the treatment of multifunctionality issues is the allocation.

Allocation means the partitioning of the inventory flows to solve the multifunctionality, allocating the amounts of individual inputs and outputs among the co-functions based on some criterion, preferably based on common physical causal relations between the co-functions, such as:

- Allocation by **mass**, where impacts during the LCIA will be calculated according to mass participation;
- Allocation by **calorific value**, where the impacts during the LCIA will be calculated according to the energy participation;

In the absence or impracticability of adopting criteria based on physical relations, other types of relation between co-functions can be adopted:

- Allocation by market value (**economic**), where the impacts during the LCIA will be calculated according to economic participation;
- Allocation by other rules, such as exergy, emergy, chemical composition, etc.

If we go back to the example of the production of wood panels and consider, for example, that in addition to the production of particleboards there is the production of wood residues for co-generation (electricity production), the procedure for allocating the multifunctional process of producing the panels could be illustrated by Fig. 3.9.

The allocation procedure in Fig. 3.9 shows the inventory values of the input flows for the multifunctional process considering the mass allocation and economic allocation. In this case, the economic allocation proved to be more relevant since a greater part of the inventory flows would be allocated to the product of interest. In the case of the choice for allocation, it will be necessary to account for the inventory flows of the multifunctional process(es) of interest; therefore, mandatory quantitative

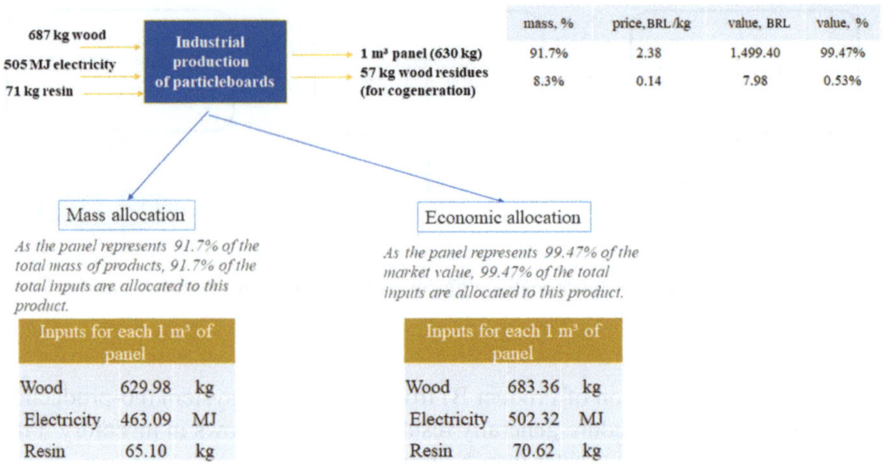

Fig. 3.9 Solution for the multifunctionality problem via allocation for the case of the production of particleboards and the production of biomass for co-generation. *Source* Author

data will be required that will be obtained in the second phase of the LCA, the LCI phase.

3.2.5 LCIA Methodology and Types of Impacts

Here, the Life Cycle Impact Assessment (LCIA) method(s) chosen for the LCA study must be addressed and the inclusion or not of the optional elements of an LCA must be defined, such as normalization, grouping and weighting.

Details on the entire LCIA process will be given in Chap. 5: LCA—Product Life Cycle Impact Assessment. However, in order to define the scope of LCA, at least the choice of impact categories and the LCIA method(s) must be justified and this decision must be based on the environmental profile of the defined product system. Environmental issues selected for the definition of cut-off criterion for environmental relevance of the system boundaries (see Sect. 3.2.3) may also serve here as a basis for choosing the categories of impact. In other words, when defining the goals and scope, it is desirable that the main environmental problems associated with the studied product system are known in order to facilitate the process of choosing the most appropriate LCIA categories and method(s). For the identification of relevant environmental problems, environmental laws and policies, as well as the review of the literature relevant to the topic, they serve as sources indicated to start collecting this information.

3.2.6 *Data Requirements, Assumptions and Limitations of LCA*

For reliable results of an LCA, it is essential to obtain high-quality data. For this purpose, the inventoried data related to the inputs and outputs for the defined product system must be collected preferably from primary sources. And information from secondary sources can also be used when the respective primary sources are not available.

Primary data, in general, are obtained from technical visits to companies. Common data collection procedures involve the following activities, which will be detailed later in Chap. 4: LCA—Life Cycle Inventory Analysis and Databases:

- Direct sampling of each analyzed process;
- Consultation of records about resource consumption and waste and emissions generation;
- Consultation of technical specifications of inputs; and
- Direct calculation through data conversions.

For secondary data, these are extracted according to the criteria:

- Reviewing of relevant literature; and
- Use of LCA database.

The establishment of quality requirements serves to specify the characteristics of the data necessary for the study of LCA and must cover according to ISO (2006a):

- **Temporal coverage**: age of the data and minimum period for the collection to occur. For example, for the case of the particleboards' production, Silva (2012) mentions that the collection period was 01 year, between 2011/2012, and that the information referring to secondary data was taken from the literature and databases of LCA dating between 2002 and 2009. The lifetime of the panel was estimated at 7 years: 3 months for seedling production; 6.5 years for wood cultivation and harvesting; and 3 months counting from the harvest in the field, the storage/drying of the logs in the field and the product manufacture in the industry until dispatch. The product's age is an important aspect, after all, the potential impacts calculated by the LCA are directly associated with the product's lifetime within the production system in question.
- **Geographical coverage**: area from which data must be collected in order to fulfill the goal of the study. It must be consistent with the defined geographical boundary;
- **Technological coverage**: specific technology or set of technologies. For example, if the study is based on the technological mix used for the studied life cycle, or if it is considered the best available technology, or if conventional technology, or if mixed technology, laboratory experimentation, use of process simulation, etc.;
- **Others**: Depending on the level of detail required for the study, items such as accuracy, completeness, representativeness of data, consistency, reproducibility,

data sources and data uncertainties may also be considered. Such requirements are described in Chap. 6: LCA—Interpretation of results.

Collecting high-quality inventory data is essential to support a proper interpretation of LCA results. For example, in LCA studies, the uncertainty analysis can be an instrument to be used in the LCA interpretation stage to analyze the quality of the inventory data, both foreground and background data, as examples which will be demonstrated by Chap. 6: LCA—Interpretation of results. Thus, it is important that all the foreground and background data used in an LCA study are methodologically consistent.

Other important elements on defining scope are the assumptions and limitations of the LCA study, as this directly affects the quality of the inventory data.

LCA assumptions involve all the assumptions/considerations taken for the LCA study. According to EC-JRC (2010), the assumptions are made based on the limitations of the study, such as the absence of databases, establishment of scenarios, among others such as the examples in Chart 3.4. However, it is important that each assumption is based on some appropriate technical-scientific justification.

Chart 3.4 Typical examples of assumptions and limitations in LCA studies

Examples of assumption	Examples of limitation
• Use of conversion factors or scale factors to consolidate inventory flows in face of the functional unit. For example, *use of indicators such as productivity, cut-off criteria, chemical composition of materials, rate of material consumption, rate of waste generation, among others*	• Information about the absence of inventory data. For example, *the atmospheric emissions of volatile organic compounds from process X were not considered due to the lack of data for the studied process*
• Definition of transport distances and type of modes for logistical operations. For example, *it was considered for the transportation of raw material X the use of a truck with 7.5 tons of payload for transporting for 100 km to the destination*	• Use of foreign data to consolidate the inventory. For example, *for process X, background data was extracted from the ecoinvent LCA database, however, to the European context due to the absence of national datasets*
• Considerations about the sources of inventory data. For example, *the electricity supply data were taken from the ecoinvent LCA database and taking Brazilian conditions into account*	• Information about the exclusion of stages of the life cycle and/or specific elementary processes. For example, *the product use stage was disregarded from the study due to the wide range of product applications, which made it difficult to model a specific use for the product in the study*
• Considerations about the methods and categories of the LCIA method. For example, *for the category of climate change impact, the midpoint-type LCIA ReCiPe (2008) method was used and for 100 years of damage*	• Information about simplifications in process modeling. For example, *the industrial production was modeled as a black-box due to the difficulty in stratifying the production process into individual elementary processes*

The limitations of LCA, on the other hand, are the stage where all the limitations/restrictions of the LCA study are clarified based, for example, on the lack of data, use of foreign impact methods not consistent with the local geopolitical context, restrictions due to the establishment of the product system's boundaries among other typical examples presented in Chart 3.4.

3.2.7 Interpretation to Be Used

The last phase of LCA is interpretation, where the results obtained during the LCI and LCIA are identified and analyzed according to the goals and scope outlined (EC-JRC 2010). The conclusions, recommendations and limitations of the study are taken and priorities must be developed and assessments made to identify opportunities to reduce the environmental burdens (Guinée et al., 2011; Thorn et al. 2011).

The elements that integrate the interpretation of results of an LCA study must be clearly defined even in the scope definition phase. In Chap. 6: LCA—Interpretation of results, further details on the subject will be provided. For now, it is important to just know and choose which of the strategies listed below will be included in the LCA study for the interpretation phase:

- **Identification of significant issues**: Based on the results of the LCI and LCIA, hotspots (critical points) must be listed and highlighted. The hotspots involve the identification of the life cycle stages and the respective elementary processes, environmental aspects and impact categories that present the greatest potentials in the studied product's life cycle. For example, in an LCA study it was found that the process of transporting raw material (life cycle stage) was a hotspot due to the use of diesel (environmental aspect) in the defined route, with an emphasis on the impacts on the potential categories of acidification and for climate change (environmental impacts). In general, the analysis of critical points must be taken by category of environmental impact studied and may also include a grouping of results based on the assessment: normalized and/or weighted impacts (if applicable).
- **Completeness check**: It aims at ensuring that all information relevant to the interpretation of LCA results is complete and available. Any missing information must be reported in the LCA report and the need for such information must be considered to satisfy the LCA goals and scope (ISO 2006b).
- **Sensitivity check**: Also known as sensitivity analysis, here we aim at determining the influence of variations in assumptions, LCI methods and LCI data on LCA results. Several modifications can be made to the product system initially defined by creating **alternative scenarios** based on modifications, for example, in rules for the solution of multifunctionality, cut-off criteria, establishment of the product system boundary (inclusion/exclusion of processes), changes in the impact categories (inclusion of more categories), exchange of the LCIA method for an alternative method, etc.

- **Consistency check**: According to ISO (2006b), it aims at "*determining whether the assumptions, methods, models and data are consistent throughout the product life cycle or between different options*". The consistency analysis must look for whether there are significant differences in data sources, data accuracy, age of data, technological scope, temporal scope and geographical scope of the LCA study. In short, the consistency analysis aims at reviewing whether the quality of LCA data is consistent with the goal and scope of the study.
- **Other checks**: ISO (2006b) mentions that the analysis of uncertainties and the analysis of data quality must supplement these three assessment checks in the LCA. Chapter 6 will give examples of LCA uncertainty analysis.

3.2.8 Other Elements of the LCA Scope Definition

Other elements related to the definition of scope are: type of critical review and type and format of the report required for the study.

The critical review is defined by ISO (2006b) as the "*process that aims at ensuring consistency between a life cycle assessment and the principles and requirements of the standards on life cycle assessment*". ISO (2006b) makes it mandatory a critical review in cases that aim at developing comparative statements between products that are disclosed to the public. For example, imagine that a study has been carried out comparing the LCA of fuel alcohol vs. gasoline, or a comparative LCA between using plastic bags vs. returnable bags, or the LCA between the disposable plastic cup vs. glass cup, or the LCA of this paper book you're reading vs. its e-book version. To answer these questions, the LCA critical review step is required, which is prepared by a panel of experts. In the academic field, in general, these critical reviews are carried out by scientific examining boards. In the case of studies carried out for private companies, consultants and external consultants specialized in LCA must be contracted by the entity interested in the publication of the study.

Finally, regarding the report format item for LCA studies, ISO (2006b) emphasizes that LCA results and conclusions must be reported to the target audience in a complete, accurate and impartial manner, taking into account all elements of Definition of Goals and Scope, LCI, LCIA and Interpretation. The study report must be transparent, impartial and consistent.

3.3 Final Considerations

In this chapter, the main elements that make up the first phase of the LCA have been presented—The Definition of Goals and Scope, its characteristics and methodological procedures. Since it refers to the overall LCA planning, the results of this phase are essential to support and guide the other LCA phases, that is, the life cycle inventory analysis (LCI), the life cycle impact assessment (LCIA) and the LCA interpretation as a whole.

The three main elements of this initial LCA phase are: the definition of the function, the functional unit and the reference flow (see Sect. 3.2.2); and the definition of product system boundaries (see Sect. 3.2.3). All other activities related to definition of scope depend directly on these two main elements being clearly established.

Exercises related to this stage of conducting an LCA are proposed below. And then, the matter will be continued through the chapter on life cycle inventory (LCI) of products.

3.4 Exercises

Question 1. What is the purpose of the Definition of Goals and Scope in LCA?

Question 2. What is functional unit and reference flow? Differentiate them.

Question 3. Give 02 product examples and complete the following chart with correct definitions for function, functional unit and reference flow.

Product	**Function**	**Functional unit**	**Reference flow**

Question 4. What is multifunctionality? How should the multifunctionality of a process be treated to avoid allocation according to ISO 14040 and 14044 guidelines?

Question 5. In the chlor-alkali chemical process, the sodium chloride solution is electrolyzed to produce chlorine gas, hydrogen gas and sodium hydroxide according to the following reaction:

$$2NaCl_{(aq)} + 2H_2O_{(l)} \rightarrow H_{2(g)} + Cl_{2(g)} + 2NaOH_{(aq)}$$

Do the mass allocation and the economic allocation of the chlor-alkali process to the product chlorine gas (Cl_2) according to the data in the following figure:

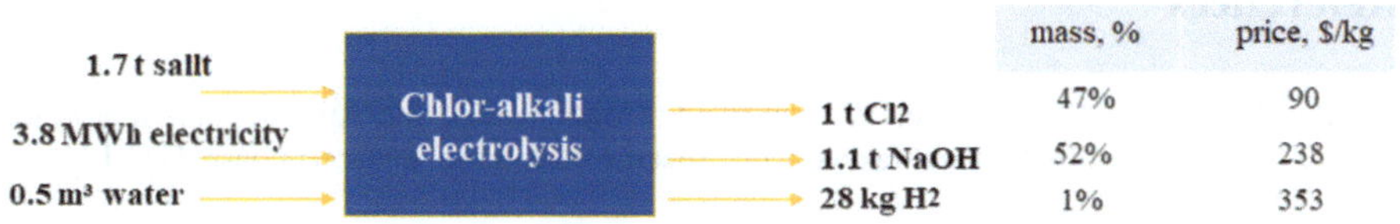

Feedback

Question 1.

Carry out an initial planning of the LCA study with all the elements necessary to provide transparent and consistent support for conducting the subsequent stages of the LCA study.

Question 2.

The functional unit mathematically expresses the function of a product, whereas the reference flow establishes an amount of product necessary to reach the functional unit.

Question 3.

Product	Function	Functional unit	Reference flow
Gasoline	Serving as fuel in transport vehicles	10,000 km route using a four-wheel vehicle at an average consumption of 10 km/l	750 kg of gasoline (density 0.75 = g/cm^3)
Air conditioner	Condition an environment at a certain temperature and for a certain time	Refrigerate a room (10 × 10) m^2 in area at a temperature of 25 °C operating 08 h a day	The production of 01 air conditioner unit with 10,000 BTUs

Question 4.

A process is said to be multifunctional when it provides more than one function – that is, it provides multiple products. In these cases, to address multifunctionality and avoid allocation, two different approaches can be employed: subdivision of multifunctional processes and system expansion (including replacing multifunctions).

Question 5.

The mass allocation and the economic allocation for the chlorine gas product will be 47% and 25%, respectively, according to the data in the following figure.

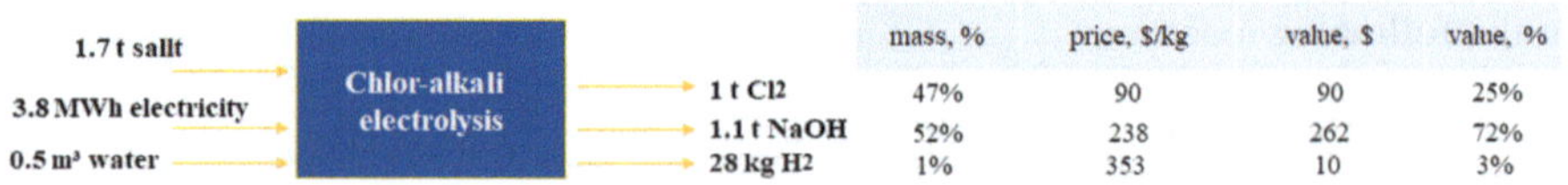

References

Brander, M., Tipper, R., Hutchison, C., Davis, G.: Consequential and Attributional Approaches to LCA: A Guide to Policy Makers with Specific Reference to Greenhouse Gas LCA of

Biofuels, 15p. Ecometrica Press, Edinburgh, UK. Technical paper, April, 2009. Disponível em: https://ecometrica.com/white-papers/consequential-and-attributional-approaches-to-lca-a-guide-to-policy-makers-with-specific-reference-to-greenhouse-gas-lca-of-biofuels

Curran, M.A.: Overview of goal and scope definition in life cycle assessment. In: Klöpler, W., Curran, M.A. (eds.) LCA Compendium—The Complete World of Life Cycle Assessment. Springer, Berkeley (2017)

Ferreira, J.V.R.: Análise do ciclo de vida dos produtos. Gestão Ambiental. Instituto Politécnico de Viseu (2004)

Guinée, J.B., et al.: Life cycle assessment: past, present, and future. Environ. Sci. Technol. **45**(1), 90–96 (2011)

Hunt, R., Franklin, E.: LCA—How it came about. Personal reflections on the origin and the development of LCA in the USA. Int. J. Life Cycle Assess. **1**(1), 4–7 (1996)

International Organization of Standardization: ISO 14040: environmental management—life cycle assessment- principles and framework. Geneva (2006a)

International Organization of Standardization: ISO 14040: environmental management—life cycle assessment- requirements and guidelines. Geneva (2006b)

Joint Research Centre of the European Commission (EC-JRC): International Reference Life Cycle Data System (ILCD) Handbook. ILCD Handbook International Reference Life Cycle Data System, European Union (2010)

Nicoletti, G.M., Notarnicola, B., Tassieli, G.: Comparative life cycle assessment of flooring materials: ceramic versus marble tiles. J. Clean. Prod. **10**, 283–296 (2002)

Ometto, A.R., et al.: Lifecycle assessment of fuel ethanol from sugarcane in Brazil. Int. J. Life Cycle Assess. **14**(3), 236–247 (2009)

Ribeiro, P. H. (2009). Contribuição ao banco de dados brasileiro para apoio à avaliação do ciclo de vida: fertilizantes nitrogenados. 343f. Tese (Doutorado) – Escola Politécnica, Departamento de Engenharia Química, Universidade de São Paulo, São Paulo (2009)

Silva, D.A.L.: Avaliação do ciclo de vida da produção do painel de madeira MDP no Brasil, 207p. Dissertação (Mestrado) – Escola de Engenharia de São Carlos, Universidade de São Paulo, São Carlos (2012)

Silva, D.A.L., et al.: Life cycle assessment of medium density particleboard (MDP) produced in Brazil. Int. J. Life Cycle Assess. **18**(7), 1404–1411 (2013)

Thorn, M.J., Kraus, J.L., Parker, D.R.: Life-cycle assessment as a sustainability management tool: strengths, weaknesses, and other considerations. Environ. Qual. Manage. **20**(3), 1–10 (2011)

Tibor, T., Feldman, I.: ISO 14000: A Guide to the New Environmental Management Standards. Times Mirror Higher Education Group, USA (1996)

United Nations Environment Programme: Life Cycle Management—A Business Guide to Sustainability. UNEP/SETAC Life Cycle Initiative, Paris (2007)

Chapter 4
LCA—Life Cycle Inventory Analysis and Database

Thiago Oliveira Rodrigues, Fernanda Belizario-Silva, Tiago Emmanuel Nunes Braga, and Marília Ieda da Silveira Folegatti Matsuura

4.1 Introduction

Life Cycle Inventory Analysis (LCI) consists of collecting, compiling and validating data regarding what enters and exits a product's life cycle. The data to be collected refer to inputs, raw materials, the product itself, co-products, by-products, solid waste, liquid effluents and gas emissions.

LCI is supported by the definitions from the previous step, covered in Chap. 3: Life Cycle Assessment (LCA)—Definition of Goals and Scope. Thus, although the data collection procedure is carried out in different ways (interviews, measurements, stoichiometric calculations, etc.), their compilation must meet the definitions agreed in the first step, aligned with the function of the product system and the respective functional unit.

The LCI's focus is on quantifying the flows that enter and exit the processes that make up a product system. Any mass or energy flow that can be measured and is part of a process within the boundaries of the study must be considered.

The main components of the LCI for products and services are presented below. It is important to note that the LCI structure discussed in this text is not oriented toward organizational LCA (LCA-O), social LCA (LCA-S) or consequential LCA. Since the

T. O. Rodrigues (✉) · T. E. N. Braga
Brazilian Institute of Information in Science and Technology—IBICT, Ministry of Science, Technology, Innovations and Communications—MCTIC, Setor de Autarquias Sul (SAUS)— Quadra 05 Lote 06 Bloco H 6th floor room 604, Asa Sul. CEP: 70910900, Brasíli, DF, Brazil
e-mail: thiagorodrigues@ibict.br

F. Belizario-Silva
Institute for Technological Research - IPT, Av. Prof. Almeida Prado 532 Cid. Universitária—Butantã, CEP: 05508-901, São Paulo, SP, Brazil

M. I. da Silveira Folegatti Matsuura
Brazilian Agricultural Research Corporation—EMBRAPA Environment, Rodovia SP-340, Km 127,5, Tanquinho Velho, Jaguariúna, SP ZIP Code: 13918-110, Brazil

J. A. de Oliveira et al. (eds.), *Life Cycle Engineering and Management of Products*,
https://doi.org/10.1007/978-3-030-78044-9_4

acknowledgments of Weidema and Ekvall (2009), attributional approach is still the traditional path to an LCI. The consequential approach has been discussed in Chap. 3 of Defining Goals and Scope and focuses on changes both inside and outside the life cycle of the product system under evaluation (Brander et al. 2009).

4.1.1 Flowchart

The basis for the development of a consistent LCI is the appropriate mapping of the product system in question, which includes:

- Identifying the existing processes within the scope of the analyzed boundary, for example, processes for the transformation of raw materials, transport operations, waste treatment processes, product use, among others. Note that these processes are not limited to the activities that take place inside the facilities that produce the final product (e.g., a factory), and it is also necessary to identify the processes that occur upstream (necessary to supply the inputs for production) and downstream (for example, use scenarios and waste disposal);
- Identifying the mass and energy flows involved in the mapped processes, that is, all the inputs and outputs of these processes;
- Understanding the relations between the processes and flows identified.

Accomplishing this mapping requires a throughout understanding of the studied product system. For this, it is recommended to consult professionals who are involved in the day-to-day production, specialists in the processes in question and the literature as well as visits to the facilities where the processes take place, as the visualization of the processes, is of great help in drawing up an appropriate mapping. In addition, visits allow understanding what kind of information is available for each of the processes (e.g., how a factory's production controls are made), which subsidizes the planning of data collection for building the inventory of the product's life cycle.

This mapping defines the types of data to be collected concerning the following sources: **primary data**, which are collected directly from the product system; **secondary data**, which are collected through databases, literature, stoichiometric calculations and consultations with specialists.

The representation of this process mapping is usually done through a flowchart. Figure 4.1 illustrates an example of a generic flowchart. It is possible to distinguish the processes that are directly related to the analyzed product (foreground system) of those that are not directly related to the product or that are common to several other processes (background system). One example of process that belongs to the foreground system is the manufacture of the product itselft, whereas electricty generation belongs to the background system (EU, 2014). Although the link between the foreground and background processes has been represented in a simplified way, it is necessary to map the inputs and outputs of each of the processes belonging to the product system to draw up its life cycle inventory correctly (in the figure, we

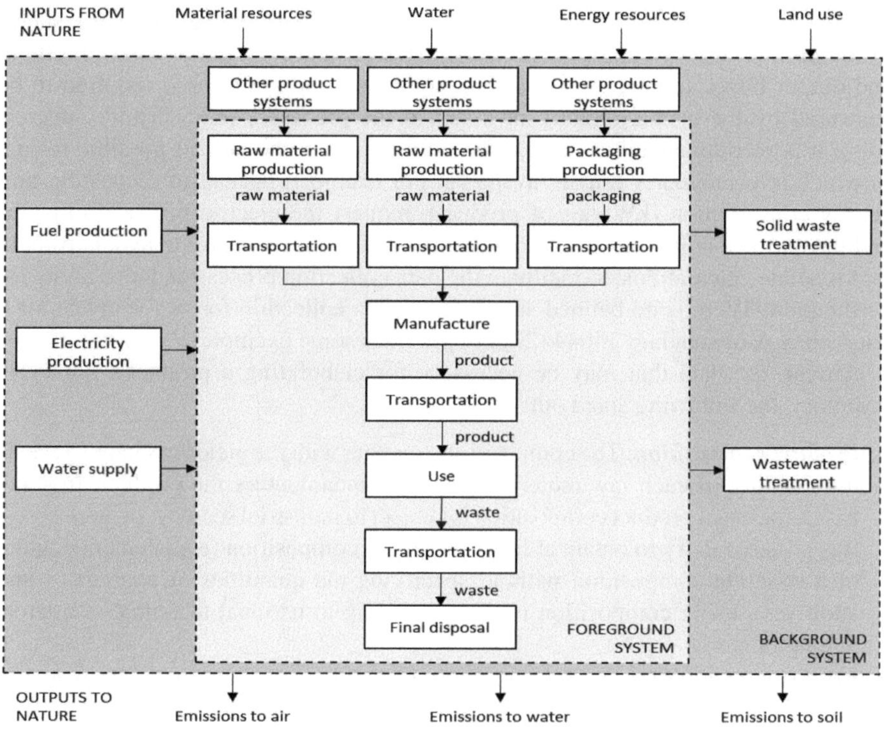

Fig. 4.1 Example of a flowchart for mapping a generic product system with the system boundary from cradle to grave. *Source* Authors

opted for the simplified representation to avoid crossing the arrows that represent the flows).

An important premise to be observed in the mapping of processes is that for preparing the inventory, it is necessary that the product system boundary begins and ends in nature; that is, all activities must be modeled until the boundary with nature is reached (ISO 2006a). For example, if any waste is destined for treatment, the type of applicable treatment (for example, landfill) must be identified and the treatment activity modeled, identifying the outputs to the nature (e.g., methane emission for the atmosphere and substances leached into the soil).

4.1.2 Data and Metadata

Once the process mapping is prepared and the product system is understood, which includes the definition of the product function and the characterization of the functional unit in relation to which the quantitative results of the LCA will be expressed, data collection must be planned for building the life cycle inventory.

For this, it is recommended to prepare a spreadsheet containing all the stages of the product life cycle, the respective processes, and for each process, the input and output flows. In cases where a data collection questionnaire is required to be answered by the agents directly involved in the processes (e.g., factory supervisors), it is recommended to request the information in the closest possible format to which information is usually available. For example, instead of requesting unit energy consumption (kWh/kg of product), request the electric bill (kWh/month) and the corresponding monthly production (kg of product/month) to avoid errors in intermediate calculations, to facilitate the data collection process and also allow for verification. There is no defined standard for data collection forms for inventories, but Annex A of standard 14044 (2006b) provides some examples.

Among the data that may be necessary for elaborating a product's life cycle inventory, the following stand out:

- ***Product composition***: The composition, together with the yield/loss index, reveals the quantity of each raw material needed to manufacture the product. In some cases, the exact product composition is subject to industrial secrecy; in these cases, it is possible to try to obtain at least part of the composition (e.g., cement content of a concrete composition without specifying the quantities of aggregates and additives), using composition ranges according to national technical standards, among others;
- ***Energy consumption***: A large part of product manufacturing processes, in several sectors, involves the consumption of electricity and/or fuels for the supply of thermal energy. Some products also require energy consumption during their stage of use, for example, vehicles and buildings. Regarding electricity, for example, the country's energy mix is usually accepted for its production; however, large consumers may choose to purchase energy from specific sources, which may be relevant to their impact results and, therefore, must be taken into account if applicable. In the case of fuels, in addition to the amount consumed, it is recommended to record their calorific value to enable the energy balance for later data validation. Another way to estimate energy consumption is to record the equipment power and its usage time, taking into account the machine's performance, or based on literature data that are deemed representative;
- ***Water consumption***: Water consumption can be obtained through the water bill, in the case of mains water; the measurement of hydrometers in the case of direct capture of water from river, lake or well; or by indirect estimates in cases where there is no measurement, such as through product moisture content. Water consumption must be broken down by source as this makes a difference in the impact assessment methods related to water, as well as its origin (e.g., hydrographic basin), as it is also a relevant data in the impact assessment phase (ISO 2014);
- ***Consumption of auxiliary material***: Depending on the characteristics of the product system, information regarding materials used as processing aids, packages, equipment maintenance material (e.g., replacement of molds and parts, lubricating oil, etc.) may be necessary. It is necessary to estimate the unit consumption

of these materials, for example, requesting the service life for a specific piece and the production corresponding to this period;

- ***Transportation***: Throughout a product's life cycle, it is common to have several transport operations. For considering them in the life cycle inventory, information is required on the type of vehicle, its load capacity, the load actually transported, fuel consumption factors and/or emission of pollutants depending on the load and distance traveled (if the vehicle returns to its origin empty, the return distance must also be counted);
- ***Emissions to water, air and soil***: Some processes generate outputs directly to nature, such as emission of pollutants into the air by combustion processes, emission of particulates from rock dismantling operations, emissions to water of certain chemical processes, among others. In the case of data from companies, it is recommended to request environmental monitoring data, which normally records the concentration of pollutants (e.g., kg/m^3 air) as well as other information necessary to convert these concentrations into unitary flows (e.g., air flow in the oven chimney, in m^3 air/h and hourly production, in kg of product/h). If this information is not available, the limits in force in the relevant environmental legislation can be used as a conservative estimate;
- ***Generation of waste***: The amount of waste can be obtained directly, for example, through waste transport certificates that declare the amount and destination; or indirectly, through loss indexes reported by the manufacturer or estimated based on literature. In addition to the quantity, it is important to record the type of destination given to the waste.

It is recommended that the data cover a sufficient period to contemplate any seasonality related to the product system. In addition, inventories must be based on representative samples of the product system intended to be depicted, for example, in the case of inventories that aim at representing a certain technology or even a certain region (e.g., inventories with national validity), enough information must be obtained from manufacturers so that general conclusions cannot be drawn from very specific cases. Those recommendations apply to product systems or parts of product systems that are already in place, for instance, factories that are already operating.

Depending on the type of LCA study, some stages of the product life cycle need to be modeled based on scenarios and assumptions: This is the case, for example, of forecasting the operating conditions of products with a long useful life such as buildings and infrastructures, which depend on user behavior and future technology scenarios that are not fully delineated at the time of elaboration of the life cycle inventory.

Not all flows and processes need to be included in the life cycle inventory. In comparative LCA studies, for example, processes common to the two compared product systems can be excluded from the inventory as they will not influence the comparison, as long as they are effectively the same. In addition, ISO 14040 (ISO 2006a) admits the application of cut-off criteria which simultaneously consider the contribution of mass flows, energy, and environmental significance, that is, in terms of their contribution to the product total environmental impact. For example, the European standard for Environmental Product Declarations for construction products, EN 15804 (DIN 2014), admits that items that contribute individually with less than 1% of the mass, energy or environmental impact of the product are disregarded as long as the sum of the disregarded items does not exceed 5% of the mass, energy and total impact. In order to do this assessment, it is necessary to carry out previous calculations with existing data (as shown in Chart 3.3 of Chap. 3: Life Cycle Assessment (LCA)—Definition of Goals and Scope and turn to expert opinions.

Another very important issue for planning life cycle inventory data collection is the Life Cycle Impact Assessment method that is intended to be used, which is established at the definitions of goals and scope stage of the LCA study (ISO 2006b). For example, if the study focuses only on global warming potential, it is necessary to collect only data on atmospheric emissions of greenhouse gases; in turn, if there is an intention to also assess other impact categories such as acidification and photochemical ozone formation potential, other pollutants emitted to the atmosphere will also need to be inventoried.

As explained in the product system mapping, there are processes which belong to the foreground system and processes which belong to the background system. It is recommended that the foreground processes are modeled with primary information collected from the agents involved in these processes. For those in the background, it is common to turn to secondary information sources such as databases of life cycle inventories, literature, among others, in order to contemplate the entire boundary of the product system in question (IBICT 2014). It is essential to observe the applicability of information from secondary sources for the LCA study, that is, whether such information can be considered representative of the studied context, especially in cases where the background processes have a great influence on the environmental impact results of the product in question.

In addition to the quantitative information necessary to compile the product life cycle inventory, information to adequately describe the information contained in the inventory is needed, i.e., period to which the data refer, conditions of boundary/limitations of the inventory, representativeness in face to the studied context, possible assumptions that have been made to calculate flows, bibliographic references used, among others. This information is called "**metadata**" (Rodrigues et al. 2016), and they are just as important for understanding and using the life cycle inventory as the quantitative information of the flows themselves. It is recommended that during the preparation of the inventory, such information is recorded (e.g., in a report) so that the application field of the resulting inventory is well defined.

4.1.3 Flows

All the data collected for a product's LCI refer to a flow. The data refer to flows of matter or energy that are moving between systems. In LCA studies, there are three basic types of flows to be collected:

- **Elementary**: Matter or energy that leaves nature (ecosphere) to the product system (technosphere) or that leaves the technosphere directly to the ecosphere, without human interference, e.g., CO_2 captured by a tree (elementary input) and CO_2 emitted by burning coal (elementary output);
- **Intermediate**: Technological flows generated by some artificial process, which occurs within a product system never reaching the ecosphere, e.g., wood chips leaving the grinding and entering the cooking to produce cellulosic mass.
- **Products**: Also technological flows, they refer to the product under assessment in a product system. They leave the technosphere boundary to enter another product system, e.g., cellulosic pulp that leaves the cellulose product system and enters the paper product system as a raw material.

There is also the **reference flow**, which is equivalent to a product flow and refers to the quantity needed to perform the function of a product defined by the functional unit (Ontologia 2019). The functional unit has already been discussed in the chapter about definition and scope.

4.1.4 Processes

Process in a product system is an activity that benefits inflows (represented by inputs and raw materials) in outflows that meet conditions for further processing. Each and every process has inputs and outputs corresponding to one or more flows.

A product system is always formed by **elementary processes**. They are the basic units of systems, the smallest element of a system that cannot be divided into smaller processes and for which data are collected.

Elementary processes can also be classified as **primary** or **secondary** depending on the level they are at. The primary ones are in the foreground system (Fig. 4.1) within the technosphere, represented by the scope of gate-to-gate (G2G) and from which it is usually possible to directly collect data through measurements and calculations. The secondary ones, on the other hand, are in the background system, outside the G2G and represented by processes that feed the foregorund system from which data are collected mainly from databases.

4.1.5 Aggregation

Although the product systems are composed by elementary processes, for several reasons, many are presented in aggregate form. This means that it is not possible to visualize the elementary processes, they are grouped in such a way that the origins and destinations of flows cannot be distinguished, and they all go in and out from a (and in a) "black box" that groups several elementary processes: an **aggregated process**.

Among the goals for aggregation, Broadbent and collaborators (2011) highlight: confidentiality, protection of data ownership, increased efficiency of data computation, guarantee of analysis efficiency, appropriate relevance, preservation of data integrity, protection of business model, increasing ease of use and even covering up processes with a significant environmental burden. The reasons, commercial or computational, must always be justified and highlighted to promote reliability and explain the lack of transparency, a key point of LCA.

The **aggregation** of processes can take two forms: **complete** or **partial**. In the complete one, it is only possible to observe the elementary and product flows entering and leaving the aggregated system boundaries. The partial one allows the identification of some intermediate flows (flows of that occur between elementary processes). The level of partial aggregation varies according to the goals, where it must be composed of at least two processes with an intermediate flow between them and at least one of them must be aggregated.

Figure 4.2 generically exemplifies the relation between primary and secondary elementary processes, aggregated processes, elementary, intermediate and product flows.

In Fig. 4.2, the processes and flows included inside the largest central rectangle (inside the technosphere) make up the first plan of the product system, known as foreground processes. Between each elementary process of this plan, there are intermediate or technological flows that are inputs for the consecutive processes until

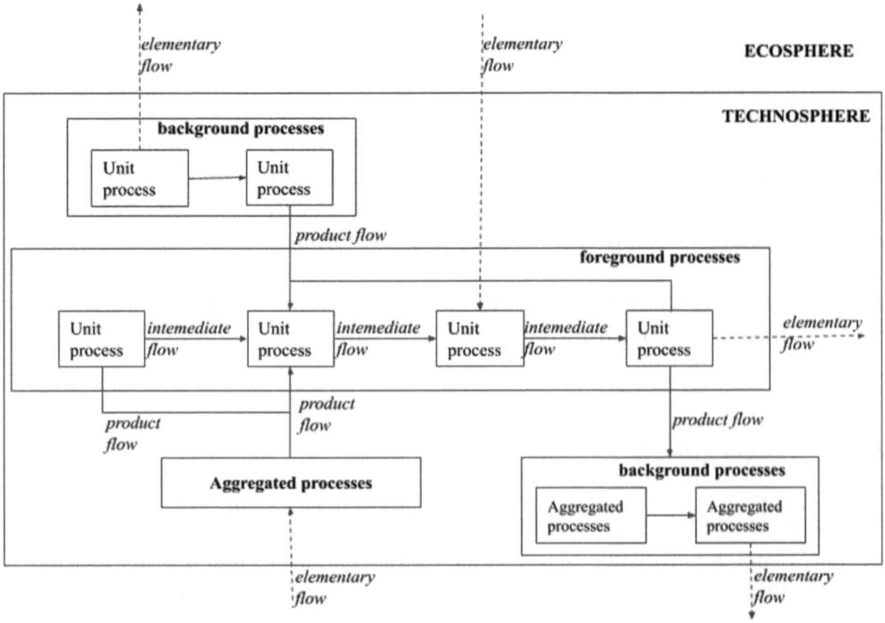

Fig. 4.2 Relations between systems (technosphere and ecosphere), processes and flows. *Source* Authors

reaching the "final" process, in that the output is a product flow which goes on to other stages such as direct consumption or as an input to another product system. In this plan elementary flows also occur, which can be input and output of an elementary process. The others smaller rectangles, above and beyond the foreground processes, represent product systems that provide product flows to the first plan. These are the background systems, composed of the second plan processes. For LCA studies in general, process data from background systems are collected in specialized LCA databases.

4.2 Data Validation

After the laborious process of data collection, some checks are necessary to guarantee their credibility and relevance. The data collected must be representative of the product system in question in order to ensure that the product's environmental profile is true to reality, does not neglect flows or processes, has a balance between what enters and leaves the system and meets a defined quality standard.

Data validation must be conducted during the data collection process (item 4.3.3.2 of ISO 14044). The standard proposes performing mass and energy balances that will attest to the validity of a unitary process's description. Therefore, if the validation points out flaws in the data collection, they must be reviewed with attention to the data quality requirements.

4.2.1 Mass and Energy Balances

The elaboration of an LCI requires a large amount of data, and depending on the scope of the LCA study and the complexity of the analyzed product system, ensuring the consistency of flows can be a challenging task. For this reason, it is essential to have data validation mechanisms. Both balances have already been dealt with previously in the mass (material) and energy balances section of Chap. 2: cleaner production (CP).

The first mechanism arises from thermodynamics: **All processes must respect the laws of conservation of mass and energy**. Therefore, one way of assessing the inventory data's consistency is checking whether the mass and energy balances are consistent, that is, "if everything that enters, leaves" (ISO 2006b). Such balances must also consider losses that occur in the product system (e.g., heat dissipation in thermal processes). It is recommended the checking of these balances to be done with the flows already unitarized, that is, already expressed in terms of the flow values that will result in the functional unit chosen for the product; thus, it is possible to identify any errors in the step of calculating unitary flows.

A subcategory of mass balances is the water balance. It is recommended the water balance of each process to be checked, including the situations in which water is consumed in chemical reactions and incorporated into the product (e.g., in the case of concrete manufacturing). Such balances can be useful even as a tool for estimating certain flows; for example, to estimate the amount of water evaporated in a process, a flow that is not normally measured.

Special attention is recommended to the units adopted in the study and to the consistency of the variables (e.g., calorific value of fuels). In the case of using life cycle inventory databases to represent certain processes, care must be taken for the calculated units to be consistent with the units in which the processes are described on this basis.

To calculate the mass, energy or water balance, the principle is the same: to assess the difference between what enters the product system (inputs) and what leaves the system (outputs). The specificities are the units and types of inputs and outputs that differ for a mass, energy or water balance. An example of water balance is agricultural crops. The entering water comes from irrigation and precipitation. Part of the water is incorporated into the plant, part percolates through the soil and part evaporates. Then, the leaving water is contained in the plant. In order to "close" the balance, it is necessary to account for everything, but knowing in fact how much water has entered the product system is a complicated task. In general, the balances are not closed 100% (mass, water or energy), but it is important to indicate the maximum achieved.

Practical examples on mass and energy balances can be reviewed in Chap. 2 about CP. The same procedures presented that there are also valid here when validating data of an LCI.

In addition to the material balances, the analysis of the inventory data quality requirements is also a relevant aspect during the data validation.

4.2.2 Data Quality Requirements

There is no definite rule for performing an LCI due to the huge variety of product systems, which can be approached in several different ways. Thus, data are presented in the most diverse ways.

The standard that rules the guidelines of an LCA study—ISO 14044—emphasizes that data in an inventory form a mixture of measured, calculated or estimated information (ISO 2006). Therefore, it is necessary to validate the data to determine the degree of interference that a data may have on the results of an LCA study and, thus, define whether there is a need to re-survey that information.

The standard lists data quality requirements (DQR) that must be met when collecting data (item 4.2.3.6). The use of DQRs promotes transparency in the construction of inventories and accomplishment of the goals and scope of the LCA study. The requirements (discussed in Sect. 1.6) are presented with a focus on their practical use during the LCI:

- Temporal coverage—period of time from which the data have been collected;

- Geographical coverage—geographic area to which the collected data refer;
- Technological coverage—determination of specific technologies or a technological mix used in the assessed system;
- Accuracy—measurement of variability of data values using statistical methods such as variance, standard deviation, etc.;
- Completeness—percentage of data actually collected;
- Representativeness—degree of significance of data within the population of interest, related to geographical, temporal and technological coverage;
- Consistency—degree of uniformity in applying the methodology in the various components of the study (qualitative);
- Reproducibility—degree of ease in carrying out a study using the data collected (qualitative);
- Data sources—origin of the collected information (measured, calculated, literature or estimates);
- Uncertainty—lack of knowledge for specific parts of the life cycle inventory.

The determination of which DQRs are important in validating LCI data is described in the standard, but how to measure them is not. Therefore, some methodological guides have been developed to guide researchers on best practices in carrying out LCA studies, including data quality assessment. An example is the *Handbook of the International Reference Life Cycle Data System,* specific to Life Cycle Inventories (EC 2010). According to the handbook, the quality of data from an LCI must be measured by representativeness (technological, temporal and geographical), adequacy and methodological consistency, completeness and precision/uncertainty.

4.2.3 Uncertainties

As with any real-world situation, the information needed to elaborate life cycle inventories has a certain level of uncertainty associated with it. There are several sources of uncertainty that can influence inventory flows, which can be divided into two major groups:

- ***Variability associated with the process***: For example, variation in the quantity of water to be added to a process depending on the moisture level of the raw material, variation of the time of growth of a tree according to environmental conditions; variation of the incorporated impact of a raw material among manufacturers (when it is not known exactly which manufacturer will be the supplier of the raw material), among others. This type of uncertainty can be described mathematically;
- ***Uncertainty***: Occurs, for example, in the case where flows are extrapolated from other processes (e.g., another manufacturer, another technology or even another location) to cover missing data; when old information is used to model the inventory without knowing whether it remains valid; when a sample is not representative of the context it portrays (e.g., using only one installation to represent a context

that has dozens of similar installations due to the difficulty of accessing a statistically representative sample or even the entire population); among others. This type of uncertainty cannot be described mathematically (at least not directly).

In general, we must try to minimize the sources of uncertainty in an LCA study and for the remaining uncertainty (since there are no exact processes), it is necessary to describe it properly. For the portion related to variability, it is recommended to calculate the flow variation parameters (minimum and maximum values, standard deviation, among others) and propagate this variability over the product's life cycle (e.g., by sum of minimums and maximums, analysis of variance, Monte Carlo simulation, among others). For the portion of uncertainty due to deficiencies in data quality, we can describe the boundary conditions/limitations of the study using the inventory metadata or adopt certain "safety factors" for the inventory flows and/or results of environmental impact in order to obtain a conservative estimate (IBICT 2014).

The Pedigree Matrix (Weidema and Wesnaes 1996) is often adopted to describe the data quality through notes referring to certain quality requirements (reliability, scope, temporal correlation, geographical correlation and technological correlation). However, it is emphasized the importance that the criteria for assigning these scores are developed sector by sector or even product by product (e.g., a 5-year-old LCA data can be considered reasonably current for certain industrial processes of the civil construction industry, but certainly not for the electronics sector). In addition, it is not recommended to associate the quality scores of the Pedigree Matrix data with the sample's variability parameters (as an "additional uncertainty"), since, according to Ciroth et al. (2013), such an approach lacks an empirical basis.

Although many LCA studies do not declare the uncertainty associated with their results either quantitatively or qualitatively (Lloyd and Ries 2007), it is considered extremely important that the assessment of uncertainties becomes an integral part of LCA studies. This aspect is particularly important in comparative LCA studies, as shown in Fig. 4.3, since decisions based only on point values (which do not necessarily correspond to average values, since for the calculation of the average it is necessary to have a representative population sample) can lead to mistaken conclusions.

The decision based on a point value can lead to the conclusion that alternative A is better than alternative B as it has a lower environmental impact. However, the maximum value that A can assume is well above the maximum value that B can assume; therefore, depending on conditions, alternative A may have a much greater impact than alternative B.

4.3 LCA Databases

Databases (DB) are data sets organized in order to generate information and allow a series of actions such as the establishment of connections between different data

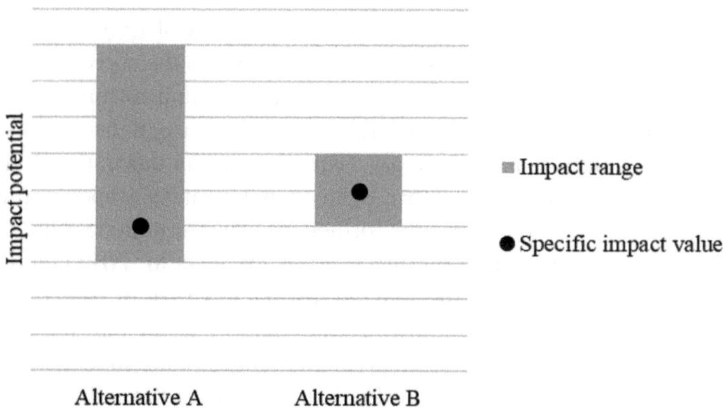

Fig. 4.3 Hypothetical comparison between two alternatives, based on the results of the potential impact of an LCA study. *Source* Authors

sets. They structure thematic data and greatly improve the efficiency of information use. In LCA, it is not different; for each type of product, there is a potential LCI that aggregates a multitude of data and metadata on its environmental profile. LCIs are organized as data sets composed mainly of flow data referring to the inputs and outputs of matter and energy that make up a product. Therefore, they are numerical values that refer to inputs, raw materials, product or process itself, emissions to air, effluents to water and solid waste. And each flow is a data set in itself, so there is a possibility to establish relations between the data sets deposited in an LCA database. Any LCI is potentially a flow of another LCI. For example, an aluminum's inventory is a flow of a bicycle's inventory.

There are LCA databases in several countries around the world. In general, they intend to represent national production systems by aggregating inventories of products produced in the country of origin. But given the obvious globalization of production systems, many databases bring with them flows that originate in other countries. If a national product's inventory does not carry the environmental loads of flows from other countries on the planet, it automatically prevents an LCA study from accurately demonstrate potential environmental impacts in case of globalized products. There is an inherent loss of reliability and transparency. Without mapping the flow chain, LCA will not address all critical points throughout the life cycle of products, processes or services. Therefore, it is desirable that LCA databases from different countries "talk" to each other, allowing access to different flows.

A challenge for the connection between different LCA databases is related to the storage format of the inventories. LCIs can be constructed in a variety of ways, including with paper cards being filled out manually. But in order to be useful in LCA studies, data collected must be compiled into digital files. These files can be spreadsheets of electronic editors or specific software for life cycle studies. Due to the amount of data and metadata collected in an LCI, files are reasonably complex. In some cases, there are files in formats exclusive to software. Failure to make these data available in open formats may affect the usefulness of LCI information, as it restricts its use to those who dominate and have access to the software.

Currently, two formats have been configured as the main formats: ecoSpold and ILCD. **EcoSpold** format is used by the largest LCA database in the world, ecoinvent, and although it is free, it is proprietary. This format is currently in the second version (Ecoinvent 2019). **ILCD,** in turn, is an open format developed initially for the European Union database and later transformed into an international one. Its creation aimed at increasing data interoperability in LCA (Wolf et al. 2011), and due to its characteristics, it was adopted as the standard format in Brazil in 2009 (Braga 2018). There are also several other lesser known formats as well as the internal formats used by the LCA software itself. One of the great challenges in the management of LCA data is related to the interoperability of LCA data between these different formats.

Silva and Masoni (2016) carried out an extensive study on national LCI databases of products in order to identify similarities and differences in their management, maintenance and use in the world. The study identified 15 relevant DBs, seven European, four Asian, three American and one from Oceania.

According to the researchers, in general, all DBs propose to support academic and business LCA studies. DBs usually cover most sectors of the economy, except for some monothematic databases dedicated to specific sectors, such as civil construction. The inventories available in these databases cover, for the most part, a cradle-to-gate approach, but there are also some LCIs that comprise the gate-to-gate one. Temporal coverage tends to vary widely with cases of LCIs dated more than 20 years old. Inventories are usually nationwide, but there are some regional cases especially when they come from DBs in countries with continental dimensions. Most banks are access free and are supported by national governments, and there are some private databases which sell user licenses. Almost all DBs include LCIs with an attributional approach in ecospold or ILCD formats.

The most representative DBs in terms of quantity of inventories and geographical coverage among those analyzed are ELCD and ecoinvent. **ELCD**, initials of European Reference Life Cycle Database, contains LCIs of European products with a focus on energy, waste management, basic materials (commodities) and transportation. There are about 2360 access free inventories in the ILCD format. ELCD was discontinued in June 2018 but has its LCIs accessible on the Life Cycle Data Network (LCDN).

Ecoinvent is a Swiss database initially developed with support from that country's government. It is currently a non-profit organization that commercializes the license to access its content. Ecoinvent's DB is now the most wide-ranging LCI bases in the world. The version, 3.7, aggregates more than 10,000 inventories of all kinds of products from different parts of the planet. The inventories are in the ecospold format, and there are examples in the attributional and consequential approach (Ecoinvent 2019).

4.3.1 SICV Brasil

Brazil is among the few countries in the world that have a database specialized in the management of LCA data. The National Bank of Life Cycle Inventories—**SICV Brasil**—is a database management system that can be created within the bank (SICV BRASIL). These bases can be thematic or institutional.

SICV Brasil is public, access free and has representative data of the reality of Brazilian agribusiness and industry (BRAGA, 2018) upon user registration. The DB is supported by the Brazilian Life Cycle Assessment Program (Programa Brasileiro de Avaliação do Ciclo de Vida- PBACV), and its management is in charge of the Brazilian Institute of Information in Science and Technology (IBICT). SICV Brasil's proposal is to organize the production of Brazilian products' LCIs in order to promote the competitiveness of the national industry in international markets (CONMETRO 2010).

The inventories currently contained in the SICV Brasil are made available in the ILCD format. The data sets are generated by researchers, members of the academy or by professionals from companies, who submit them to the database. Figure 4.4 illustrates the path between submission and publication of the inventory at the Brazilian bank. As soon as an inventory is submitted, it goes through the verification of the minimum requirements defined in the Qualidata guide (Rodrigues et al. 2016). The guide assesses 42 requirements that assist in determining whether an LCI is eligible for SICV Brasil or not. They are requirements on general data of an inventory, methods, processes, flows and review. 30 are mandatory, 10 are recommended, and two are optional.

After the requirements analysis, the inventories are technically reviewed by specialists in the technologies that they represent. The specialist also validates methodological aspects of applying the LCA standards. If necessary, the LCI is sent back to the author for adjustments. After this process and readjustments by the authors, the inventory is made available to the public on the SICV Brazil platform.

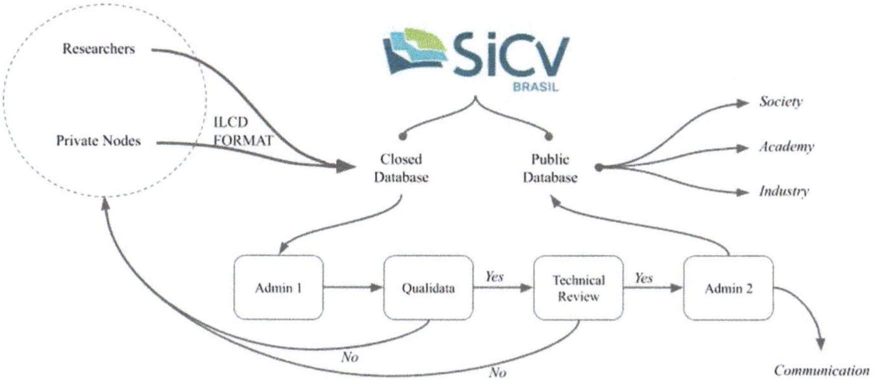

Fig. 4.4 Process for publishing inventories at SICV Brazil. 2019 *Source* SICV (2019)

4.3.2 GLAD

With the increase in the number of international LCA databases, the challenge of building interoperability mechanisms between datasets produced and made available in different formats has become even more urgent. LCA studies are invariably global, in the sense that they will have at least one product flow that originates in another country. Therefore, it is essential to create a mechanism for making these sets of data from around the world available.

It was in this context that the Global LCA Data Access Network (GLAD) emerged in, as an initiative led by the UN Environment, assumed by several other countries, including Brazil (GLAD 2019). GLAD aims at improving the accessibility and interoperability of LCA DBs around the world. For this, it has a platform on which to search, from a central point, for inventories of products of interest. The search is carried out in the entire network of connected databases, and although it is an open-access platform, the availability of the selected LCI depends on the source DB. On the platform, it is possible to access the metadata of each inventory with the description of the product system and technology. Each database that connects to GLAD is called a knot.

GLAD is a secretariat by the United Nations Environment Organization (UN Environment) and has been developed by a group of government institutions from 14 countries. Ibict is one of the main actors in the design and maintenance of GLAD, and SICV Brazil is one of the knots in the network. Currently, GLAD has eight knots, LCA DBs from the US (3), Germany, Japan, the European Union and Brazil. With the development, by GLAD, of a universal converter between LCA formats, the network tends to strengthen and become a reference in the search and availability of LCIs.

4.4 Aggregated Process Inventory

Chart 4.1 compiles the results of an aggregated process inventory to produce one cubic meter of wood particleboard panel. It is a product widely used in the furniture industry to replace solid wood.

In this case, the inventory is for the process and not for the life cycle because it does not include the phases after the factory gate: distribution, use and disposal. And it is aggregated because there is no way to distinguish the elementary processes. Therefore, the input flows are mostly products except for cooling water, which is obtained without any technological interference and directly from the ecosphere. In turn, the output flows are all elementary except for the flow of wood particleboard panel product, which is the reference flow of the process.

Chart 4.1 presents five columns that detail information about each flow. The first column contains the names of the flows. The nomenclature of flows is a critical point

Chart 4.1 Aggregated process inventory to produce one cubic meter of wood particleboard panel

	Flow	Compartments	Flow property	Unit	Quantity
Inputs	Ammonium sulfate	Product flow	Mass	kg	1.38
	Diesel	Product flow	Mass	kg	1.72
	Electricity, medium voltage	Product flow	Energy	kWh	140.83
	Heavy fuel oil	Product flow	Mass	kg	13.7
	Lubricating oil	Product flow	Mass	kg	0.02
	Paraffin	Product flow	Mass	kg	5.47
	Wood log, *Eucalyptus spp.*, from sustainable forest management with bark	Product flow	Volume	m^3	1.42
	Urea–formaldehyde resin	Product flow	Mass	kg	71.7
	Water, cooling, unspecified source	Elementary flow/resource/water	Volume	m^3	0.09
	Wood fuel, coniferous, wet, measured as solid wood with bark	Product flow	Volume	m^3	0.11
Outputs	Carbon dioxide, biogenic	Elementary flow/emissions to air/unspecified	Mass	kg	23.92
	Carbon dioxide, fossil	Elementary flow/emissions to air/unspecified	Mass	kg	48.09
	Carbon monoxide, biogenic	Elementary flow/emissions to air/unspecified	Mass	kg	0.31
	Carbon monoxide, fossil	Elementary flow/emissions to air/unspecified	Mass	kg	0.01
	Formaldehyde	Elementary flow/emissions to water/unspecified	Mass	kg	7.3E-5
	Formaldehyde	Elementary flow/emissions to air/unspecified	Mass	kg	0.15
	Hydrocarbons, aliphatic, alkanes, unspecified	Elementary flow/emissions to air/unspecified	Mass	kg	1.64E-3
	Methane, fossil	Elementary flow/emissions to air/high population density	Mass	kg	1.69E-3

(continued)

Chart 4.1 (continued)

Flow	Compartments	Flow property	Unit	Quantity
Nitrogen oxides	Elementary flow/emissions to air/unspecified	Mass	kg	0.18
Volatile non-methane organic compounds, unspecified origin	Elementary flow/emissions to air/unspecified	Mass	kg	0.36
Oils, unspecified	Elementary flow /emissions to soil /industrial	Mass	kg	0.02
Wood particleboard panel, indoor use	Product flow	Volume	m^3	1
Particulate matter < 2.5 μm	Elementary flow/emissions to air/unspecified	Mass	kg	0.02
Particulate matter > 10 μm	Elementary flow/emissions to air/unspecified	Mass	kg	0.08
Particulate matter > 2.5 and < 10 μm	Elementary flow/emissions to air/unspecified	Mass	kg	0.08
Sulfur dioxide	Elementary flow/emissions to air/high population density	Mass	kg	1.31
Sulfur trioxide	Elementary flow/emissions to air/unspecified	Mass	kg	0.02
Suspended solids, unspecified	Elementary flow/emissions to water/unspecified	Mass	kg	0.02
Wood waste, untreated	By-product	Mass	kg	97.2
Residual water from the production of wood particleboard panel	Elementary flow/emissions to water/unspecified	Volume	m^3	6E-3
Water	Elementary flow/emissions to air/unspecified	Mass	kg	0.33
Water/m^3	Elementary flow/emissions to water/unspecified	Volume	m^3	0.049
Mixture of sawdust, pure	By-product	Mass	kg	0.39

Source SICV Brasil (2019)

for LCA studies. The interoperability promoted by GLAD can be greatly compromised by different names for the same flow. The user is able to access an inventory through GLAD but may experience problems in its use as the system may not recognize flows due to the difference in nomenclature. This situation is not usually a problem for elementary flows, which are almost always named in the same way with respect to the nomenclature developed by the International Union of Pure and Applied Chemistry (IUPAC). The same is true for some product flows that are almost ubiquitous in product systems, such as diesel or electricity.

But for some flows created in very specific product systems, the software may not recognize it, and it will not be possible to send it to the database that will relate it to its respective environmental impacts. Thus, for software, this flow would have no influence on the product's environmental profile. For example, if the paraffin input flow is not recognized by an LCA software, it will not be able to connect it to the database and will not bring out the impacts caused by paraffin (production and use). An attempt to harmonize nomenclature systems has been proposed by the European Commission Institute for the Environment and Sustainability (IES/EC) by publishing a manual with procedures for creating names of flows (ECa 2010).

The second column presents the compartments and sub-compartments from which the flows originate and to which they go. They define as precisely (as possible) where the flow has a direct influence. In some cases, there may be an influence on more than one compartment, such as formaldehyde: emission to water and to air. The ILCD manual (ECa 2010) also provides guidelines for the compartmentalization of flows and processes. The types of flows (elementary or product) are also presented in this column as they indicate the type also indicates the origin or destination of each flow: A product flow comes from or goes to a product system, and an elementary flow will come from or will go to ecosphere.

The third and fourth columns provide information on the units of the flows. In general, flows are presented in mass and energy. For some cases, there are flows that are presented in volume or area. Although it is customary for certain products, such as wood log and water in cubic meters, it is necessary to have the information in kilograms to be able to make the mass balance.

The last column refers to the input and output values for each flow. These values are related to the quantities demanded by the reference flow of the inventory's main product. In the case of Chart 4.1, these are the quantities of matter and energy needed to produce one cubic meter of wood particleboard panel.

It is important to note that Chart 4.1 is not exhaustive for information relevant to an inventory. Each of the flows can bring more information about uncertainty (e.g., standard deviation), about the collection way (measured, calculated, literature or estimate), about mathematical parameters (e.g., flow $a = 0.2 \times$ flow b) and even comments that bring more transparency about how the flow has been achieved. Furthermore, there is still a series of relevant metadata as discussed in item 3 of this chapter. Such metadata will not appear in the flowchart but must be available to complete the inventory. Examples of indispensable metadata are the year or period to which the wood particleboard panel production process refers, technological description of the process, geographical representativeness, cut-off criteria, type of allocation, contacts of those responsible for developing the LCI, among others.

4.5 Final Considerations

This chapter highlighted the main steps in the Life Cycle Inventory Analysis phase and the important support of databases. LCI is the most laborious step in an LCA study, dependent on the definitions of goals and scope (Chap. 3: Life Cycle Assessment (LCA)—Definition of Goals and Scope), and is a subsidy for the Life Cycle Impact Assessment (LCIA), which will be addressed in Chap. 5: LCA—Product Life Cycle Impact Assessment.

LCI is a very critical stage of the study as it is different from the other phases, and it depends a lot on the data holders. First plan or foreground data are only accessible by raising awareness among the various actors involved in the technosphere. If the importance of the study and the LCI methodology are not clear to everyone, data collection may be compromised and affect the credibility of LCA.

Standards and methodological guides rule the data collection process, but there is no common method for all possible product systems. What does not escape the rule is that every product system follows the law of mass and energy conservation, therefore, all matter and energy that comes in has to come out in some way.

LCI alone brings a lot of information about a product's environmental profile on aspects such as consumption of renewable and non-renewable energy, consumption of material of renewable or non-renewable origin, water consumption, air emissions, effluents to water, inert and dangerous solid waste. Thus, the results of the inventory themselves support decision making.

A well-conducted inventory will certainly bring transparency and reliability for the environmental performance of a product or service. LCIA stage will only be able to point out a product system's environmental impacts based on an inventory aligned to the requirements of ISO standards, oriented to completeness, to a data quality standard, attentive to the uncertainties inherent to the process and supported by robust databases.

4.6 Exercises and Case Studies

Question 1: The data collected in an inventory goes through some validation processes to ensure representativeness and suitability to the purpose. The ISO 14044 standard lists data quality requirements (DQRs) that must be considered. List and define the DQRs present in the standard:

Feedback: Temporal coverage, geographic coverage, technological coverage, representativeness, precision, uncertainty, completeness, reproducibility, consistency and data sources.

Question 2: In the flowchart below, identify the types of flows listed, that is, elementary flow, product flow or intermediate flow.

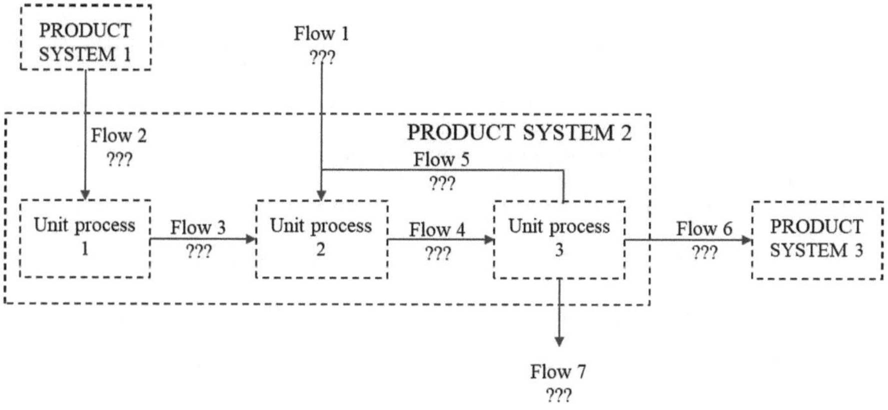

Feedback: ***Flow 1****—elementary/****Flow 2****—product/****Flow 3****—intermediate/****Flow 4****—intermediate/****Flow 5****—intermediate/****Flow 6****—product/****Flow 7****—elementary.*

Question 3: What is the importance of Life Cycle Inventory Databases for LCA studies?

Feedback: LCI databases (DBs) are important because they provide data from background (or second plan) processesfrom which the collection of primary data is practically impossible (time, resource). Thus, DBs significantly reduce the costsof an LCA study.

References

Braga, T.E.N.: Modelo conceitual para gestão da informação tecnológica no Programa Brasileiro de Avaliação do Ciclo de Vida. 2018. 219 f. Tese (doutorado em Ciência da Informação) – Universidade de Brasília, Faculdade de Ciência da Informação, Madrid (2018). Disponível em: http://repositorio.unb.br/handle/10482/34223. Acesso em: 10 abr 2019

Brander, M., Tipper, R., Hutchison, C., Davis, G.: Consequential and Attributional Approaches to LCA: A Guide to Policy Makers with Specific Reference to Greenhouse Gas LCA of Biofuels, 15p. Ecometrica Press, Edinburgh, UK. Technical paper, April, 2009. Disponível em: https://ecometrica.com/white-papers/consequential-and-attributional-approaches-to-lca-a-guide-to-policy-makers-with-specific-reference-to-greenhouse-gas-lca-of-biofuels

Ciroth, A., et al.: Empirically based uncertainty factors for the Pedigree matrix in ecoinvent. Int. J. Life Cycle Assess. 1–11 (2013)

CNT: A fase p7 do Proconve e o impacto no setor de transporte, 20p. CNT – SEST/SENAT, Brasília (2012)

CONMETRO. Resolução no 04, de 15 de dezembro de 2010. [S.l: s.n.] (2010). Disponível em: http://www.inmetro.gov.br/legislacao/resc/pdf/RESC000236.pdf. Acesso em: 28 abr 2016
DIN. DIN EN 15804: Sustainability of construction works—environmental product declarations—core rules for the product category of construction products (2014)
EC—European Commission: General Guide for Life Cycle Assessment—Detailed Guidance, 417p. EC-JRC-IES, Ispra, Italy (2010). Disponível em: https://eplca.jrc.ec.europa.eu/uploads/ILCD-Handbook-General-guide-for-LCA-DETAILED-GUIDANCE-12March2010-ISBN-fin-v1.0-EN.pdf
ECa—European Commission: Specific Guide for Life Cycle Inventory Data Sets, 142p. EC-JRC-IES, Ispra, Italy (2010). Disponível em: http://publications.jrc.ec.europa.eu/repository/handle/JRC48182
Ecoinvent: Data provider toolkit—ecoSpold2. Disponível em: https://www.ecoinvent.org/data-provider/data-provider-oolkit/ecospold2/ecospold2.html
Ecoinvent: Ecoinvent 3.5. Disponível em: https://www.ecoinvent.org/database/ecoinvent-35/ecoinvent-35.html
ELCD—European Reference Life-Cycle Database. Disponível em: https://eplca.jrc.ec.europa.eu/ELCD3/
GLAD—Global LCA Data Access Network. Disponível em: https://www.globallcadataaccess.org/about
ISO 14040: Environmental Management - Life Cycle Assessment - Principles and Framework. 20p. ISO, Geneva (2006a)
ISO 14044: Environmental management — Life cycle assessment — Requirements and guidelines, 46p. ISO, Geneva (2006b)
ISO 14046—Environmental Footprint—Water Footprint—Principles, Requirements and Guidelines, 33p. International Organization for Standardization, Genebra (2014)
LCDN—Life-Cycle Data Network. Disponível em: https://eplca.jrc.ec.europa.eu/LCDN/datasetList.xhtml;jsessionid=56EB113CFCF1F802BBB17A2AC1D3BEB5
Lloyd, S.M., Ries, R.: Characterizing, propagating, and analyzing uncertainty in life-cycle assessment. J. Ind. Ecol. **11**(1), 161–181 (2007)
Ontologia: Avaliação do Ciclo de Vida – Ibict. Disponível em: http://ontologia.acv.ibict.br/search?browse-all=yes
Rodrigues, T.O., Sugawara, E.T., Silva, D.L.A., Matsuura, M.I.S.F., Braga, T.E.N., Ugaya, C.M.L.: Guia Qualidata: requisitos de qualidade de conjuntos de dados para o banco nacional de inventários do ciclo de vida, 49p. Ibict, Brasília (2016). Disponível em: http://acv.ibict.br/wp-content/uploads/2017/05/Qualidata.pdf
SICV BRASIL – Banco Nacional de Inventários do Ciclo de Vida. Disponível em: http://sicv.acv.ibict.br/
Silva, D.A.L., Masoni, P.: Diálogos Setoriais Brasil e União Europeia: análise crítica das principais políticas de gestão, manutenção e uso de bancos de dados internacionais de inventários do ciclo de vida de produto. Instituto Brasileiro de Informação em Ciência e Tecnologia- Ibict, 139p. Brasília (2016)
UNEP—United Nations Environment Programme: Global Guidance Principles for Life Cycle Assessment databases—A Basis for Greener Processes and Products, 160p. UNEP-SETAC, Paris, France (2011). Disponível em: https://www.lifecycleinitiative.org/wp-content/uploads/2012/12/2011%20-%20Global%20Guidance%20Principles.pdf
UNIÃO EUROPEIA: Manual do Sistema ILCD : Sistema internacional de referência de dados do ciclo de produtos e processos : Guia geral para avaliações do ciclo de vida : orientações detalhadas. Trad.: Instituto Brasileiro de Informação em Ciência e Tecnologia. Ibict, Brasília (2014). Disponível em: http://acv.ibict.br/documentos/publicacoes/869-manual-do-sistema-ilcd-sistema-internacional-de-referencia-de-dados-do-ciclo-de-produtos-e-processos-guia-geral-para-avaliacoes-do-ciclo-de-vida-orientacoes-detalhadas/

Weidema, B.P., Ekvall, T.: Guidelines for applications of deepened and broadened LCA: consequential LCA. Chapter for CALCAS project, 19p. Deliverable D18 (2009). Disponível em: https://lca-net.com/files/consequential_LCA_CALCAS_final.pdf

Weidema, B.P., Wesnaes, M.S.: Data quality management for life cycle inventories—an example of using data quality indicators. J. Clean. Prod. **4**(3–4), 167–174 (1996)

Wolf, M.A., Düpmeier, C., Kusche, O.: The international reference life cycle data system (ILCD) format—basic concepts and implementation of life cycle impact assessment (LCIA) method data sets. In: EnviroInfo 2011: Innovations in Sharing Environmental Observation and Information, Proceedings of the 25th International Conference Environmental Informatics, 5–7 Oct 2011, Ispra, Italy. Disponível em: http://enviroinfo.eu/sites/default/files/pdfs/vol7233/0809.pdf

Chapter 5
LCA—Product Life Cycle Impact Assessment

Ana Laura Raymundo Pavan and Natalia Crespo Mendes

5.1 Introduction

In this chapter, the life cycle impact assessment (LCIA) phase will be addressed, which is fundamental in quantifying environmental impacts according to the goal and scope of each study, as we saw in Chap. 3: Life Cycle Assessment (LCA)—Definition of Goals and Scope. Thus, LCIA is defined as the stage that "*aims at understanding and assessing the magnitude and significance of a product system's potential environmental impacts throughout the life cycle*" (ABNT 2009a).

The purpose of the impact assessment phase is, therefore, to convert life cycle emissions and resource consumption into indicators related to the impact categories and areas of protection (AoPs), which will be discussed below. As previously mentioned, LCA studies are constructed based on the function of the product system; thus, LCIA has a relative approach based on a functional unit. This is one of the main differences between LCIA and other techniques such as, for example, the environmental impact assessment (EIA) and the environmental performance assessment. Another particular characteristic of LCIA is that its results are estimates of potential impacts, which, according to Hauschild and Huijbregts (2015), can be seen as relative performance indicators guiding comparisons and system improvements, although they cannot be monitored in the real world. For this purpose, in LCIA, the assessment is based on scientifically accepted environmental models adopted to operate within the restrictions of the LCA.

In the next sections, the methodological elements that make up the LCIA (Fig. 5.1) according to ISO 14040 (ABNT 2009a) will be addressed, which may be mandatory and optional. In addition, not only will be presented the fundamental principles of LCIA, its phases and state of the art, but also a brief discussion of the existing impact categories.

A. L. R. Pavan (✉) · N. C. Mendes
Katholieke Universiteit Leuven (KU Leuven), Kasteelpark Arenberg 44, 3001 Leuven, Belgium

J. A. de Oliveira et al. (eds.), *Life Cycle Engineering and Management of Products*,
https://doi.org/10.1007/978-3-030-78044-9_5

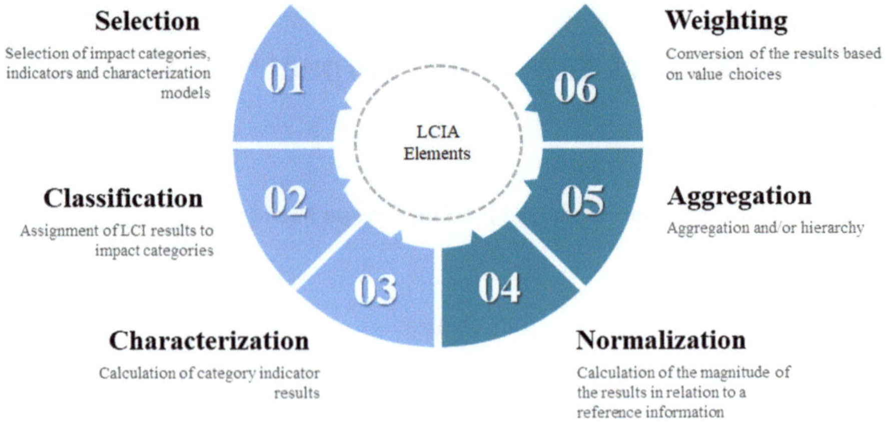

Fig. 5.1 Life cycle impact assessment: mandatory (in light blue) and optional (in dark blue) methodological elements. *Source* Authors

Although the normative guidelines of ISO 14040 (ABNT 2009a) and ISO 14044 (ABNT 2009b) provide a standardization and a general guide for LCIA, they do not provide technically detailed recommendations on indicators and models to be used, for example. Thus, the LCIA phase is in constant debate and improvement by the scientific community in order to obtain methodological advances in terms of modeling, indicators and analysis of uncertainties.

5.2 Mandatory Elements of the Life Cycle Impact Assessment

5.2.1 Selection of Impact Categories, Category Indicators and Characterization Models

The first step of LCIA consists of selecting which impact categories will compose the study (in a manner consistent with the goal and scope of LCA, seen in Chap. 3: Life Cycle Assessment (LCA)—Definition of Goals and Scope) and, therefore, what will be the characterization models and indicators of each category used in the quantification of environmental impacts. This step aims at ensuring that the impact categories are identified and that LCA user explicitly selects and justifies the relevant categories for his particular study.

It must be noted that such categories of environmental impact are classes that represent relevant environmental issues for the assessment of the environmental performance of products and decision making (ABNT 2009b). For each impact

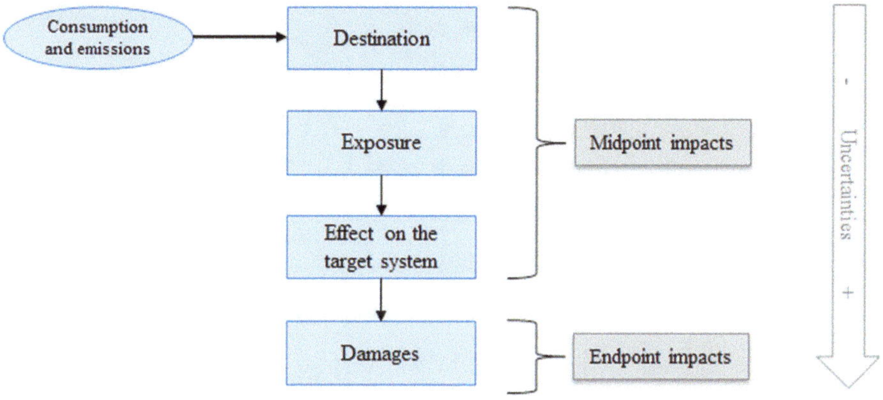

Fig. 5.2 General structure of a cause and effect chain in the life cycle impact assessment. *Source* Authors

category, there is a system of physical, chemical and biological processes with environmental effects that connect the results of the Life Cycle Inventory—LCI phase, seen in Chap. 4: LCA—Life Cycle Inventory Analysis and databases, to the category indicators and category end points.[1] This system is called cause and effect chain or environmental mechanism (ABNT 2009a).

Figure 5.2 shows the general scheme of the cause and effect chain logic, which begins with the elementary flow data (consumption and emissions). For the development of the environmental mechanism, not only the properties of the substances are considered but also the characteristics of the receiving environment, exposure to pollutants and the potential effect. It is also possible to observe in Fig. 5.2 different levels of impacts usually classified as intermediate and final, referring respectively to primary effects and damage caused to humans and ecosystems.

These terms will be detailed in item 5.4 of this chapter. According to ISO 14040 (ABNT, 2009a), at this stage, not only the environmental mechanism of the categories must be described but also the characterization models that correlate the results of the LCI to the defined indicators.

In addition, to facilitate understanding, Fig. 5.3 summarizes the procedures that make up the first stage of LCIA. For the choice of the characterization method and model, the main aspects to be considered, according to the ISO 14040 standard, refer to scientific and technical validity as well as the basis on a specific environmental mechanism. In this example, we see that for the acidification impact category, the selected indicator refers to base saturation and for eutrophication, the concentration of phosphorus. Following the logic of the cause and effect chain, the approaches for characterizing these two categories aim at tracing the destination of emissions (acidifying and eutrophic), the degree to which a receiving environment is exposed, the effect of this exposure and the severity of the effect. In the example given in

[1] They refer to the attribute or aspect of the natural environment, human health or resources that identifies an environmental issue that deserves attention (ABNT 2009b).

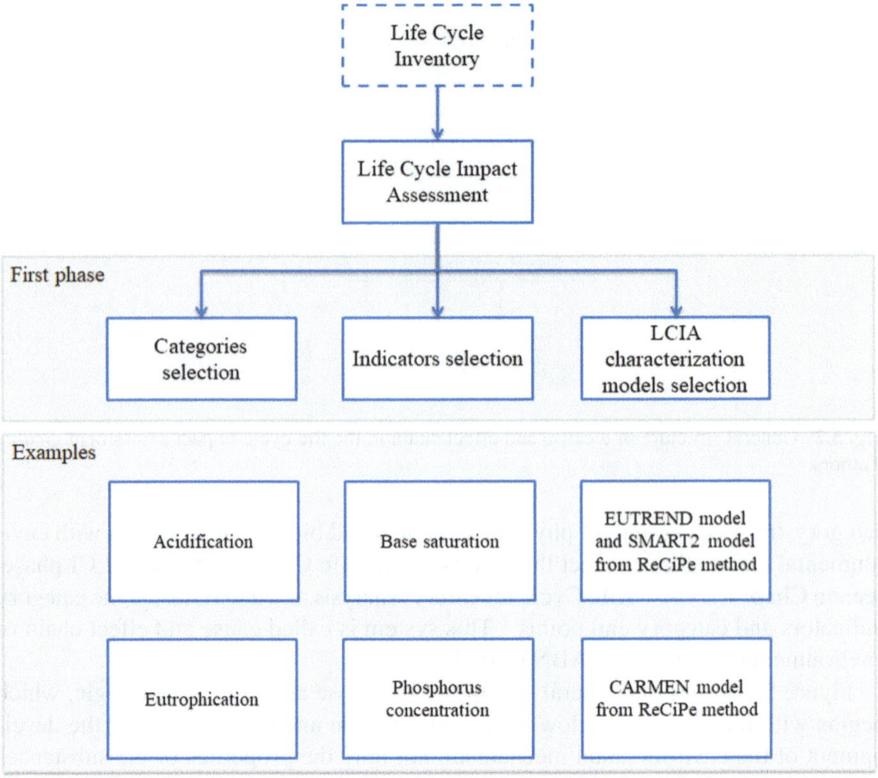

Fig. 5.3 First stage of the life cycle impact assessment: selection of impact categories. *Source* Authors

Fig. 5.3, the selected method is called ReCiPe composed of different characterization models, as shown in the figure.

We can find a large number of methods and characterization models incorporated in LCA softwares and available in literature. They differ in specific aspects such as the number of impact categories available, the number of substances included, the environmental models developed for the characterization phase and the level used in the modeling (intermediate or final).

Therefore, it is important that at this stage, a review of the scientific guidelines on the subject is carried out. It is worth mentioning that, in addition to checking the technical validity of the methods, the LCA user must also verify the geographic applicability and assumptions of the model, so that his choice is aligned with its goal and scope. More details on the different existing LCIA methods will be covered in the following topics.

Some practical difficulties at this stage, such as the choice between the different indicators and characterization models available for each impact category, are mainly due to the lack of standardization in the LCA literature. Despite efforts to develop a methodological consensus (Jolliet et al. 2018), these issues can influence comparability between LCA studies, data collection efforts and the quality of results (Reap et al. 2008).

5.2.2 *Classification*

In the classification step, all elementary flows compiled in the inventory must be assigned to the impact categories selected in the previous step (5.2.1. Selection of impact categories, category indicators and characterization models) (EC-JRC 2010). Each elementary flow can be assigned to one or more impact categories according to the environmental impacts they can cause.

According to Hauschild and Huijbregts (2015), when working with a standard list of elementary flows for which characterization factors were previously derived, the work to be performed as part of the classification step is significantly reduced. Various LCA guides, tools and software support this step. Thus, most of the inventory results have already been pre-classified for pre-selected impact categories. Also according to these authors, for the impact categories already recognized in the field of LCA, with well-defined environmental mechanisms and with a limited number of elementary contributing flows (such as climate change, acidification and eutrophication), the standard classification lists of elementary flows are highly similar among the different models of LCIA.

In case there is no pre-classification for inventory results, the LCA user can follow the recommendations of ISO 14044 (ABNT 2009b) to perform the classification step of his study. Initially, the attribution of LCI results to the impact categories must consider the following cases:

- Attribution of LCI results that are exclusive to an impact category;
- Identification of LCI results related to more than one impact category, including a distinction between parallel mechanisms (e.g., sulfur dioxide—SO_2—is distributed between the categories human health impact and acidification) and attribution to serial mechanisms (e.g., nitrogen oxides—NO_x—can be classified to contribute to both ozone formation at ground level and acidification).

The cases in which LCI results related to more than one impact category are identified, that is, when some environmental aspects contribute to different environmental problems, they may generate confusion or even lead to possible "double count" problems (Hauschild and Huijbregts 2015). For example, nitrogen oxides (NO_x) can contribute to acidification and eutrophication. Duplication occurs when the same NO_x emission is erroneously attributed to two or more categories instead of properly distributing it. This type of error may occur either due to user failure or also due to failures in the models used to divide the emissions between the impacted compartments (terrestrial, aquatic or air environment) (Guinée and Heijungs 1993). Duplication increases the impact of a specific load and compromises the results of a study generating erroneous conclusions during the interpretation phase. Therefore, it is important to understand the environmental mechanism of impact categories and appropriate classification of models.

5.2.3 Characterization

Once the classification step is completed, environmental impacts are quantified, that is, the conversion of LCI results to common units and aggregation of results within the same impact category.

The environmental impact of each emission or consumption of resources is modeled quantitatively according to the environmental mechanism through mathematical models called characterization models. Such models reflect the sum of environmental processes related to the characterization of impacts and are used to calculate the characterization factors. Thus, each elementary flow assigned to an impact category must be multiplied by a corresponding characterization factor. For example, all greenhouse gases can be expressed in terms of equivalent CO_2, multiplying the relevant LCI results by characterization factors and combining the resulting impact indicators to provide a general indicator of the potential impact of global warming (Huijbregts et al. 2016).

The ISO 14044 standard (ABNT 2009b) requires the characterization models used for each category indicator to be internationally accepted and scientifically valid. The characterization models are in constant development and for each impact category there may be several models with different sets of characterization factors available. Consequently, the same LCA study may present different results for a given impact category depending on the model used, which may increase the uncertainty of the results. It is also important to clarify the definition of the LCIA method, which in this publication is considered to be the collection of models for characterizing different impact categories and is sometimes given a name or trademark.

Figure 5.4 summarizes the structure of LCIA based on its mandatory elements and exemplifies some of the categories of environmental impact assessed in this step. As previously discussed, there are several characterization models available varying in terms of the indicator used and the definition of the characterization factors. In the figure below, there are some examples of indicators and characterization factors for the categories of global warming, ozone depletion, human toxicity and eutrophication. Note that for each impact category, there are characterization factors that express the impact of each inventory flow in relation to a reference substance. This level of detail will depend on each category and characterization model used and is something that must be carefully observed and reported in the LCIA results.

Further details on the different impact categories and the existing characterization models will be presented in Sect. 5.4. With regard to the characterization factors, Udo de Haes et al. (2002) proposed a generic structure for calculation in which the characterization factor is the product of a destination factor (considering the receiving environment), an exposure factor (for the exposure of the receiving environment) and an effect factor (expresses the ability of a substance to cause disease or toxic effects

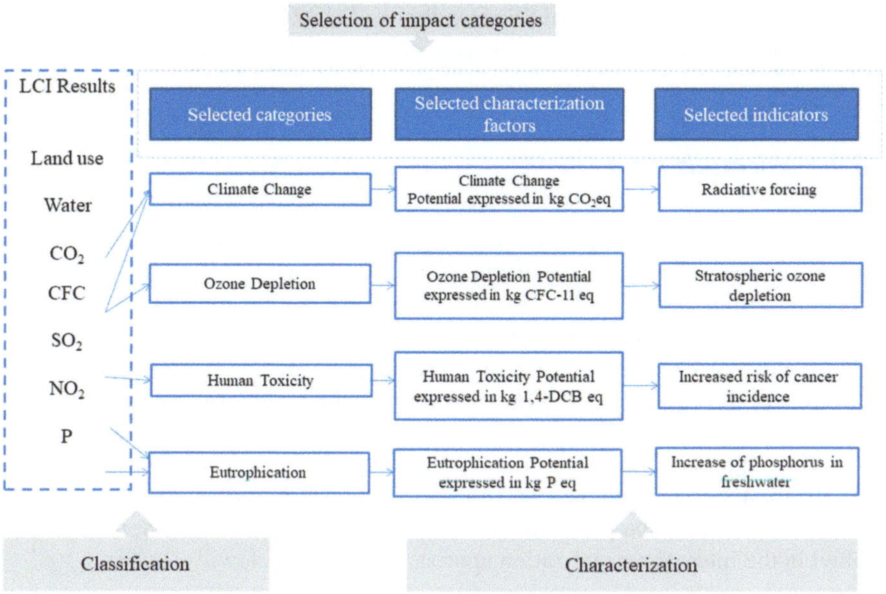

Fig. 5.4 Structure of the LCIA based on its mandatory elements. *Source* Authors

to exposed ecosystems). This generic structure has been applied to most impact categories related to emissions with modifications that vary according to each impact category (Hauschild and Huijbregts 2015). Taking the case of human toxicity as an example, the fate and exposure factors are combined to reflect the ingested fraction of a substance, representing the mass fraction of the substance that affects the population (Rosenbaum et al. 2008).

In the end, these factors are used in the total quantification of the impact potential, in general, through Eq. 5.1:

$$I = \sum i \sum x FC_{x,i} \times E_{x,i} \tag{5.1}$$

where:

I	Impact potential;
$FC_{x,I}$	Characterization factor of substance x emitted to compartment i;
$E_{x,I}$	Flow registered in the LCI, that is, emission of substance x to compartment i.

5.3 Optional Elements of the Life Cycle Impact Assessment

The LCA user also has optional elements of the LCIA to complement his study, they are: normalization, grouping and weighting. These elements are provided by ISO

14044 (ABNT 2009b) and the ILCD Handbook (EC-JRC 2010), and their need for application is defined according to the goal and scope of each study.

5.3.1 *Normalization*

The normalization stage consists of converting the results obtained for each impact category into a unit common to all the categories selected in the study. From this procedure, a scale capable of expressing the magnitude of the impact scores obtained in the characterization stage is established (see Sect. 5.2.3. Characterization).

Normalization can be performed by two approaches according to its purpose and the source of the reference data: internal normalization and external normalization (Norris 2001).

The first approach, internal normalization, is so named because it uses data and normalization factors from the system under study and is applied to obtain comparable scales and to solve problems related to the plurality of units. Some methods applied in the internal normalization approach are listed below:

- Division by baseline: The results of the characterization are divided by a value from a baseline selected among the alternatives. Examples and further information can be found in the work of Hauschild and Huijbregts (2015).
- Division by maximum value: The results of the characterized indicators are divided by the highest score obtained among the alternatives of each impact category. Examples and further information can be found in Norris and Marshal (1995).
- Division by summation: The results of the characterized indicators are divided by the sum of the scores obtained for all alternatives. More information about this methodological procedure is also presented in the study by Norris (2001).

External standardization, in turn, analyzes the significance of the results of the category indicators and the impacts are normalized with external references and, therefore, independent of the object of study of LCA. In this case, normalization factors can be an ideal value for that category, defined based on public policies or scientific criteria, for example. This procedure is also known as *distance to target*. Another option of external normalization is done by dividing the results characterized by an estimate of the total or *per capita* equivalent emissions associated with an entire geographic region, as shown in Eq. 5.2 (Prado et al. 2017).

$$\mathrm{NI}_{a,i} = \frac{\mathrm{CI}_{a,i}}{\mathrm{NR}_i} \tag{5.2}$$

where:

$\mathrm{NI}_{a,I}$ the normalized impact per year of the alternative a of category i;
$\mathrm{CI}_{a,I}$ the characterized impact of the alternative a of impact category i;

NR_i the normalization reference representing a specific geographic region for impact category i on physical units (per year) corresponding to the characterized impact $CI_{a,i}$.

Pizzol et al. (2017) listed the following existing methods for performing external normalization:

- Global normalization: The results of the characterized indicators of the system(s) under study are divided by the results of the characterized indicators of the total activities underway in the world throughout the reference duration (assumed balance between consumption and production). For more information and examples, see Sleeswijk et al. (2008).
- Normalization based on production: The results of the characterization of the system(s) under study are divided by the characteristic indicators associated with all territorial activities in a region or country, including its exports, but excluding its imports, accounting for all flows that occur within the geographic limits of that region or country throughout the reference duration. Several studies present details for applying this method, such as Bare et al. (2006).
- Normalization based on consumption: The results of the characterization of the system(s) under study are divided by the results of the characteristic indicators associated with the total territorial consumption of a region or country, including its imports, but excluding its exports. Thus, it accounts for the environmental flows of all the processes necessary to support the consumption activities of that region or country throughout the reference period including those that occur outside its physical or geographic limits. Examples and description of the methods can be found in Dahlbo et al. (2013).
- Carrying capacity (planetary boundaries): Most recent approach in which the characterized indicator results are divided by the normalization references that represent the supportability of the reference system for each impact category (Bjørn and Hauschild 2015). More information on the challenges of implementing the concept of planetary limits and example of application can be found in the study by Uusitalo et al. (2019).

The current recommendation of the scientific community is to prioritize external normalization approaches in studies that apply normalization, that is, approaches in which the reference system is independent or not directly related to the alternatives assessed in the study. In practice, it is also important to note that there are data gaps in the current external normalization references, which can compromise the results (Verones et al. 2017).

5.3.2 Grouping

In the grouping stage, the LCIA user can group and rank the impact categories selected in his study as long as he maintains consistency between the units. This can

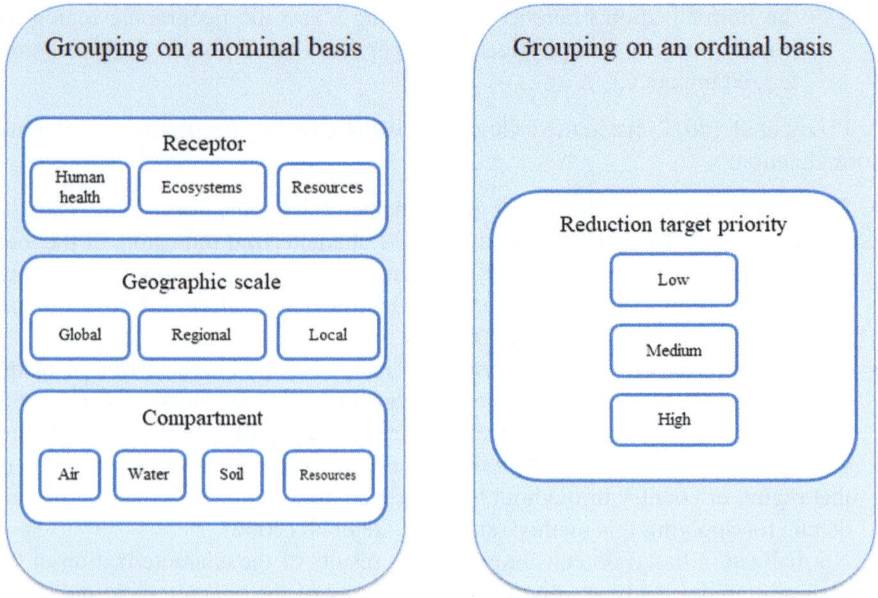

Fig. 5.5 Examples of types of grouping. *Source* Authors

be done in two ways: by classification, by grouping on a nominal basis (e.g., global versus regional) and ranking, by grouping on an ordinal basis (e.g., high, medium and low priority). Figure 5.5 presents examples for the two forms of grouping.

5.3.3 Weighting

The weighting stage allows the definition of factors which indicate the relevance that the selected impact categories have within a given LCA study. This stage, therefore, aims at resolving situations in which there is a conflict of choice (trade-offs) between the environmental impacts that are measured in different biophysical units and, therefore, are incomparable. It is considered by many to be a controversial stage, sometimes being criticized in part for being a process based on value choices and not scientifically based and for requiring the incorporation of social, political and ethical values (Finnveden et al. 2006).

However, weighting can facilitate the aggregation of indicators and the interpretation of LCIA, in addition to reflecting the preferences of the decision-maker(s) and stakeholders. The weights applied must represent an assessment of the relative importance of impacts according to the specific value of the choices, reflecting the preferences, for example: of people, specialists or organizations, in relation to time (present versus future impacts), geography (local versus global), urgency, political

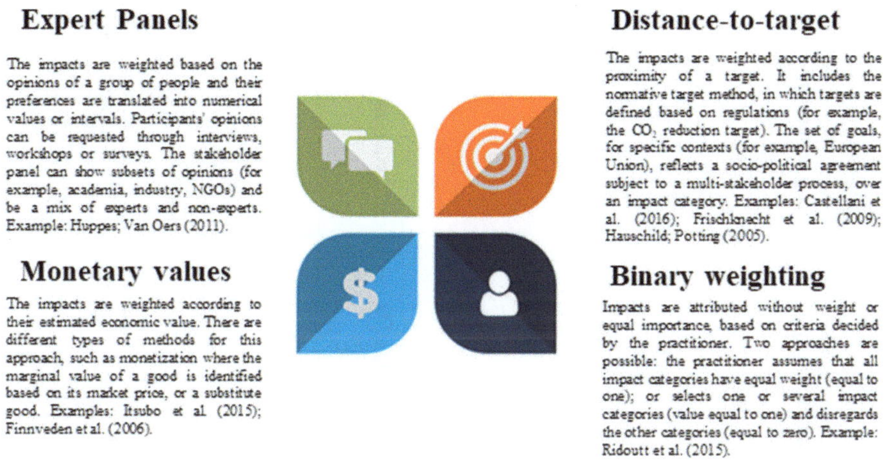

Fig. 5.6 Four different methodological approaches for the weighting stage. *Source* Authors

agendas or cost. Thus, with weighting, the results can be added in impact categories in order to get to a single score indicator (Pizzol et al. 2017).

Not only are there values involved in choosing weighting factors but also in choosing the type of weighting method to be used. Weighting methods use data and methodological procedures taken from different scientific disciplines that may and must be assessed, and value choices can be identified and clarified. In Fig. 5.6, some examples of methodological approaches are described, namely: panel of experts, monetary weighting, distance to target and binary weighting. Bibliographical references to access more details of each of these approaches are also found in Fig. 5.6.

5.4 Main Impact Categories and Methods of LCIA

As previously discussed, LCIA structures the various effects caused to the environment resulting from human actions in different impact categories according to its environmental mechanisms. Next, it is presented a summary of the categories commonly assessed in LCIA and some of their characteristics. It is emphasized that, often, different methodological approaches can be found for the same impact category, and there is no intention here to analyze and recommend which one to adopt. The following detailed information, in Chart 5.1, is based on the LCIA ReCiPe method (Goedkoop et al. 2009) as an example.

It is also important to highlight that there are two methodological approaches according to the levels of impact assessment:

- Intermediate impacts (midpoint): Modeling that links the results of the LCI to the impact categories but does not reach the end of the environmental damage

Chart 5.1 Examples of impact categories, their indicators and characterization factors

Impact category	Foundation	Characterization factor	Indicator	Unit
Climate change	Impact related to the increase in terrestrial temperature in the lower atmosphere caused by the increase of greenhouse gases. These gases absorb the radiation emitted by the Earth's surface, causing it to heat up	Global warming potential	Infrared radiative forcing	kg CO_2 equivalent
Ozone depletion	Impact related to decreased levels of ozone (O_3) in the stratosphere caused by emissions of substances such as chlorofluorocarbons (CFCs). This causes a higher incidence of ultraviolet radiation on the Earth's surface	Ozone depletion potential	Stratospheric ozone depletion	kg CFC-11 equivalent
Terrestrial acidification	Related to the release of hydrogen ions (H^+) in terrestrial and aquatic ecosystems. The reaction of substances such as sulfur dioxide (SO_2), for example, with atmospheric water vapor, gives rise to sulfuric (H_2SO_4) and nitric (HNO_3) acids which, therefore, results in changes in the natural and built environment	Terrestrial acidification potential	Increase of protons in soils	kg SO_2 equivalent

(continued)

Chart 5.1 (continued)

Impact category	Foundation	Characterization factor	Indicator	Unit
Eutrophication in fresh water	Impact related to the addition of nutrients to aquatic surfaces that causes an increase in the production of biomass and consequent reduction in the concentration of dissolved oxygen	Potential for eutrophication in fresh water	Increase of phosphorus in fresh water	kg P equivalent
Human toxicity (cancer)	Impacts related to human exposure to toxic substances, especially through ingestion and inhalation, with carcinogenic effects	Marine eutrophication potential	Increased risk of cancer incidence	kg 1.4-DCB equivalent
Human toxicity (non-cancer related)	Impacts related to human exposure to toxic substances	Human toxicity potential	Increased risk of incidence of other diseases	kg 1.4-DCB equivalent
Formation of particulate material	The exposure of particulate materials is associated with several adverse health effects and reduced life expectancy	Potential for formation of particulate material	PM2.5 increase	kg PM2.5
Terrestrial ecotoxicity	Related to the emission of toxic substances in the soil	Terrestrial ecotoxicity potential	Weighted increase of risk in natural soils	kg 1.4-DCB equivalent
Ecotoxicity in fresh water	Related to the emission of toxic substances in water	Potential for ecotoxicity in fresh water	Weighted increase of risk in fresh water	kg 1.4-DCB equivalent

Source Adapted from Huijbregts et al. (2016)

assessment. It can be said that this modeling has a lower degree of uncertainty but requires greater efforts in the interpretation of results.

- Environmental damage (endpoint): It covers the entire environmental mechanism referring to the damage caused to human health, ecosystems and natural resources. The characterization of damages allows the assessment of the environmental contribution of each of the intermediate categories in relation to one or different damage categories. In this case, it can be said that it presents a higher degree of uncertainty, but results are easily interpreted.

Chart 5.2 presents a summary of the main LCIA methods used in LCA studies and which of these approaches fit.

As seen in Chart 5.2, most of the LCIA methods have been developed by foreign institutions with a scope restricted to specific countries or continents. Even when they have a global application scope, these methods are often based on a set of data and parameters from specific regions. In order to promote the regionalization of characterization methods and models for application in Brazil, several studies have been developed for each impact category in search of a more consistent model with the country's specificities. In this sense, the publication by Ugaya et al. (2019) gathers information on the challenges and opportunities for the development and application of regionalized parameters. This publication also summarizes the recommendations of characterization models for the categories of impact of water scarcity, abiotic resources, biotic resources, resource accounting method (RAM), ecosystem services, eutrophication and acidification for the Brazilian context.

In Chart 5.2, it is also possible to observe that some models mentioned have a combined approach of intermediate and final impacts, that is, midpoint and endpoint. The impact assessment at the level of environmental damage (or endpoint) includes specific damages related to a wider area of protection and the characterization considers the entire environmental mechanism up to its final point. Similar to intermediate impact modeling, damage modeling must also have a scientific basis. Damage modeling is available at a disaggregated level by impact categories (e.g., climate change), which can be aggregated at more comprehensive levels in the so-called areas of protection (AoP). The definition of these areas aims at protecting the values considered important for society. For example, AoP "human health" uses aggregated impacts of morbidity and mortality as an indicator to measure damage to human health.

After a process of revising the AoPs, a new division into two comprehensive systems has been defined by Verones et al. (2017): (i) natural systems and, (ii) anthropic systems; with three possible value[2] assignments. Natural systems go beyond the concept of ecosystems, also including immaterial goods such as natural heritage, while anthropic systems are related to values related to man such as socioeconomic assets and cultural heritage. Thus, six potential areas of protection are identified for consideration in LCIA (Fig. 5.7).

[2] Here considered as aspects that society considers a target of protection.

Chart 5.2 Examples of LCIA methods and their characteristics

Method	Origin	Approach	Application scope	References
CML 2002	CML—Netherlands	Midpoint	Most categories are global in scope, but the categories acidification and formation of photo-oxidants are created specifically for Europe	Guinée et al. (2002)
EDIP97; EDIP2003	DTU—Denmark	Midpoint	The first version is global in scope while the 2003 version has been developed for application in Europe	Wenzel et al. (1997), Hauschild and Wenzel (1998), Hauschild and Potting (2005), Potting and Hauschild (2005)
Impact 2002 +	EPFL—Switzerland	Combined (midpoint e endpoint)	Europe	Crettaz et al., (2002), Jolliet et al. (2004), Payet (2002), Pennington et al. (2005), Pennington et al. (2006), Rochat et al. (2006), Rosenbaum (2006), Rosenbaum et al. (2007)
ReCiPe	PRé, CML, RUN, RIVM—Netherlands	Combined (midpoint e endpoint)	Categories such as climate change, ozone depletion and depletion of fossil resources for example are global in scope, while categories such as ecotoxicity, ionizing radiation and human toxicity for example have been developed for Europe	De Schryver et al. (2009), Huijbregts et al. (2005a, b), Struijs et al. (2008), Van Zelm et al. (2007a, b) Wegener Sleeswijk et al. (2008)

(continued)

Chart 5.2 (continued)

Method	Origin	Approach	Application scope	References
TRACI	US EPA—USA	Midpoint	Categories such as ozone depletion, global warming and depletion of fossil fuels are global in scope while others, such as acidification and human health for example, have been developed for the USA	Bare et al. (2003), Hertwich et al. (1997), Hertwich et al, (1998), Hertwich et al. (1999); Hertwich et al. (2001), Norris (2002)
USEtox	Life Cycle Initiative Program	Midpoint	Global	Rosenbaum et al. (2008), Hauschild et al. (2008)

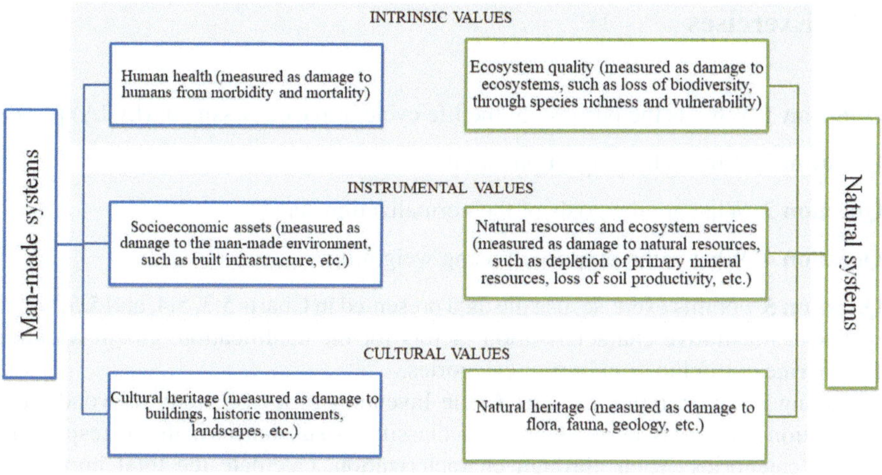

Fig. 5.7 Classification of AoPs according to their origin in anthropic or natural systems and their values. *Source* Based on Verones et al. (2017)

Finally, it is emphasized that the aggregation in single scores by AoP can facilitate the decision-making process and the communication of results but, at the same time, it can decrease transparency in relation to uncertainties and trade-offs between impact categories.

5.5 Final Considerations

In this chapter, the elements that make up the LCIA (mandatory and optional), its characteristics and methodological procedures were presented. As they refer to the phase in which the magnitude and significance of potential environmental impacts is assessed, the results of the LCIA are essential to support the Engineering and Life Cycle Management (LCEM) of products and must be interpreted accordingly with the goal and scope of the LCA study.

It is noteworthy that LCIA is highly interdependent with scientific advances inside and outside the LCA field and the approaches to modeling impacts are constantly being revised and improved. In this way, the user must remain attentive to discussions and recommendations on topics such as category indicators, scope of AoP coverage, documentation of the characterization models and their uncertainties, so that decision making through LCA is consistent and scientifically based.

Finally, considering the limitations that involve the quantification of potential environmental impacts and the discussions of trade-offs that permeate it, the studies must be guided especially in terms of transparency and ensure that all environmentally relevant information and indicators are included.

5.6 Exercises

Question 1. What is the purpose of the life cycle impact assessment (LCIA) phase?

Question 2. What is the impact potential?

Question 3. What are the goals of the normalization stage?

Question 4. What is the purpose of using weighting?

Question 5. For this exercise, use the data presented in Charts 5.3, 5.4, and 5.5, which represent illustrative characterization factors for the acidification, global warming and marine eutrophication impact categories.

Taking as an example, the Life Cycle Inventory of a case study of wood panel production in Brazil (Charts 5.6 and 5.7), classify the substances in the corresponding impact categories. Then, through characterization, calculate the total impact for

Chart 5.3 Characterization factors for the terrestrial acidification category referring to the emission to air

Substances	Characterization factor (kg SO_2-eq/kg substance)
NO_x (nitrogen oxides)	0.35
NH_3 (ammonia)	1.9
SO_2 (sulfur dioxide)	1

Chart 5.4 Characterization factors for the global warming category referring to the emission to air

Substances	Characterization factor (kg CO_2-eq/kg substance)
CO_2 (non-fossil carbon dioxide)	1
CO_2 (fossil carbon dioxide)	1
CH_4 (non-fossil methane)	84
CH_4 (fossil methane)	85

Chart 5.5 Characterization factors for marine eutrophication category

Emission compartment	Substance	Characterization factor (kg N-eq/kg substance)
Fresh water	NH_3 (ammonium)	0.30
Air	NH_3 (ammonium)	0.09
	NOx (nitrogen oxides)	0.04
	NO_2 (nitrogen dioxide)	0.04

Chart 5.6 LCI of the forest stage

Eucalyptus wood produced in Brazil					
Inputs			Outputs		
Material	Quantity	Unit	Material	Quantity	Unit
Lubricating oils	1.00×10^{-2}	kg	Wood (Eucalyptus ssp. from sustainable forest management)	1.00 (ou 687 kg)	m^3
Phosphate fertilizer, as P_2O_5	7.36×10^{-1}	kg	*Emissions to the air*		
Ammonium sulfate, as N	5.85×10^{-2}	kg	Sulfur dioxide	8.00×10^{-3}	kg
Urea, as N	3.10×10^{-2}	kg	Ammonia	6.33×10^{-2}	kg
Tree seedlings	6.38	Units	Glyphosate	1.87×10^{-1}	kg
Diesel	3.35	kg	Xylene	1.76×10^{-5}	kg
Potassium chloride, such as K_2O	1.14	kg	Methane, fossil	5.27×10^{-5}	kg
Carbon dioxide, in the air	9.16×10^{2}	kg	Nitrogen oxides	1.16×10^{-1}	kg
Land occupation	1.95×10^{2}	m^2*year	NMVOC, of unspecified origin	7.80×10^{-2}	kg
Sulfluramid	7.70×10^{-2}	kg	Carbon monoxide, fossil	5.40×10^{-2}	kg
Land transformation	30.00	m^2	Nitrogen dioxide	7.00×10^{-3}	kg
			Carbon dioxide, fossil	15.10	kg
			Emissions to the soil		
			Oils, unspecified	2.00×10^{-3}	kg

NMVOC Non-methane volatile organic compounds. *Source* The SICV Brasil database (https://sicv.ibict.br/Node/)

each impact category and identify the environmental[3] aspects that make the greatest contribution to each.

Feedback

Question 1.

The life cycle impact assessment phase introduces knowledge about the environment and resources to help the interpretation of results of the inventory according to the goals of the study. In this phase, the information present in the inventory is used in the quantitative modeling of the impact potentials for different impact categories.

[3] Element of the activities, products or services of an organization that can interact with the environment. (NBR ISO 14,040: 2009, p. 2).

Chart 5.7 LCI of the wood panel industrial production stage

Wood panel industrial production					
Inputs			Outputs		
Material	Quantity	Unit	Material	Quantity	Unit
Ammonium sulfate	1.38	kg	Panel	630	kg
Lubricants (oil and grease)	1.80×10^{-1}	kg	*Emissions to the air*		
Paraffin emulsion	5.47	kg	Ash (from wood waste)	3.90×10^{-1}	kg
UF resin	7.17×10^{1}	kg	Carbon dioxide (from fossil fuel)	4.8×10^{1}	kg
Water	9.04×10^{1}	kg	Carbon monoxide	1.90×10^{-1}	kg
Wood (logs)	6.87×10^{2}	kg	VOC, unspecified	3.60×10^{-1}	kg
Energy consumption			Formaldehyde	1.50×10^{-1}	kg
Diesel	1.72	kg	NMVOC	9.48×10^{-4}	g
Eletricity	5.07×10^{2}	MJ	Methane (non-fossil)	1.69	g
Fuel oil	1.37×10^{1}	kg	Nitrogen oxides	1.80×10^{-1}	kg
Wood waste	3.85×10^{1}	kg	Sulfur oxides	1.32	kg
			Emissions to the water		
			Ammonia	1.21×10^{-1}	g
			BOD	6.16×10^{-1}	g
			Formaldehyde	7.29×10^{-1}	g
			Suspended solids	2.44×10^{1}	g
			Emissions to the soil		
			Lubricant Waste	1.59×10^{1}	g
			Wood waste	9.72×10^{1}	kg

VOC volatile organic compounds; *NMVOC* non-methane VOC; *BOD* biochemical oxygen demand.
Source Silva et al. (2013)

Question 2.

Impact potential is a quantitative expression which represents the potential contribution of the life cycle of a product or process to an environmental impact category. The impact potential is usually aggregated over time and space and does not represent a real impact.

Question 3.

The normalization stage aims at facilitating the comparison of contributions for different impact categories, integrating them on a common scale and, thus, preparing them for the weighting stage. This stage allows the verification of impact potentials calculated for the product relating them to impact potentials of a known activity, also called reference system. In addition, it provides a didactic and more easily understandable presentation of the LCIA results.

Question 4.

Weighting is used to classify normalized impact scores according to their relative importance. Weighted impact scores are obtained by multiplying the normalized impact score by a weighting factor. Therefore, weighting can facilitate the aggregation of indicators and the interpretation of LCIA.

Question 5.

The first step is to identify, with the help of Charts 5.3, 5.4 and 5.5, in which impact categories the substances contribute. Thus, it turns out that:

Emissions to the air:

- Ammonia contributes to the marine acidification and eutrophication categories.
- Sulfur dioxide contributes to acidification.
- Nitrogen oxides have effects on marine acidification and eutrophication.
- Carbon dioxide is classified in the global warming category.
- Methane (fossil and non-fossil) contributes to the global warming category.
- Nitrogen dioxide contributes to the marine eutrophication category.

Emissions to the water:

- Ammonia contributes to the marine eutrophication category.

The characterization stage takes place as in Charts 5.8, 5.9 and 5.10.

Thus, the biggest contribution in the global warming category comes from the carbon dioxide emitted in industrial production.

In the case of the acidification category, the ammonia (NH_3) emitted into the air in the forest phase is the major responsible for the impact.

For the marine eutrophication category, nitrogen oxides (NO_x) emitted into the air in the forest phase are the ones which most contribute to the potential impacts.

Chart 5.8 Calculation of the potential impacts of global warming

Global warming	Quantity (kg)	CF (kg CO_2-eq/kg GHG)	Impact characterization
Methane, CH_4, fossil (forest phase)	5.27×10^{-5}	85	4.48×10^{-3}
Methane, CH_4, non-fossil (industrial production)	1.69×10^{-3}	84	1.42×10^{-1}
Carbon dioxide CO_2, fossil (forest phase)	15.1	1	15.1
Carbon dioxide, CO_2, fossil (industrial production)	48	1	48
Total			63.25 kg CO_2-eq

Chart 5.9 Calculation of the potential impacts of terrestrial acidification

Terrestrial acidification	Quantity (kg)	CF (kg SO_2-eq/kg)	Impact characterization
Ammonia, NH_3 (emission to the air in the forest phase)	6.33×10^{-2}	1.9	1.20×10^{-1}
Sulfur dioxide, SO_2 (emission to air in the forest phase)	8.00×10^{-3}	1.0	8.00×10^{-3}
Nitrogen oxides, NO_x (emission to air in the forest phase)	1.16×10^{-1}	3.5×10^{-1}	4.06×10^{-2}
Nitrogen oxides, NO_x (emission to air in the industrial phase)	1.80×10^{-1}	3.5×10^{-1}	6.30×10^{-2}
Total			2.32×10^{-1} kgSO_2-eq

Chart 5.10 Calculation of the potential impacts of marine eutrophication

Marine eutrophication	Quantity (kg)	CF (kg N-eq/kg)	Impact characterization
Ammonia, NH_3 (emitted into the air in the forest phase)	6.33×10^{-2}	9×10^{-2}	5.70×10^{-3}
Ammonia, NH_3 (emitted into fresh water in the industrial phase)	1.21×10^{-4}	3×10^{-1}	3.63×10^{-5}
Nitrogen oxides, NO_x (emitted into the air in the forest phase)	1.16×10^{-1}	4×10^{-2}	4.64×10^{-3}
Nitrogen oxides, NO_x (emitted into the air in the industrial phase)	1.80×10^{-1}	4×10^{-2}	7.20×10^{-3}
Nitrogen dioxide, NO_2 (emitted into the air in the forest phase)	7.00×10^{-3}	4×10^{-2}	2.80×10^{-4}
Total			1.79×10^{-2} kg N-eq

References

Associação Brasileira de Normas Técnicas – ABNT: NBR ISO 14040: Gestão Ambiental - Avaliação do Ciclo de Vida - Princípios e Estrutura: ABNT (2009a)

Associação Brasileira de Normas Técnicas – ABNT: NBR ISO 14044: Gestão Ambiental - Avaliação do Ciclo de Vida - Requisitos e Orientações. Brasil: ABNT (2009b)

Bare, J., Gloria, T., Norris, G.: Development of the method and US normalization database for life cycle impact assessment and sustainability metrics. Environ. Sci. Technol. **40**, 5108–5115 (2006)

Bare, J.C., Norris, G.A., Pennington, D.W., McKone, T.T.: The Tool for the reduction and assessment of chemical and other environmental impacts. J. Ind. Ecol. **6**(3–4) (2003)

Bjørn, A., Hauschild, M.: Introducing carrying capacity-based normalisation in LCA: framework and development of references at midpoint level. Int. J. Life Cycle Assess. **20**, 1005–1018 (2015)

Castellani, V., Benini, L., Sala, S., Pant, R.: A distance-to-target weighting method for Europe 2020. Int. J. Life Cycle Assess. **21**, 1159–1169 (2016)

Crettaz, P., Rhomberg, L., Brand, K., Pennington, D.W., Jolliet, O.: Assessing human health response in life cycle assessment using ED10s and DALYs: carcinogenic effects. Int. J. Risk Anal **22**(5), 929–944 (2002)

Dahlbo, H., Koskela, S., Pihkola, H., et al.: Comparison of different normalised LCIA results and their feasibility in communication. Int. J. Life Cycle Assess. **18**, 850–860 (2013)

De Haes, H.A.U., Finnveden, G., Goedkoop, M., Hauschild, M.Z., Hertwich, E., Hofstetter, P., Klöpffer, W., Krewitt, W., Lindeijer, E., Jolliet, O., Mueller-Wenk, R., Olsen, S., Pennington, D., Potting, J., Steen, B. (eds.): Life cycle impact assessment: striving towards best practice. SETAC Press, Pensacola (2002). ISBN 1-880611-54-6

De Schryver, A.M., Brakkee, K.W., Goedkoop, M.J., Huijbregts, M.A.J.: Characterization factors for global warming in life cycle assessment based on damages to humans and ecosystems. Environ. Sci. Technol. **43**(6), 1689–1695 (2009)

Finnveden, G., Eldh, P., Johansson, J.: Weighting in LCA based on ecotaxes: development of a mid-point method and experiences from case studies. Int. J. Life Cycle Assess. **11**, 81–88 (2006)

Frischknecht, R., Steiner, R., Jungbluth, N.: The ecological scarcity method—eco-factors 2006. A method for impact assessment in LCA. Federal Office for the Environment (FOEN), Bern (2009)

Guinée, J., Heijungs, R.: A proposal for the classification of toxic substances within the framework of life cycle assessment of products. Chemosphere **26**, 1925–1944 (1993)

Guinée, J.B., Gorrée, M., Heijungs, R., Huppes, G., Kleijn, R., De Koning, A., Van Oers, L., Wegener Sleeswijk, A., Suh, S., Udo De Haes, H.A., De Bruijn, J.A., Van Duin, R., Huijbregts, M.A.J.: Handbook on life cycle assessment: operational guide to the ISO Standards. Series: Eco-efficiency in industry and science. Kluwer Academic Publishers, Dordrecht (2002). Paperback, ISBN 1-4020-0557-1

Goedkoop, M., et al.: ReCiPe 2008: a life cycle impact assessment method which comprises harmonized category indicators at the midpoint and the endpoint level, 1st ed. Report I: Characterisation. Ruimte em Milieu Ministerie van Volkshuisvesting, Ruimtelijke Ordening en Milieubeheer (2009)

Hauschild, M.Z., Huijbregts, M.A.J. (eds.): Life cycle impact assessment. In: LCA Compendium—The Complete World of Life Cycle Assessment, 349p. Springer, Berlin (2015). ISBN 978-94-017-9743-6

Hauschild, M., Potting, J.: Spatial differentiation in life cycle impact assessment—the EDIP2003 methodology. Environmental News no. 80. The Danish Ministry of the Environment, Environmental Protection Agency, Copenhagen (2005)

Hauschild, M.Z., Wenzel, H.: Environmental assessment of products, vol. 2—Scientific Background, 565 pp. Chapman & Hall, London (1998). ISBN 0412 80810 2

Hauschild, M., Huijbregts, M., Jolliet, O., Macleod, M., Margni, M., Payet, J., Schuhmacher, M., Van De Meent, D., McKone, T.: Building a consensus model for life cycle impact assessment of chemicals: the search for harmony and parsimony. Environ. Sci. Technol. **42**(19), 7032–7037 (2008)

Hertwich, E.G., Pease, W.S., Koshland, C.P.: Evaluating the environmental impact of products and production processes: a comparison of six methods. Sci. Total Environ. **196**, 13–29 (1997)

Hertwich, E.G., Pease, W.S., McKone, T.E.: Evaluating toxic impact assessment methods: what works best? Environ. Sci. Technol. **32**, A138–A144 (1998)

Hertwich, E., McKone, T., Pease, W.: Parameter uncertainty and variability in evaluative fate and exposure models. Risk Anal. **19**, 1193–1204 (1999)

Hertwich, E.G., Matales, S.F., Pease, W.S., McKone, T.E.: Human toxicity potentials for life cycle analysis and toxics release inventory risk screening. Environ. Toxicol. Chem. **20**, 928–939 (2001)

Huijbregts, M.A.J., Struijs, J., Goedkoop, M., Heijungs, R., Hendriks, A.J., Van De Meent, D.: Human population intake fractions and environmental fate factors of toxic pollutants in life cycle impact assessment. Chemosphere **61**(10), 1495–1504 (2005a)

Huijbregts, M.A.J., Rombouts, L.J.A., Ragas, A.M.J., Van De Meent, D.: Human toxicological effect and damage factors of carcinogenic and non-carcinogenic chemicals for life cycle impact assessment. Integr. Environ. Assess. Manag. **1**(3), 181–244 (2005b)

Huijbregts, M.A.J., Steinmann, Z.J., Elshout, P.M.F., Stam, G., Verones, F., Vieira, M.D.M., Zijp, M., Van Zelm, R.: ReCiPe 2016: a Harmonized Life Cycle Impact Assessment Method at Midpoint and Enpoint Level—Report 1: Characterization, 194p. National Institute for Public Health and the Environment, Bilthoven, The Netherlands (2016)

Huppes, G., Van Oers, L.: Background review of existing weighting approaches in life cycle impact assessment (LCIA). Joint Research Centre—Institute for Environment and Sustainability (2011)

Joint Research Centre of the European Commission (EC-JRC): General guide for Life Cycle Assessment - Detailed guidance. ILCD Handbook International Reference Life Cycle Data System, European Union (2010)

Jolliet, O., Anton, A., Boulay, A.-M., Cherubini, F., Fantke, P., Levasseur, A., McKone, T.E., Michelsen, O., Milà I Canals, L., Motoshita, M., et al.: Global guidance on environmental life cycle impact assessment indicators: impacts of climate change, fine particulate matter formation, water consumption and land use. Int. J. Life Cycle Assess. (2018). https://doi.org/10.1007/s11367-018-1443-y

Jolliet, O., Müller-Wenk, R., Bare, J.C., Brent, A., Goedkoop, M., Heijungs, R., Itsubo, N., Peña, C., Pennington, D., Potting, J., Rebitzer, G., Stewart, M., Udo de Haes, H., Weidema, B.: The LCIA midpoint-damage framework of the unep/setac life cycle initiative. Int. J. Life Cycle Assess. **9**(6), 394–404 (2004)

Norris, G.A.: The requirement for congruence in normalization. Int. J. Life Cycle Assess. **6**, 85–88 (2001)

Norris, G.: Impact characterisation in the tool for the reduction and assessment of chemical and other environmental impacts: methods for acidification, eutrophication, and ozone formation. J. Ind. Ecol. **6**(3/4), 83–105 (2002)

Norris, G.A., Marshall, H.E.: Multiattribute decision analysis method for evaluating buildings and building systems. Building and Fire Research Laboratory, National Institute of Standards and Technology, Gaithersburg (1995)

Payet, J.: Assessing Toxic Impacts on Aquatic Ecosystems in LCA, 232p. Thesis 3112, Ecole Polytechnique Fédérale de Lausanne (2004)

Pennington, D.W., Margni, M., Amman, C., Jolliet, O.: Multimedia fate and human intake modeling: spatial versus non-spatial insights for chemical emissions in Western Europe. Environ. Sci. Technol. **39**(4), 1119–1128 (2005)

Pennington, D.W., Margni, M., Payet, J., Jolliet, O.: Risk and regulatory hazard-based toxicological effect indicators in life-cycle assessment (LCA). Hum. Ecol. Risk Assess. **12**(3), 450–475 (2006)

Pizzol, M., Laurent, A., Sala, S., Weidema, B., Verones, F., Koffler, C.: Normalisation and weighting in life cycle assessment: quo vadis? Int. J. Life Cycle Assess. **22**, 853–866 (2017)

Potting, J., Hauschild, M.: Spatial differentiation in life cycle impact assessment—the EDIP2003 methodology. Environmental News no. 80. The Danish Ministry of the Environment, Environmental Protection Agency, Copenhagen (2005)

Prado, V., Wender, B.A., Seager, T.P.: Interpretation of comparative LCAs: external normalization and a method of mutual differences. Int. J. Life Cycle Assess. **22**, 2018–2029 (2017). https://doi.org/10.1007/s11367-017-1281-3

Reap, J., et al.: A survey of unresolved problems in life cycle assessment. Int. J. Life Cycle Assess. **13**(5), 374–388 (2008)

Ridoutt, B., Fantke, P., Pfister, S., et al.: Making sense of the minefield of footprint indicators. Environ. Sci. Technol. **49**, 2601–2603 (2015)

Rochat, D., Margni, M., Jolliet, O.: Continent-specific intake fractions and characterization factors for toxic emissions: does it make a difference? Int. J. Life Cycle Assess. **11** (1), 55–63 (2006)

Rosenbaum, R.: Multimedia and food chain modelling of toxics for comparative risk and life cycle impact assessment. Thesis 3539, Ecole Polytechnique Fédérale de Lausanne (2006)

Rosenbaum, R., Margni, M., Jolliet, O.: A flexible matrix algebra framework for the multimedia multipathway modeling of emission to impacts. Environ. Int. **33**(5), 624–634 (2007)

Rosenbaum, R.K., Bachmann, T.M., Gold, L.S., Huijbregts, M.A.J., Jolliet, O., Juraske, R., Köhler, A., Larsen, H.F., Macleod, M., Margni, M., Mckone, T.E., Payet, J., Schuhmacher, M., Van de meent, D., Hauschild, M.Z.: USEtox—the UNEP SETAC toxicity model: recommended characterisation factors for human toxicity and freshwater ecotoxicity in life cycle impact assessment. Int. J. Life Cycle Assess. **13**(7), 532–546 (2008)

Sleeswijk, A.W., Van Oers, L.F.C.M., Guinée, J.B., et al.: Normalisation in product life cycle assessment: an LCA of the global and European economic systems in the year 2000. Sci. Total Environ. **390**, 227–240 (2008)

Struijs, J., Van Wijnen, H.J, van Dijk, A., Huijbregts, M.A.J.: Ozone depletion. In: Goedkoop, M., Heijungs, R., Huijbregts, M.A.J., De Schryver, A., Struijs, J., Van Zelm, R. (eds.) ReCiPe 2008: a Life Cycle Impact Assessment Method which Comprises Harmonised Category Indicators at the Midpoint and the Endpoint Level, 1st edn, pp. 37–53. Pre Consultants, CML University of Leiden, Radboud University, RIVM, Bilthoven (2009)

Ugaya, C.M.L., Almeida Neto, J.A., Figueiredo, M.C.B.: Recomendação de Modelos de Avaliação de Impacto do Ciclo de Vida para o Contexto Brasileiro. Relatório da Rede de Pesquisa de Avaliação do Impacto do Ciclo de Vida - RAICV. Brasília:IBICT/MCT (2019)

Uusitalo, V., Kuokkanen, A., Grönman, K., Ko, N., Mäkinen, H., Koistinen, K.: Environmental sustainability assessment from planetary boundaries perspective—a case study of an organic sheep farm in Finland. Sci. Total Environ. **687**, 168–176 (2019)

Van Zelm, R., Huijbregts, M.A.J., Van Jaarsveld, H.A., Reinds, G.J., De Zwart, D., Struijs, J., Van De Meent, D.: Time horizon dependent characterization factors for acidification in lifecycle impact assessment based on the disappeared fraction of plant species in European forests. Environ. Sci. Technol. **41**, 922–927 (2007a)

Van Zelm, R., Huijbregts, M.A.J., Harbers, J.V., Wintersen, A., Struijs, J., Posthuma, L., Van De Meent, D.: Uncertainty in msPAF-based ecotoxicological freshwater effect factors for chemicals with a non-specific mode of action in life cycle impact assessment. Integr. Environ. Assess. Manag. **3**(2), 203–210 (2007b)

Verones, F., et al.: LCIA framework and cross-cutting issues guidance within the UNEP- SETAC life cycle initiative. J. Clean. Prod. **161**, 957–967 (2017)

Wegener Sleeswijk, A., Van Oers, L., Guinée, J., Struijs, J., Huijbregts, M.A.J.: Normalisation in product life cycle assessment: an LCA of the global and European economic systems in the year 2000. Sci. Total Environ. **390**(1), 227–240 (2008)

Wenzel, H., Hauschild, M.Z., Alting, L.: Environmental assessment of products. Vol. 1—Methodology, tools and case studies in product development, 544p. Chapman & Hall,London (1997). ISBN 0412 80800 5

Chapter 6
LCA—Interpretation of Results

Marcella Ruschi Mendes Saade, Vanessa Gomes, and Maristela Gomes da Silva

6.1 Introduction

As LCA gains ground as a support for decision making both in the industrial sphere and in policy making, careful interpretation of the final results becomes crucial to determine their reliability and communicate them in an accurate, complete and fair manner (Skone, 2000).

Most LCA studies currently available point very defined deterministic values (indicated in Fig. 6.1 as μ_1; μ_2 and μ_3) as results, which suggest a false sense of certainty but actually cloud the decision-making process (Pomponi et al. 2017).

Exclusively based on such deterministic values, the analyst would see only the tip of the iceberg and possibly conclude that "alternative 1 has a greater impact than alternatives 2 or 3" ignoring the fact that the latter have much larger standard deviations (and consequently variances), while the probabilistic distribution of alternative 1 points to a much narrower range of results variation around the central value. From this perspective, alternative 1 becomes equally or more attractive than the others and the importance of analyzing the uncertainties involved in the assessment then becomes very evident.

When performing the inventory analysis and life cycle impact assessment, the analyst makes estimates, assumptions and subjective methodological decisions. All these considerations must be communicated, and their effects measured, in order to

M. R. M. Saade (✉)
Institute of Structural Design, Graz University of Technology, Technikerstraße 4 / IV, A-8010 Graz, Austria

V. Gomes
Campinas State University, Saturnino de Brito, 224 - Cidade Universitária, CEP: 13083-889, Campinas, SP, Brazil

M. Gomes da Silva
Federal University of Espírito Santo, Avenida Fernando Ferrari, 514 – Goiabeiras, Vitória, ES ZIP Code: 29075-910, Brazil

J. A. de Oliveira et al. (eds.), *Life Cycle Engineering and Management of Products*,
https://doi.org/10.1007/978-3-030-78044-9_6

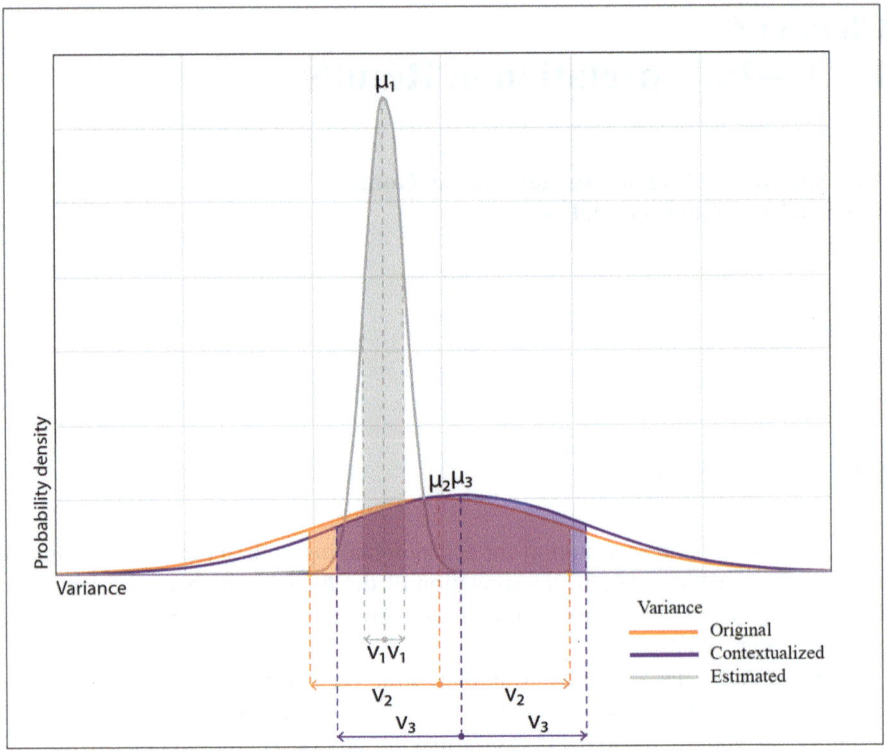

Fig. 6.1 Probability of occurrence of deterministic value and variance of LCA results for three hypothetical alternatives. *Source* Authors

ensure the correct understanding of the results. Despite being known as the fourth stage of an LCA (see Chap. 3: Life Cycle Assessment (LCA)—Definition of Goals and Scope), in fact, the interpretation activities permeate the study throughout its conduction (ABNT 2009b). Precisely because of its broad and iterative nature, it is common for LCA studies not to present this stage separately.

The interpretation activities recommended in ABNT NBR ISO 14044: 2009 include: (i) identification of significant issues; (ii) data assessment through completeness, sensitivity and consistency checks, among others; and (iii) conclusions and recommendations (Fig. 6.2). Although dealt with in the standard, this LCA stage is notably less systematized than the others. Interpretation is also often understood as a qualitative assessment (Klöpffer and Grahl 2014) when, in fact, it represents the opportunity for expanded and in-depth observation of the results obtained in the LCA. However, one frequently encounters studies without any reference to the uncertainties of the result or consistency of methodological choices with the goal and scope previously outlined.

Given the important role of the interpretation phase within the LCA technique, this chapter expands and discusses the guidelines established in the specific international

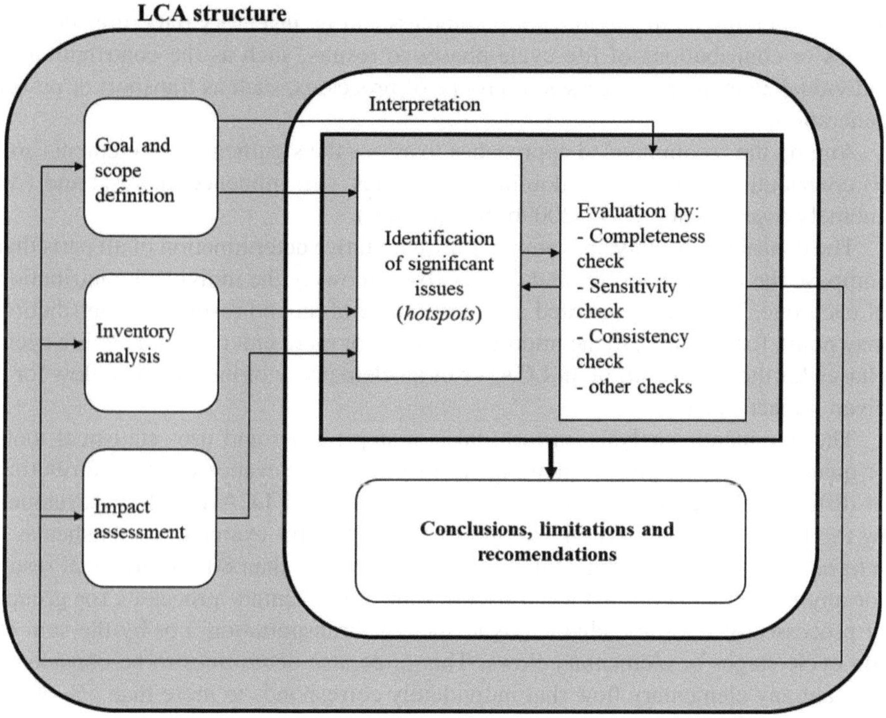

Fig. 6.2 Relation between elements of the interpretation phase with the other LCA phases. *Source* Adapted from ABNT NBR ISO 14044:2009 (ABNT 2009b)

normalization (ABNT 2009a,b) and highlights the mathematical robustness that must accompany the conclusions formulated in this stage.

6.2 Identification of Significant Issues

Performing an LCA uses a huge amount of data, which makes it impossible to verify all the parameters involved. The *identification of significant issues* (or "hotspots analysis") aims at optimizing the mathematical work to be conducted in the interpretation stage. The information from the inventory analysis and the impact assessment is reviewed to detect the parameters that present a significant quantitative difference in relation to the others (Klöpffer and Grahl 2014). This procedure (see Fig. 6.2) allows to filter and guide subsequent checks concentrated on the parameters and choices with the greatest contribution to the results obtained.

ABNT NBR ISO 14044: 2009 does not determine the types of elements to be assessed or contribution values which characterize criticality. It is up to the LCA analyst to define them. Significant parameters can be inventory data, impact categories or contributions of life cycle phases to results, such as the contribution of individual elementary processes or groups of processes, such as transport or power generation.

Among the recommended approaches to assess the significance of elements are: (i) contribution analysis, (ii) dominance analysis, (iii) influence analysis and (iv) anomaly assessment (ABNT 2009b; Skone 2000).

The contribution analysis is simply the quantitative determination of all parts that compose the result (Klöpffer and Grahl 2014), showing the individual contribution of each one. Typically presented as a percentage of an end result, the contribution may point, for example, to the importance of one or more unit processes or life cycle phases for the final result of an LCA, or of an elementary or intermediate flow for a given impact category.

The dominance analysis, in turn, takes a step further and uses statistical tools or quantitative or qualitative rankings to identify the relevance of the contribution of different elements. The "Guide for Interpretation of LCA Results," developed by the European Commission (Zampori et al. 2016), for example, establishes as a *relevance criterion* the participation equal to or greater than 80% in the final result for any impact category achieved by the sum of elementary processes (or groups of processes of the same characteristic such as "transportation") or by the sum of life cycle stages or elementary flows. The guide also determines to be considered relevant any elementary flow that individually corresponds to more than 5% of the final result of an impact category.

The influence analysis determines the influence degree; that is, the effect that a change in a certain element causes on the results of the inventory or of the impact assessment combined with the level of control (by the LCA analyst, manufacturer or responsible for the process) on the element in question. ABNT NBR ISO 14044: 2009 gives the following qualitative classification as an example:

A. flow on which you have significant control and whose change implies a great improvement in the result;
B. flow on which you have little control and whose change implies some improvement;
C. flow on which you have no control. The information obtained from the influence analysis can assist investigations that intend to minimize the impacts of a given production process.

Finally, the assessment of anomalies depends on the LCA analyst's previous experience to detect unexpected deviations in the results and carefully examine their relevance in the final result. Anomalies can represent errors of calculation or representativeness, and their identification can feed strategies for improvement in the conduct of the study. The limitations and assumptions established during the LCA scope definition phase (see Chap. 3: Life Cycle Assessment (LCA)— Definition of Goals and Scope) can also represent sources of anomalies at the end of the study, which will need to be reviewed during LCA interpretation.

Charts 6.1 and 6.2 show examples of structuring inventory flows (Chart 6.1) for later contribution analysis (Chart 6.2) conducted in a hypothetical study (James and Galatola 2015). Examples of the other approaches described here can be found in annex III of ABNT NBR ISO 14044: 2009.

Chart 6.1 LCIA results for the "climate change" category, expressed in gCO_{2eq}

Inventory flow	Substance 1 (gCO_{2eq})	Substance 2 (gCO_{2eq})	Substance 3 (gCO_{2eq})	Substance 4 (gCO_{2eq})	Substance 5 (gCO_{2eq})	Total (gCO_{2eq})
Process A	249	85	6	45	5	390
Process B	1100	600	500	450	50	2700
Process C	300	250	20	30	430	1030
Process D	60	30	20	10	5	125
Process E	64	1	1	1	1	68
Process F	15	10	8	5	3	41
Other processes	15	10	8	5	3	41
Total	1803	986	563	546	497	4395

Source Adapted from James and Galatola (2015)

Chart 6.2 Contribution analysis and flow dominance of substances 1–5 (highlighted in italics) most relevant to the final result of the climate change category

Inventory flow	Substance 1 (%)	Substance 2 (%)	Substance 3 (%)	Substance 4 (%)	Substance 5 (%)	Total (%)
Process A	*5.7*	1.9	0.1	1.0	0.1	8.9
Process B	*25.0*	*13.7*	*11.4*	*10.2*	1.1	61.4
Process C	*6.8*	*5.7*	0.5	0.7	*9.8*	23.4
Process D	1.4	0.7	0.5	0.2	0.1	2.8
Process E	1.5	0.0	0.0	0.0	0.0	1.5
Process F	0.3	0.2	0.2	0.1	0.1	0.9
Other processes	0.3	0.2	0.2	0.1	0.1	0.9
Total	41.0	22.4	12.8	12.4	11.3	100

Source Adapted from James and Galatola (2015), adopting the relevance criteria suggested by Zampori et al. (2016)

6.3 Verification

The function of the verification aims at ensuring reliability to the LCA results and, particularly, to the significant parameters identified. The verification result must provide a clear picture of the study's outcomes to any interested parties. ABNT NBR ISO 14044: 2009 recommends the use of three different techniques: (i) completeness verification, (ii) consistency verification and (iii) sensitivity verification.

In the completeness verification, the analyst confirms that all information and data relevant to interpreting the study results are available and complete. If any information considered relevant – due to the identification of significant issues or previous knowledge of the process and its – components is missing, its necessity must be assessed to satisfy the proposed goal of the study. If possible, data from the inventory analysis and/or from the impact assessment must be supplemented or optimized and the calculations fed back iteratively. Alternatively, goal and scope may be adjusted to consider the limitations found, which means that, in some cases, the expectation of quality or completeness of LCA will need to be reduced to ensure the consistency of the study (Klöpffer and Grahl 2014).

The consistency verification assesses whether all the assumptions made and whether the methods, quantity and quality of data used in the study were consistent with the previously defined goal and scope. ABNT NBR ISO 14044: 2009 suggests the analyst to consider four main aspects:

- Are the differences in data quality in a given product system or between different product systems consistent with the expectation outlined in the goal and scope of the study?
- Have the regional and/or time differences, if any, been consistently applied to the same—or between different—product system(s)?
- Have the allocation criteria and the system boundaries been consistently applied to all product systems?
- Have the elements of the impact assessment been consistently applied?

ABNT NBR ISO 14,044: 2009 cites as examples of inconsistency the differences in: (i) data source, for example: alternative A is based on literature data, while in alternative B primary data have been used; (ii) accuracy of the data used in each alternative; (iii) technological boundary, for example: data for alternative A are experimental, while for alternative B they are based on technology applied on a large scale; (iv) temporal coverage, referring to e.g. the age of the data collected for each alternative or process; and (v) geographical coverage, for example: data collected representing local production for an alternative or process, while the other alternatives or processes use national production data.

Finally, the sensitivity verification assesses whether the uncertainty related to the critical elements identified at the beginning of the interpretation stage affects the reliability of the conclusions and recommendations made (Skone 2000). In this step, the uncertainty or sensitivity analyzes results are included in the inventory or in the impact assessment. As they are often confused, it is worth mentioning that the uncertainty analysis aims at understanding the *variability* of the final results, while the sensitivity analysis aims at understanding the *influence of the input parameters* on the final results.

ABNT NBR ISO 14044:2009 forecasts the performance of "other verifications" (Fig. 6.2) that may supplement this stage. Sections 4 and 5 further explore the concepts of uncertainty and sensitivity analysis, statistical procedures that provide robustness for verification.

6.4 Uncertainty Analysis

Uncertainties are inevitable in a life cycle approach and neglecting them in the analysis result may lead to biased or even incorrect conclusions. Considering uncertainties becomes even more important in assessments which include systems with high complexity and a long life cycle. Additionally, LCA has been increasingly used to support decision-making processes. The quality of the information underpinning such process must be known so that reliable decisions can be made.

There are basically two major groups of uncertainty sources: random and epistemic. Both are often present in LCA. Epistemic uncertainty results from the imperfect description of reality in the models used to represent it, whether due to insufficient data, inaccuracies of measurement or estimate, or ignorance of the real value. Random uncertainty, on the other hand, refers to the variability (geographical, temporal, technological) in the information observed due to the randomness of the underlying phenomenon (Igos et al. 2019).

As for the type of uncertainty, in turn, there are several classifications in literature (for example, review Sect. 4.2.3. Uncertainties of Chap. 4: LCA—Life Cycle Inventory Analysis and databases), but the most frequently used categorizes them into parameter, model and choice uncertainties (Huijbregts 2001), also called "scenario uncertainties." The first group includes inaccurate, incomplete or outdated measurements, or the absence of data in the inventory and models. Model uncertainties are introduced, for example, by using simplified or linear models and by disregarding geographical and temporal information in the inventory analysis. Finally, the uncertainties resulted from choices (or scenarios) refer to normative decisions made by the analyst as in the definition of functional units, allocation criteria and the adoption of characterization, weighting and impact assessment methods.

Each modeled environmental aspect introduces an inherent uncertainty and each choice leads to variability in the analysis result. Knowing the sources and types of uncertainties is important to determine the best way to address them (Heijungs and Huijbregts 2004; Gervasio 2018).

Different statistical approaches can be used to consider uncertainties in LCA (Heijungs and Huijbregts 2004): (i) analysis of scenarios, in which different sets of data, models and/or normative choices are contrasted; (ii) sampling methods, such as Monte Carlo simulation; (iii) analytical methods based on mathematical expressions;

and (iv) non-traditional methods, such as *fuzzy* sets, Bayesian methods and neural networks, among others.

Among the quantitative approaches to analyze and propagate uncertainties in LCA, the use of stochastic modeling (through probabilistic distributions) is clearly dominant, present in around 67% of published studies (Lloyd and Ries 2007), being Monte Carlo simulation the most frequently used sampling approach. The aim is, basically, to propagate the uncertainty of each parameter that composes the model to obtain the distribution that represents the uncertainty in the complete model and, then, to estimate the range of uncertainty linked to the final result.

The first step is to identify all parameters of the model that may influence the analysis result. Assuming that the uncertainties in the model parameters can be represented by probability distributions, they are characterized, that is the probability distribution that best fits each input is selected, based on measurements, literature data, expert judgment or the analyst's experience. The analyst estimates or assumes, based on the best information available, that the values of each input variable follow a certain probabilistic distribution.

The choice for one of the several types of probabilistic distribution (Chart 6.3) depends on the knowledge available about the variable at hand. The simplest probabilistic distribution is the uniform distribution, which requires only the upper and lower limits to be described. The triangular distribution, in turn, is a continuous probability distribution with a lower limit a, upper limit b; and mode c, in which $a < b$ and $a \leq c \leq b$. Three-point estimates are widely used in engineering in situations of scarce data and excessively expensive or unviable surveys, but with a relation between the data reasonably known to the point of allowing the analyst to estimate

Chart 6.3 Most common probabilistic distributions and respective descriptive parameters required for modeling uncertainty

Type of distribution	Parameters required	Graphic representation
Uniform	Maximum and minimum values (A and B)	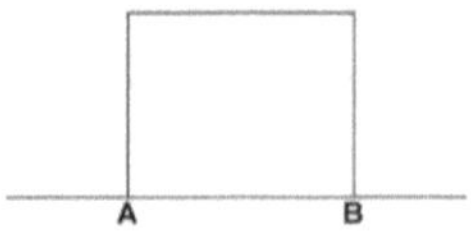
Triangular	Maximum and minimum values and mode (A)	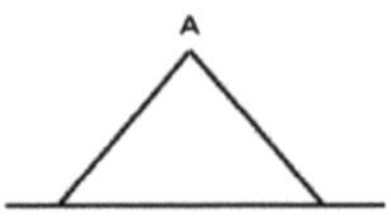
Normal	Arithmetic mean and standard deviation	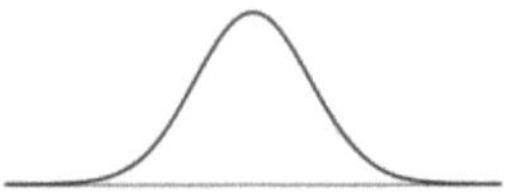
Log-normal	Geometric mean and standard deviation	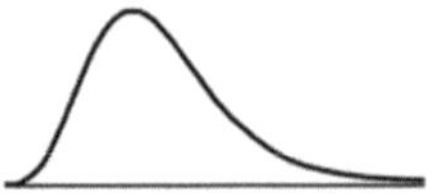

the mode (most frequent value) based on his experience. Triangular distributions therefore offer some extra information relatively to uniform distributions.

The normal distribution is another continuous probabilistic distribution described by its mean (which coincides with mode and median) and standard deviation. The normal distribution is particularly useful due to the central limit theorem which basically states that when the number of observations is sufficiently large, the means of random variables observation samples converge to a normal distribution, regardless of independent distributions.

In the log-normal distribution, the logarithm of the random variable is normally distributed. The distribution is described by the mean and standard deviation of the random variable logarithms and is useful for describing natural phenomena whose growth is often driven by the accumulation of a series of small percentage changes until they become additives on a logarithmic scale. For this reason, this is the default distribution adopted by the international database "Ecoinvent" whose pertinence has been recently validated by Qin and Suh (2017).

At this point, it is possible to estimate a range of values (uncertainty) for each input variable. Once the most appropriate distribution of each "input" has been selected, the uncertainty of these "inputs" is propagated in the result. Suppose an analyst wishes to measure the uncertainty of the result obtained for the climate change category. The elementary flows of greenhouse gas emissions in their inventory have a probabilistic distribution already identified. Then, a value taken randomly from each distribution referring to the emission of greenhouse gases is multiplied by the corresponding characterization factor and added to the other products of other greenhouse gases, which, in turn, is multiplied by the corresponding characterization factor and added to the other products of other greenhouse gases. This calculation provides a deterministic value of global warming potential (GWP). In Monte Carlo simulation, this procedure is repeated iteratively, normally 100 to 10,000 times, obtaining a sampling of solutions (Ang and Tang 2007) with different results for the global warming potential and which indicates the probability of occurrence of each result within a certain interval (Fig. 6.3). Each blue column in the histogram in Fig. 6.3 represents the probabilistic distribution function (PDF), that is, probability of occurrence of a given GWP value. The orange curve represents the corresponding Cumulative Distribution Function (CDF) and illustrates, in an accumulated manner, the probability of occurring values equal to or less than those described on the x-axis.

The usefulness of Monte Carlo simulation lies in the premise that a sufficient number of attempts are made to make statements about data distribution and the uncertainty related to them, based on the calculation of the corresponding statistics (Pomponi et al. 2017). As the iterations progress, the simulated values accumulate in a sample of increasing size. The more iterations that occur, the closer the sample—and the corresponding statistics—will get to the population and its statistics.

Adopiting distributions which describe the dataset well makes it possible to reduce the number of iterations required in the Monte Carlo simulation to achieve convergence of the estimated sample means. From a very high number of iterations (10^7, for example), the central limit theorem described above prevails. Thus, the type of data distribution adopted and the number of data points in the original sample make

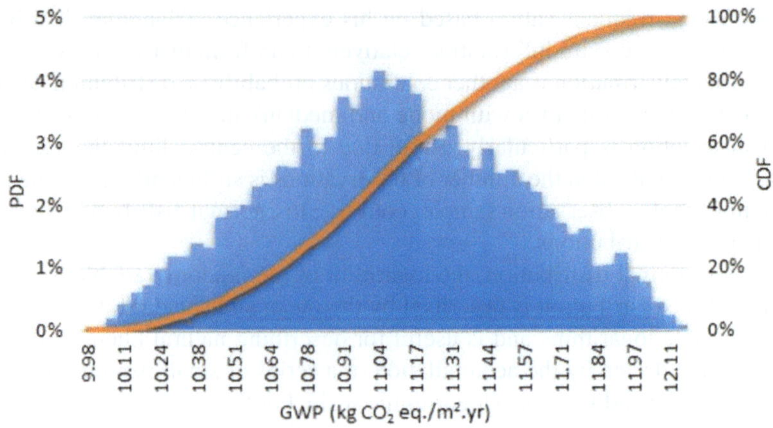

Fig. 6.3 Propagation of uncertainty of the climate change category result of an LCA of buildings, using Monte Carlo simulation. *Source* Gervasio (2018)

little difference (Pomponi et al. 2017), but at the expense of additional computational effort, which becomes the practical limitation to running simulations.

Basically, what defines the number of necessary iterations is the fulfillment of a certain convergence criterion: the calculated values of the sample mean stabilize within a maximum tolerated percentage error margin for a specified confidence level. It is possible to verify the convergence rate by plotting the estimated mean versus the number of required iterations. Estimated means will converge to a specific value (true mean), and additional iterations will no longer change it significantly.

Figure 6.4a illustrates the most common graphic representation of LCA uncertainty analysis using the so-called error bars to indicate only the range of results. But

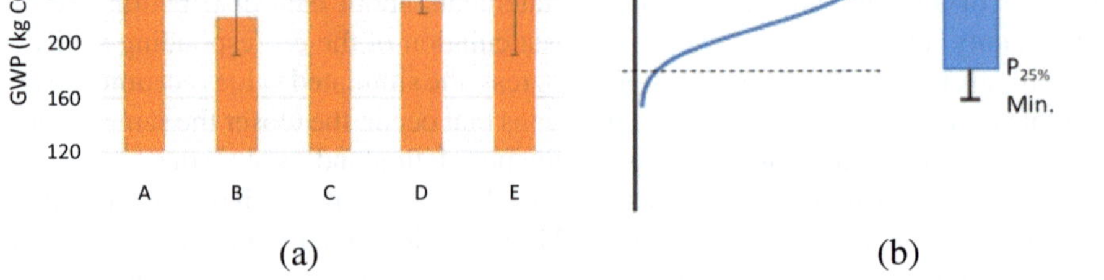

Fig. 6.4 Uncertainty ranges in the global warming potential (GWP) values for different alternatives. *Source* Authors

the bars can provide a more complete representation, and also indicate the probability of occurrence of the values obtained by the uncertainty propagation procedure (Fig. 6.4b). The boxplot in Fig. 6.4b offers a more complete representation, for it also summarizes the statistical parameters which describe the probabilistic distribution function of the results.

The example illustrated in Fig. 6.4a points out that alternatives A and C obvioulsly have a greater impact than B and D. The distinction between the other alternatives, in turn, is less clear: although D has a higher deterministic value and a more concentrated distribution around the median than B, when considering the distribution of the probability of results at the upper end, these alternatives get closer. Based exclusively on the deterministic value, alternative E could initially be understood as less impacting than C, but there is the possibility of occurrence of maximum values equivalent to those of C.

Figure 6.4 indicates results of the so-called absolute uncertainty of data. When there are overlaps in the distributions of two alternatives (such as between C and E), the comparisons become less obvious. Therefore, one must assess the uncertainty of difference between the alternatives, by measuring the probability of one being more impactful than the other. Through Monte Carlo simulations, for example, it is determined how many times the iterations show product E as the alternative of lower impact to estimate the probability of E being better than C. An example of this type of analysis has been described by Cherubini et al. (2018), who assessed the uncertainty of the results of a comparison between two types of swine production considering the variability coming from methodological choices (scenario uncertainty). The shaded region (<1.0) in Fig. 6.5 corresponds to the probability of reference scenario production to be less impacting than the alternative scenario.

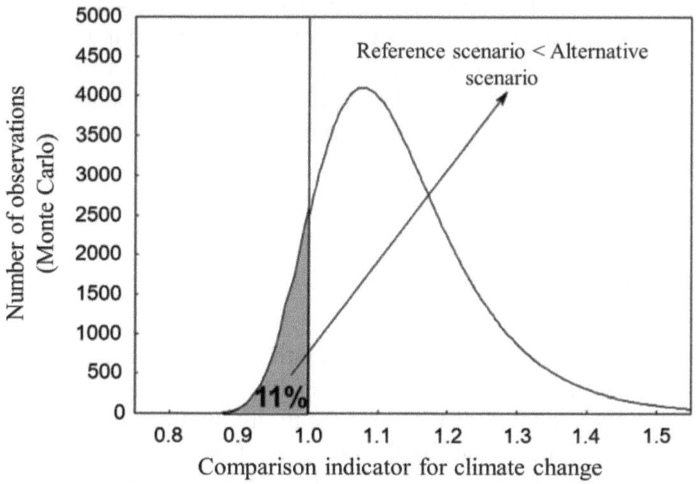

Fig. 6.5 Example of comparative analysis of scenario uncertainty. *Source* Adapted from Cherubini et al. (2018)

The main limitation for carrying out more complex analyzes and using more sophisticated models, such as intervals, fuzzy sets and stochastic modeling, is clearly the computational capacity available. Recent advances in computer technology explain the strengthening of the "processing uncertainties" trend. Even so, such approaches require a large amount of data to characterize uncertainty, which is directly related to computational strength and, consequently, to the required processing time.

In Monte Carlo simulation, the greater the number of iterations, the better the sample representation, however—as mentioned earlier—the greater the computational effort required. To reduce computational complexity, sampling methods that allow fewer samples with the same quality of result can be used, such as the "Latin Hyper Cube" (Gervasio 2018), quasi-Monte Carlo sequences, multidimensional Fourier transform (FAST) (Iooss and Lemaître 2015), or that allow to concentrate on the distribution area with the highest probability of occurrence (Gervasio 2018).

The analyst will also face the need for a combination of tools. LCA platforms such as *SimaPro*, *GaBi* and *OpenLCA* have a built-in Monte Carlo module, which includes basic simulations. These simulations are fed by the variance (or standard deviation) and the type of distribution of each flow inside a unitary process. The variance is determined by the result of the Pedigree matrix (described in Chap. 4: LCA—Life Cycle Inventory Analysis and databases) plus the so-called basic variance—which refers to the variability of the sample of collected data. More sophisticated analyzes, however, require the use of external software, whether commercial such as *Oracle Crystal Ball*® and *Palisade @Risk*®, or free such as *SimLab*, developed by the European Commission's *Joint Research Center* (SimLab 2008).

At this point, the analyst already has enough elements to perform a more advanced sensitivity analysis such as, for example, the global sensitivity analysis (GSA), in which sensitivity measures are considered and estimated based on the results of the Monte Carlo simulation to identify the most influential inputs in the model's result (Cucurachi et al. 2016). This allows to focus on the needs and strategies for improving the quality of corresponding data and establishes the attention points to be observed in order to complete and communicate the results of the analysis carried out (Gervasio 2018).

6.5 Sensitivity Analysis

While the uncertainty analysis focuses on quantifying the uncertainty of the model's output (Saltelli et al. 2008), the sensitivity analysis studies how this output uncertainty can be associated and distributed among different sources of uncertainty in the model's inputs (Saltelli et al. 2000). Thus, sensitivity analyzes are used to determine the input variables that most influence the model's output behavior. In these circumstances, it assists in verifying and understanding the model, simplifying it and prioritizing factors that require better measurement or estimation to reduce the output variance.

Before proceeding to the application of sensitivity analysis in LCA, it is important to synthesize the theoretical principles involved and the characteristics of the different methods available so that the analyst gains autonomy in choosing the most appropriate way to deal with the problem and the model in question.

There are basically two major groups of methods for performing sensitivity analysis: local and global analyzes. The first sensitivity analysis focused on studied the effect that small disturbances around nominal values (the mean or median of a random variable, for example) of input variables (inputs) would have on a model's result (output). To overcome the limitations of local methods, a new class of methods was developed in the late 1980s: the "global sensitivity analysis," so called because it considers the entire range of inputs variation (Saltelli et al. 2000). The model's output overall variability is usually measured by its variance. The methods of global sensitivity analysis therefore determine how much each input variable contributes to the variance of the result (Fig. 6.6) and can be interpreted as an extension of the propagation of uncertainty (Groen et al. 2017).

Ideally, the uncertainty and sensitivity analyzes should be carried out in parallel (Saltelli et al. 2008). However, in the current practice of analyzing numerical models, the analyzes end up being performed sequentially: first, the uncertainty is quantified and then proceeded to the sensitivity analysis, especially in the case of global sensitivity analysis (Fig. 6.6). In LCA, the difficulties of data processing and integration of

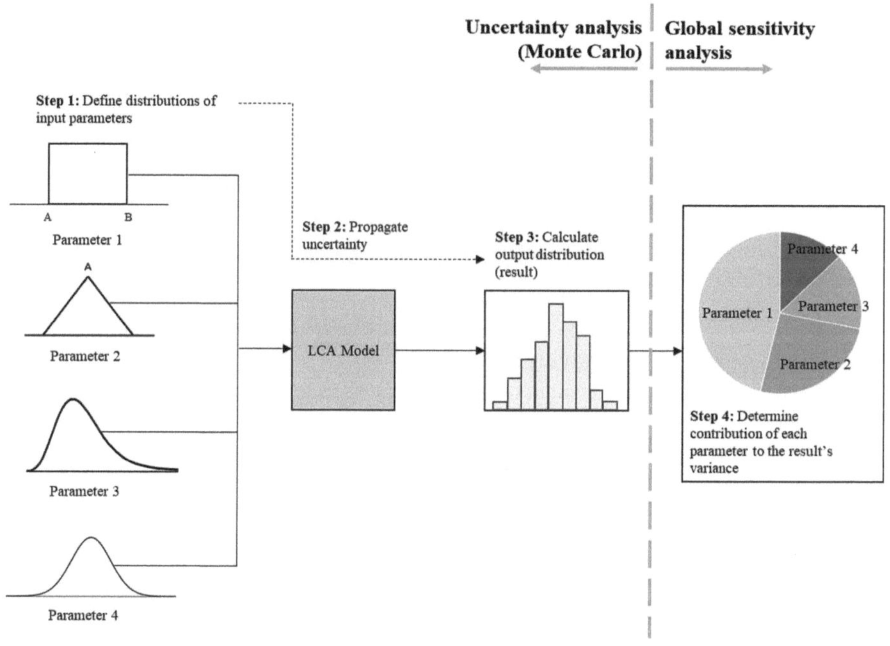

Fig. 6.6 Structure of an LCA global sensitivity analysis. *Source* Adapted from Groen et al. (2017) and Saltelli et al. (1999)

tools inherent to the uncertainty analysis popularized the performance of sensitivity analysis independently when—for example—changing an input data in isolation, the impact assessment method or even the allocation criteria to estimate the influence that such parameters, considered scenarios or normative choices would have on the final result of the analysis.

ABNT NBR ISO 14044:2009 recommends performing a sensitivity analysis without identifying specific methods to conduct it, limiting itself to requiring a comparison of the results obtained with the different methodological choices, with the sensitivity to the choices made indicated by the difference in values achieved. Thus, the methods most frequently applied for sensitivity analysis in LCA studies are statistically simple.

Most studies use local sensitivity analyzes, mainly those called "disturbance analysis" or "one-at-a-time" (OAT), or scenario analysis. In the first case, the value of one input variable is changed one at a time, while the others remain constant (Iooss and Lemaître 2015) to test its effect on the model's output. Figure 6.7 shows an example of OAT sensitivity analysis, in which each input has been individually varied within a range of − 10% to + 10% in order to identify the sensitivity of the result (climate change indicator) to the input variables. Mathematically, this method discretizes the spaces of different inputs, and through the repetition of OAT experiments, it estimates corresponding elementary effects from which sensitivity indices are derived for each of the input variables.

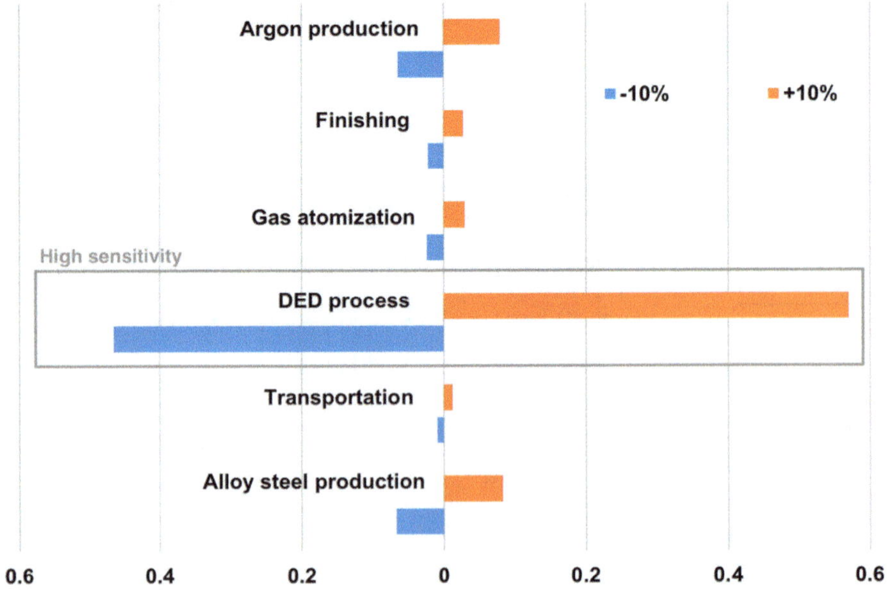

Fig. 6.7 Example of sensitivity analysis "one-at-a-time" (OAT) performed in an LCA of additive manufacturing process (3D printing) of a steel gear. *Source* Adapted from Liu et al. (2018)

Figure 6.7 shows, therefore, the effect of a 10% increase or decrease in the quantity of each element of the inventory of the additive manufacturing of a steel gear (indicated on the ordinate axis) on the climate change indicator result. In scenario analysis, in turn, the analyst modifies a base modeling scenario to assess possible differences induced in the results; for example, by choosing a certain method for impact assessment or for solving a multifunctionality problem. For proposing new scenarios, there may also be a change in more than one input parameter simultaneously, a characteristic that differentiates this type of analysis from local sensitivity analyzes. For being the procedure indicated by ABNT NBR ISO 14044:2009 for carrying out the only mandatory sensitivity analysis (referring to the choice of impact distribution criteria) and for dispensing the use of a complementary statistical package, the scenario analysis has become very common in LCA. However, it has some limitations such as the obvious impossibility of detecting sensitivity to individual aspects of the analyzed scenarios; and the difficulty of generalization as the analyzes are restricted to the sets of aspects included in the scenarios created in each study.

In line with ABNT NBR ISO 14044:2009, Fig. 6.8 illustrates the LCA results of eight concrete mixtures for three different impact distribution methods—avoided impact approach (columns), mass-based allocation (circles) and economic value (squares)—adopted in the modeling of blast furnace slag, a steel industry's byproduct used in the manufacture of cement and in the dosing of concrete mixtures (Silva et al. 2018).

Five impact categories have been considered: acidification potential (AP), eutrophication potential (EP), global warming potential (GWP), ozone layer depletion potential (ODP) and photochemical ozone creation potential (POCP). The change in the impact distribution criterion chosen implied a great variation in the final result, showing the LCA modeling output's sensitivity to this specific methodological choice. In these circumstances, the conclusions drawn must stick to those

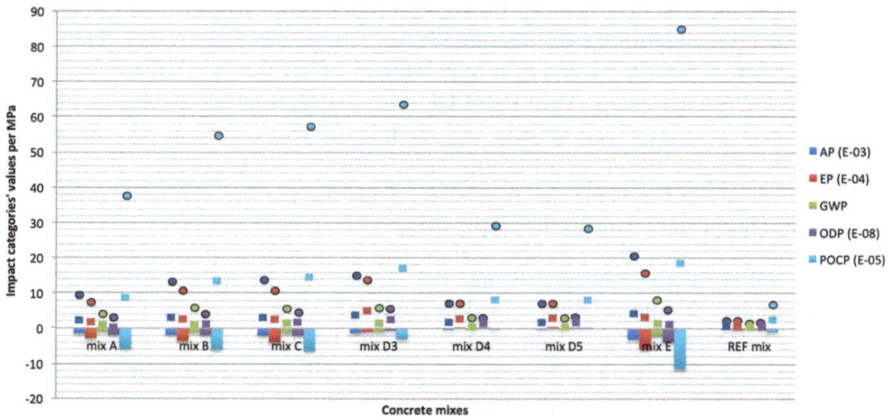

Fig. 6.8 LCA results' sensitivity of different concrete mixtures to the choice of impact distribution method in the modeling of blast furnace slag. *Source* Silva et al. (2018)

that would remain unchanged if the analyst opted for a different distribution criterion, and the sensitivity to each aspect is considered explicitly stated.

Although simple sensitivity analyzes, such as disturbance or scenario analyzes, are useful to preliminarily identify the most relevant parameters and scenarios, they disregard uncertainty and offer very limited insight of model's inputs and no information on the interactions among them.

Driven by advances in computing and numerical methods fields, the use of global sensitivity analysis is also increasingly present in LCA literature especially from 2017 on. An example is the work by Guimarães et al. (2019), which details the construction of the scenarios that configured the probabilistic distributions used in the propagation of uncertainties by Monte Carlo simulation, followed by global sensitivity analysis in specific software. The analysis result, which focused on a non-residential building, pointed to the dominance of the contribution of the uncertainties modeled in the operational phase (99.2%) to the total variance of the case study, in relation to the construction stages (0.79%) and end of life (~ 0%) of the building.

For allowing the individual contribution assessment of each input variable to the model's global variance, this type of analysis is particularly useful in LCAs with a large number of choices to consider. Recent advances in tools and computational capacity have substantially reduced the time and expertise required from the LCA analyst, who must be encouraged to deepen sensitivity analyzes. They are essential to support the conclusions and allow LCA to be used to its maximum capacity to support decision-making processes.

6.6 Exercises

1. According to the international standard ABNT NBR ISO 14044 (2009b), what are the four steps that make up the interpretation phase from which conclusions and recommendations are elaborated?
2. Chart 6.4 shows the results of a hypothetical LCIA for the "Climate Change" ('GWP') impact category, in kg of CO_2eq, considering two elementary flows: carbon dioxide (CO_2) emission and methane (CH_4) emission. Do the contribution analysis filling in all empty cells with the corresponding contribution percentages and indicate the elementary flow and the life cycle phase that you would direct for a sensitivity verification.
3. The GWP values defined in Chart 6.4 above have been used to characterize the climate change category considering a 100-year timescale. From the data listed in Chart 6.5, do a sensitivity analysis by changing a timescale of the characterization factor to 20 years. From your calculations, would you say that this hypothetical LCA result is sensitive to the timescale of the characterization factor? Justify.
4. There are some methods for obtaining characterization factors for climate change. For this reason, the Intergovernmental Panel on Climate Change (IPCC)

Chart 6.4 LCIA results for the "Climate Change" category

Alternative A LCIA Data for 'GWP'	Raw material extraction		Manufacturing		Use		Final disposal		Total	
	(CO_2eq)	(%)	(CO_2eq	(%)	(CO_2eq)	(%)	(CO_2eq)	(%)	(CO_2eq)	(%)
CO_2	5,050,500		2,000,060		1,020,000		895,421		8,965,981	
Methane (CH_4)	255,276		179,844		73,080		18,012		526,302	
Total	5,305,776		2,179,904		1,093,080		913,523		9,492,283	100

Source Adapted from Skone (2000)

Chart 6.5 Sensitivity of the LCI results for the "Climate Change" category in relation to the characterization factors

Alternative A LCIA Data for 'GWP'	Total				
	kg emitted	Characterization factors (100 years)	kg CO_2eq (100 years)	Characterization factors (20 years)	kg CO_2eq (20 years)?
CO_2	8,965.981	1	8,965,981	1	
Methane (CH_4)	18,796.5	28	526,302	84	
Total	–	–	9,492,283	–	

publishes an interval of values for each greenhouse gas in each timescale considered. (4a) Considering the interval from 28 to 36 for the 100-year-scale characterization factor of methane gas (CH_4), determine the uncertainty (also in an interval form) of the result of the climate change category presented in exercise 2 (9,492,283 kg CO_2eq). (4b) Suppose you need to do an uncertainty analysis of the results obtained for the climate change category, considering solely this variation source and having only the intervals of the characterization factors for each greenhouse gas at hand. What kind of probabilistic distribution would you use for the propagation of uncertainties via Monte Carlo simulation? What are the parameters needed to feed this simulation?

5. Differentiate local and global sensitivity analysis citing the main advantages and disadvantages of each approach.

Feedback

1. Identification of significant issues; completeness verification; sensitivity verification; consistency verification.
2. Elementary flow of CO_2, extraction of raw material phase.

Alternative A LCIA Data for 'GWP'	Extraction of raw material		Manufacturing		Use		Final disposal		Total	
	(CO_2eq)	(%)	(CO_2eq)	(%)	(CO_2eq)	(%)	(CO_2eq)	(%)	(CO_2eq)	**(%)**
CO_2	5,050,500	95.2	2,000,060	91.7	1,020,000	93.3	895,421	98	8,965,981	94.5
Methane (CH_4)	255,276	4.8	179,844	8.3	73,080	6.7	18,012	2	526,302	5.5
Total	5,305,776	**55.9**	2,179,904	**23**	1,093,080	**11.5**	913,523	**9.6**	9,492,283	**100**

3.

Alternative A LCIA Data for 'GWP'	Total				
	kg emitted	Characterization factors (100 years)	kg CO_2eq (100 years)	Characterization factors (20 years)	**kg CO_2eq (20 years)?**
CO_2	8,965,981	1	8,965,981	1	**8,965,981**
Methane (CH_4)	18,796.5	28	526,302	84	**1,578,906**
Total	-	-	9,492,283	-	**10,544,887**

4 (a) Interval: 9,492,283–9,642,628 kg CO_2eq.
(b) Uniform. Maximum and minimum values.

5. Local sensitivity analyses assess the effect that small disturbances around the nominal values of input variables have on a model's result. Global sensitivity analyses consider the entire range of variation of the input variables, measuring the output variability by the variance. Therefore, it determines how much each input variable contributes to the result's variance.

References

Ang, A., Tang, W.: Probability Concepts in Engineering. Emphasis on Applications to Civil and Environmental Engineering, 2nd ed. Wiley, New York (2007)

Associação Brasileira de Normas Técnicas – ABNT: NBR ISO 14040: Gestão Ambiental - Avaliação do Ciclo de Vida - Princípios e Estrutura: ABNT (2009a)

Associação Brasileira de Normas Técnicas – ABNT: NBR ISO 14044: Gestão Ambiental - Avaliação do Ciclo de Vida - Requisitos e Orientações. Brasil: ABNT (2009b)

Cherubini, E., Franco, D., Zanghelini, G.M. et al.: Uncertainty in LCA case study due to allocation approaches and life cycle impact assessment methods. Int. J. Life Cycle Assess **23**, 2055–2070 (2018). https://doi.org/10.1007/s11367-017-1432-6

Cucurachi, S., Borgonovo, E., Heijungs, R.: A protocol for the global sensitivity analysis of impact assessment models in life cycle assessment. Risk Anal. **36**(2), 357–377 (2016)

Gervasio, H.: Sustainable design of buildings, EUR 29324 EN, Publications Office of the European Union (2018). ISBN 978-92-79-92882-6, https://doi.org/10.2760/356384

Groen, E.A., Bokkers, E.A.M., Heijungs, R., De Boer, I.J.M.: Methods for global sensitivity analysis in life cycle assessment. Int. J. Life Cycle Assess. **22**, 1125–1137 (2017)

Guimaraes, G.D., Saade, M.R.M., Zara, O.O.D.C., Gomes Da Silva, V.: Scenario uncertainties assessment within whole building LCA. In: SBEGraz—Sustainable Built Environment Conference 2019, 2019, Graz. Sustainable Built Environment D-A-CH Conference 2019 Transition Towards a Net Zero Carbon Built Environment. UGraz, Graz (2019)

Heijungs, R., Huijbregts, M.A.J.: A review of approaches to treat uncertainty in LCA. In: iEMSs 2004 International Congress, Osnabruck, Germany, p. 8 (2004)

Huijbregts, M.: Uncertainty and variability in environmental life cycle assessment. Universiteit van Amsterdam (2001)

Iooss, B., Lemaître, P.: A review on global sensitivity analysis methods. In: Meloni, C., Dellino, G. (eds.) Uncertainty Management in Simulation Optimization of Complex Systems: Algorithms and Applications. Springer, New York (2015)

Igos, E., Benetto, E., Meyer, R., Baustert, P., Othoniel, B.: How to treat uncertainties in life cycle assessment studies? Int J Life Cycle Assess **24**, 794 (2019). https://doi.org/10.1007/s11367-018-1477-1

James, K., Galatola, M.: Screening and hotspot analysis: procedure to identify the hotspots and the most relevant contributions (in terms of, impact categories, life cycle stages, processes and flows) Versão 4.0, 24 April 2015. Disponível em: https://webgate.ec.europa.eu/fpfis/wikis/display/EUENVFP/Documents+of+common+interest?preview=%2F66618509%2F101648875%2FHotspot+Analysis+version+4.0.docx. Acesso em 10 de Julho de 2019

Klöpffer, W., Grahl, B.: Life cycle interpretation, reporting and critical review. In: Life Cycle Assessment (LCA): A Guide to Best Practice, pp. 329–356. Wiley, New York (2014)

Liu, Z., Jiang, Q., Cong, W., Li, T., Zhang, H.-C.: Comparative study for environmental performances of traditional manufacturing and directed energy deposition processes. Int. J. Environ. Sci. Technol. **15**(11), 2273–2282 (2018)

Lloyd, S.M., Ries, R.: Analyzing uncertainty in life cycle assessment: a survey of quantitative approaches. J. Ind. Ecol. **11**(1), 161–181 (2007)

Pomponi, F., D'amico, B., Moncaster, A.M.: A method to facilitate uncertainty analysis in LCAs of buildings. *Energies* **10**, 524 (2017). https://doi.org/10.3390/en10040524

Qin, Y., Suh, S.: What distribution function do life cycle inventories follow? Int J Life Cycle Assess **22**, 1138–1145 (2017)

Saltelli, A., Tarantola, S., Chan, K.P.S.: A quantitative model-independent method for global sensitivity analysis of model output. Technometrics **41**, 39–56 (1999)

Saltelli, A., Chan, K., Scott, M. (eds.).: Sensitivity Analysis. Wiley Series in Probability and Statistics. Wiley, New York (2000)

Saltelli, A., Ratto, M., Andres, T., Campolongo, F., Cariboni, J., Gatelli, D., Saisana, M., Tarantola, S. (eds.).: Introduction to sensitivity analysis. In: Global Sensitivity Analysis, pp. 1–51. The Primer. Wiley, New York (2008)

Silva, M.G., Gomes, V., Saade, M.R.M.: The contribution of life-cycle assessment to environmentally preferable concrete mix selection for breakwater applications. Ambiente Construído **18**(2), 413–429 (2018)

Skone, T.J.: What is life cycle interpretation? Environ. Prog. **19**(2), 92–100 (2000)

Zampori, L., Saouter, E., Castellani, V., Schau, E., Cristobal, J., Sala, S.: Guide for interpreting life cycle assessment result; EUR 28266 EN; 2016. https://doi.org/10.2788/171315

Chapter 7
Theory and Practice on Social Life Cycle Assessment

Alessandra Zamagni, Laura Zanchi, Silvia Di Cesare, Federica Silveri, and Luigia Petti

7.1 Introduction

In 1992, in Rio de Janeiro, the United Nations have declared "sustainability" as the guiding principle for the twenty-first century, and the term has become popular also thanks to the Bruntland Report of the World Commission on Environment and Development drafted in 1987 (Klöpffer 2002, 2008). In this report, the definition of sustainable development ("the development that meets the needs of present without compromising the ability of future generation to meet their own needs") has been introduced for the first time, and the responsibility of humankind towards the future generation has been emphasized (Klöpffer 2002).

Since then, the world has gained a deeper understanding of the interconnected challenges we face and has recognised that sustainable development has to embrace several sustainability pillars: from the three fundamental pillars related to environmental, economic and social aspects to pillars concerning, e.g. institutional, cultural and technological (Sala et al. 2013) ones.

Thus, the sustainability concerns embrace a wide range of aspects, including the social ones. The latter have been taken up by several policy frameworks such as the United Nations' Sustainable Development Goals, national initiatives and in standardisation frameworks such as the ISO 26000 (ISO 26000 2010). Social aspects are characterised by particular features: they are *bipolar* (they refer both to individual and collective levels), *reflexives* (our perceptions and interpretations of the objective social conditions change the behaviour of individuals and community), and they are *immaterial* (the social phenomena are difficult to grasp and analyse quantitatively)

A. Zamagni (✉)
Ecoinnovazione Srl Spin-Off ENEA, Via della Liberazione, 6, 40128 Bologna, Italia
e-mail: a.zamagni@ecoinnovazione.it

L. Zanchi · S. Di Cesare · F. Silveri · L. Petti
Department of Economic Studies (DEc), University "G. D'Annunzio" Chieti-Pescara, Viale Pindaro 42, 65127 Pescara, PE, Italy

J. A. de Oliveira et al. (eds.), *Life Cycle Engineering and Management of Products*,
https://doi.org/10.1007/978-3-030-78044-9_7

(Lehtonen 2004). In addition, they are perceived and evaluated differently by different stakeholders' categories, and might evolve rapidly over time (Grieβhammer et al. 2006).

Specific methodologies are needed to quantify social impacts experienced by different stakeholders, as a key step then for identifying and defining plans and actions to provide adequate solutions. Methodologies and frameworks have been developed for evaluating the social impacts of projects and organisations, such as the social impact assessment (SIA) and the corporate social responsibility (CSR). In particular, the CSR has become an important component in the management of relationships between companies and communities, the public, employees and shareholders, since companies that successfully pursue a strategy of seeking profits while solving social needs may earn better reputation and gain a competitive advantage over companies esteemed socially irresponsible (Fet 2006; Cochran 2007).

At product level, the social life cycle assessment (hereinafter S-LCA) has emerged as a methodological approach aimed at evaluating social and socio-economic aspects of products (and recently also of organisations: for more details, please refer to Martinez-Blanco et al. 2015) and their potential positive and negative impacts along their life cycle. It builds upon the life cycle concept, which is considered to provide a valuable support in integrating sustainability into design, innovation and evaluation of products and services, thanks to its systemic approach. S-LCA is a relatively new discipline and expanding field of research that completes environmental LCA (ISO 14040; 14044) and life cycle costing—LCC (Swarr et al. 2011), the latter for the determination of the most cost-effective option along the whole life cycle.

In June 2009, the scientific field of S-LCA issued a major contribution thanks to the publication of the "guidelines for social life cycle assessment of products" by the UNEP/SETAC task force (UNEP/SETAC 2009), whose updated and revised version has been recently published (UNEP 2020) (not available at the time of writing this chapter), followed by the publication of the Methodological Sheets for the Subcategories of Social LCA (UNEP/SETAC 2013). They aimed to provide practical guidance for conducting S-LCA case studies by making available information on data, examples of inventory indicators, units of measurement and data sources. According to the Guidelines, "A social and socio-economic Life Cycle Assessment (S-LCA) is a social impact (and potential impact) assessment technique that aims to assess the social and socio-economic aspects of products and their potential positive and negative impacts along their life cycle encompassing extraction and processing of raw materials; manufacturing; distribution; use; re-use; maintenance; recycling; and final disposal". (UNEP/SETAC 2009, p. 37).

During the second International Seminar in Social Life Cycle Assessment (2010, Montpellier), the previous definition of S-LCA was fine-tuned: first, social LCA is seen as a method (and not as a technique); in addition, "**social LCA may be used either to analyze the social effects caused by the functioning of chains of products compared with the situation where the chain does not exist or to look at the difference between the potential variants of one chain of products and even to look at the differences between two scenarios delivering the same service**" (Macombe et al. 2011). S-LCA assesses social aspects/impacts of all life cycle phases, from cradle to grave, and it has been developed for including a great number of impacts that vary from those concerning workers (accidents, remuneration, working conditions) and local communities (toxic pollutants, human rights abuses), to the consequences on society (corruption, payment of taxes) (Grieβhammer et al. 2006; Jørgensen et al. 2008).

7.2 The Object of Assessment in Social LCA

S-LCA evaluates social *risks*, *effects*, *performances* and *impacts*. Social *impacts* are consequences of positive or negative pressures on social areas of protection (AoP) (i.e. well-being of stakeholders), which can be caused by, e.g. a specific behaviour of one or more stakeholders, which for example causes effects related to changes in life expectancy, health and social status. Social *effects* measure the effect of an activity on stakeholders but an intermediate level, as the entire causal relationship is not identified. Social *performances* are neither social effects nor social impacts of changes, but "[…] features of a situation in a relevant organization (or features of the value chain of organizations shaping the life cycle), referring to social issues" (Macombe et al. 2013, p. 205). Finally, a social *risk* measures the likelihood of negative effects only (damage, injury and loss) that may be avoided through pre-emptive action.

These different objects of the assessment are inherently related to the geographic and cultural context where they unfold, even if culture can cross geography. Cultural indicators are also present in S-LCA subcategories (e.g. cultural heritage, respect of indigenous rights); however, performing an S-LCA does not always guarantee the inclusion of cultural values because the supporting data are often associated only with the presence or absence of national and international policies, agreements, standards and reports.

Impacts, effects, performances and risks are evaluated at different levels, namely (i) *micro* (products/services/technologies); (ii) *meso*, which includes "groups of related products and technologies, baskets of commodities (e.g. the product folio of a company), a municipality, a household" (Guinée et al. 2011: 93); (iii) *macro*, i.e. economies of states or other geographical/political entities and eventually the world. Micro-level type of studies are currently more common in the scientific literature, if not dominant, and only a few examples of macro-level studies have been published so far (Pelletier et al. 2018). It is important to point out that the level of assessment does not correspond to the level at which data are collected. In fact, in many cases, S-LCA studies at micro-level are carried out using data at company, regional and state level. Company-level data are site-specific, and as such are the most representative of the product system at hand, despite being more difficult to collect; regional- and country-level data provide average information on a given territory or sector, which can be used to address the data gaps in S-LCA, at least for the most remote life cycle stages.

7.3 Which Uses of Social LCA Results?

The ultimate goal for conducting a S-LCA is to promote improvement of social conditions and of the overall socio-economic performance of a product throughout its life cycle for all of its stakeholders, while it does not have the purpose to provide information on whether or not a product should be produced. S-LCA can assist in

- Enhancing social performance of the concerned companies by helping them to build a targeted strategy for future development of social policies;
- Managing social risk due to the identification of the social hot spots;
- Bringing structure, credibility and consistency to supply chain materiality assessment;
- Quantifying and qualifying social performance, complementing other CSR approaches;
- Increasing knowledge about the social issues along the value chain.

In addition to the educational and informative role, S-LCA main function is to support the decision-making process within organisations, by providing a strategic and managerial vision of the social sustainability of products and organisations. Companies can take advantage from the evaluation provided by the S-LCA and choose alternatives that present the most favourable social consequences. From this point of view, it can be stated that the S-LCA creates a positive effect, precisely because it allows the choice of alternatives that otherwise would not have been chosen. Furthermore, the S-LCA can also create indirect positive effects through the encouragement of socially responsible practices and behaviours on the part of companies, which consequently affect the market itself and its actors (Jørgensen et al. 2010).

7.4 The Social LCA Methodology

S-LCA has been developed mirroring the LCA framework, as described in the ISO 14040 and 14044, which implies the following considerations:

- It is structured along four main phases, namely goal and scope definition, social life cycle inventory, social life cycle impact assessment and interpretation;
- It is based on the concept of functional unit, i.e. product, service and organisation systems are defined based on the function they deliver. Consequently, based on the function, it is possible to compare different systems;
- It adopts the same modelling principles as attributional LCA, namely (i) linearity (the double the quantity considered, the double the impact; (ii) "ceteris paribus assumption", i.e. other things being equal or held constant, according to which the product, service or organisation system are considered to work under the hypothesis of isolation, without reacting to the effects of the surrounding context in which it is embedded;

- It focuses on routine functioning, i.e. exceptional situations that might occur in the functioning of the product, service or organisation system are not accounted for;
- Facts and values are both present and part of the assessment.

One reason for adopting the same framework is that there is the possibility to use S-LCA integrated /combined with other life cycle-based methods such as environmental LCA and LCC, for a life cycle sustainability assessment (Kloepffer 2008).

In the following paragraphs, the life cycle phases are analysed in details, together with further explanations about the key characteristics of S-LCA.

7.4.1 Goal and Scope Definition

Like in LCA, in the goal and scope phase of S-LCA, the following aspects are defined: (i) the goal of the study, i.e. the goal pursued and the intended use; (ii) the scope, i.e. the extent of the subject matter that it deals with, which in this case means the depth and breadth of S-LCA.

More specifically, when defining the goal, a set of guiding questions can be used: *What do we want to assess? Which decisions do the study intend to support? What are the potential improvement opportunities that are being sought through the knowledge that will be produced by the study?* Examples of intended applications and uses are, e.g.: the identification of social hot spots along the life cycle; the evaluation of potential social risks the value chain is exposed to; the quantification of potential social performances/impacts and the identification of improvement strategies; comparison of social performances/impacts of different products; reporting and labelling. It is important to identify also to whom the results will be communicated, namely the target audience. This can be represented by consumers, governments, NGOs, shareholders, product designers or trade unions and workers' representatives, and each of them requires a tailored way for communicating the outcomes.

The definition of the goal is the basis then for setting the depth and breadth of the study, namely the scope. This is the step in which the key methodological aspects of the overall study are defined, in a way similar to an LCA study:

- Functional unit definition and related reference flow;
- The product system and the system boundaries;
- The activity variables, i.e. those variable(s) that measure the magnitude of a process in the product system. They are useful to represent the relative significance of each unit process in the whole system. Activity variables are defined in the goal and scope phase, but then applied during the inventory phase of the study;
- Stakeholders to be included in the evaluation and a strategy for involving them;
- The key elements needed for evaluating the social impact assessment: type of impact assessment method, impact categories and subcategories included;
- Social indicators to be used in the assessment;

- Data collection strategies and quality requirements;
- Allocation procedures, assumptions and value choices, limitations, communication strategies (including requirements for a critical review, if needed).

While S-LCA structure strictly mirrors the LCA's one, however, there are three key important differences in the setting of the scope, out of all the aspects that have to be defined in this step:

- The definition of system boundaries;
- The definition of stakeholders;
- The identification of social indicators.

Regarding the definition of the system under study and its ***boundaries***, it is important to consider that in S-LCA, the social life cycle is still defined by processes (social impacts arise as a consequence of interactions among stakeholders driven by activities) but also by socio-economic mechanisms. The latter consist of relations among stakeholders involved in the life cycle, as a result of the activities carried out. According to this perspective, stakeholders are the key elements for defining the system under study and its boundaries.

Thus, as shown in Fig. 7.1, system boundaries in S-LCA should take into account

- The causal relationships that connect the level of two activities within the domain of technology, i.e. processes needed for delivering the output of the system;
- Social relations (second layer defined by dotted lines), driven by stakeholders: at each step of the system life cycle, there are different actors who might be potentially affected. These in turns have different kind of relationships (e.g. commercial, administrative) with other actors who might not be strictly connected with the technological life cycle, but this interaction gives rise to social effects that affect the system under study.

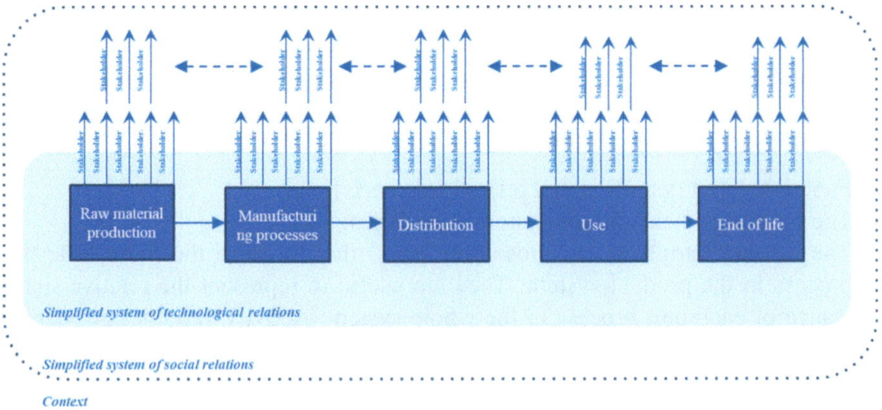

Fig. 7.1 System boundaries in S-LCA. *Source* Scorelca (2017)

As far as ***stakeholders*** are concerned, they are a cluster of actors who are expected to have shared interests, and currently in S-LCA they are grouped into six main categories, namely worker, consumer, local community, society, value chain actors not including consumers and children (UNEP 2020). These categories are not exhaustive per se, and other could be added when deemed relevant: as such, the current structure of the guidelines, while keeping its applicability, is not static, but can be broadened and enriched, building upon the knowledge developed with the applications. Stakeholders categories can be linked to different socially relevant themes or attributes (called "***subcategories***"), such as child labour, fair salary, fair competition and health and safety, to mention just a few. The relation between stakeholders and subcategories is one-to-many, as represented in Chart 7.1, i.e. one stakeholder category can be affected by more than one social theme, and usually, one social theme can be of interest for one specific stakeholder category only.

The quantification of a social effect, risk and performance or impact is expressed by means of ***social indicators***. Several and different sets of social indicators exist, developed within different methodological frameworks, for different purposes and with a different resolution, and their selection represents a challenging issue. The relevance is often mentioned as the criterion for indicators selection, but the rationale behind the choice is often not provided. Most of the studies rely on the indicators proposed in the UNEP/SETAC methodological sheets (UNEP/SETAC 2013) and approach them on the basis of data availability, while a few stress the need of introducing additional indicators or stakeholder groups specific for their case studies. However, in most of the cases, the addition of other stakeholder groups (and related indicators) relies upon the author's perception of what matters, while a sound and reproducible approach is neither presented nor its relevance is discussed (Zanchi et al. 2018).

Finally, an important aspect in the goal and scope is the definition of the social impact assessment method, which is described in Sect. 7.4.3, and might require (depending on the method chosen) the definition also of the reference scale used for the assessment and the weighting method.

7.4.2 Social Life Cycle Inventory

The inventory phase of S-LCA relies on the ISO definition (ISO 14040 2006), and therefore, it consists of the collection of data to model the product system. However, some differences with environmental LCA exist. First, inventory data in environmental LCA are related to physical quantities linked to unit processes (e.g., energy consumption for injection moulding), and can be measured, calculated or estimated. In S-LCA data are often qualitative and reflect the capability of a company to manage social aspects of concern. Thus, they do not have a direct link with physical quantities. Secondly, social impacts of an activity also depend on its geographic location and social context. A parallel can be done considering the geographic scale of environmental impact categories; the magnitude of a regional-scale impact category (e.g.

Table 7.1 Relation between stakeholders and subcategories

Stakeholder categories	Worker	Local community	Value chain actors (not including consumers)	Consumer	Society	Children
Subcategories	1. Freedom of Association and Collective Bargaining 2. Child labour 3. Fair salary 4. Working hours 5. Forced labour 6. Equal opportunities/ discrimination 7. Health and safety 8. Social benefits/social security 9. Employment relationship 10. Sexual harassment 11. Smallholders including farmers	1. Access to material resources 2. Access to immaterial resources 3. Delocalisation and migration 4. Cultural heritage 5. Safe and healthy living conditions 6. Respect of indigenous rights 7. Community engagement 8. Local employment 9. Secure living conditions	1. Fair competition 2. Promoting social responsibility 3. Supplier relationships 4. Respect of intellectual property rights 5. Wealth distribution	1. Health and safety 2. Feedback mechanism 3. Consumer privacy 4. Transparency 5. End-of-life responsibility	1. Public commitments to sustainability issues 2. Contribution to economic development 3. Prevention and mitigation of armed conflicts 4. Technology development 5. Corruption 6. Ethical treatment of animals 7. Poverty alleviation	1. Education provided in the local community 2. Health issues for children as consumers 3. Children concerns regarding marketing practices

Source UNEP (2020, page 23)

eutrophication) also depends on the existing state of environmental departments and receivers. A third clear difference lies in the fact that inventory data in a social analysis need to be specified in relation to different stakeholders (e.g. local community and consumer) and contexts.

The Guidelines and the Methodological Sheets for Subcategories in S-LCA(UNEP 2020) currently provide a structured framework to implement the social data inventory according to a cradle-to-grave perspective. The S-LCA inventory indicators are classified according to subcategories and stakeholders, as represented in Fig. 7.2.

Subcategories have been retrieved from international instruments, CSR initiatives, model legal framework and social impacts assessment literature in order to define inventory indicators with proper references to international instruments.

Overall, inventory data are classified in generic or site-specific data. The first are typically country-level data that could be retrieved from government and intergovernmental documents and that could be useful to carry out a screening of high-risk regions in case of organisation with global supply chain. The second group is

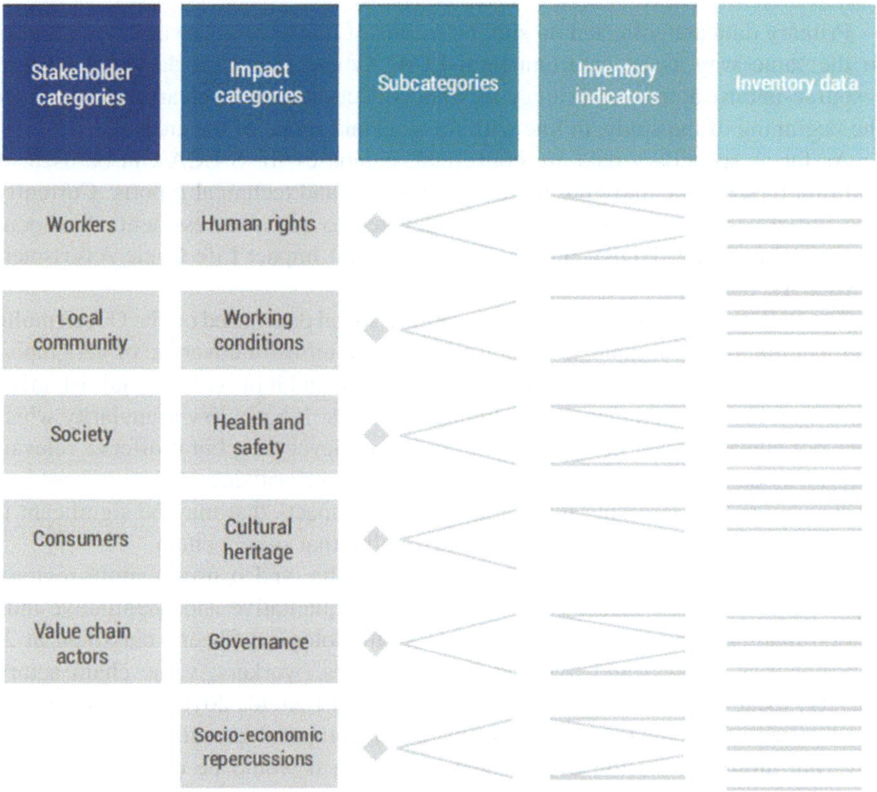

Fig. 7.2 Structure of the S-LCA framework in the S-LCA guidelines. *Source* UNEP (2020)

represented by information that can be collected as primary data by an organisation in one or more of its plants; this data collection could be particularly time and cost-consuming, and for this reason, it could proceed in areas highlighted as hot spots.

Life cycle inventory (LCI) in the S-LCA also involves the collection of ***activity variable*** information. The activity variable measures the magnitude of a process in the product system, and it reflects the share of a given activity associated with each unit process. It is thus useful to give an idea of the relative significance of each unit process in the whole system. The most common used activity variables are the number of worker hours necessary at each unit processes to provide the input to the final product or the added value; however, also, other variables can be chosen.

When the significance of unit processes has been defined, thus, data collection has been somehow prioritised, and it is necessary then to proceed with the data collection. Different types of data can be used in S-LCA, which can be overall classified according to

- The type of data, i.e. qualitative vs quantitative vs semi-quantitative data;
- Their resolution, i.e. primary and generic data.

Primary data are collected on site, by means of interviews, surveys, audit results in the same way as an environmental LCA. Considering that data collection is resource-intense activity, a strategy for data collection shall be clearly defined since the beginning of the study, in line with the goal and scope of the analysis.

As far as secondary data are concerned, databases for S-LCA can be used, in addition to scientific and grey literature, statistics and technical reports. Currently, two initiatives exist regarding database devoted to social assessment: the Social Hotspot Database (SHDB)[1] and the Product Social Impact Life Cycle Assessment (PSILCA)[2] database.

The SHDB provides sector-country-specific social data based on the GTAP multi-regional input–output table; it is comprehensive in terms of coverage of geographic contexts and sectors (113 countries and 57 sectors), and it provides social risk information on 22 social themes and including 89 issues. It has a low granularity which does not allow to cover process-level or company-level data, but it offers a relevant way to model product category supply chains by prioritising hot spots based on worker hours and assessing the potential social impacts that may be significant in particular countries and for specific sectors within that supply chain.

The PSILCA has been developed by GreenDelta, and it uses a multi-regional input/output database called Eora. It includes 88 qualitative and quantitative indicators, mainly inspired by UNEP/SETAC Methodological Sheets, classified in 23 subcategories (topics) and five stakeholder groups (workers; value chain actors; society; local community; consumers) (Ciroth and Eisfeldt 2016). Both databases can be used within the Open LCA software tool and provide useful information for covering those remote life cycle stages for which it would be difficult to collect

[1] https://www.socialhotspot.org/about-shdb.html#.

[2] https://psilca.net/.

primary data and also for performing a preliminary social risk assessment; however, primary data are key for a reliable and robust assessment of the social impacts of products and organisations.

7.4.3 Social Life Cycle Impact Assessment

Social impact assessment is the phase aimed at understanding and evaluating the magnitude and significance of the potential social impacts of a product system throughout the life cycle of the product. There are two main types of impact assessment in S-LCA:

- *Reference scale assessment* (also referred to as Type I) is aimed at evaluating social performances and/or risks, and it is based upon the concept of subcategories and their relation to the affected stakeholders;
- *Impact pathway assessment* (also referred to as Type II) is aimed at evaluating potential social impacts, and it is based upon the identification and measurement of causal links in impact pathways.

In the reference scale assessment, a key aspect is the choice of the referencing system, i.e. reference scales of different levels, each of which corresponds to a performance reference point (PRPs). PRPs are thresholds, targets or objectives that set different levels of social performance or social risk, which allow to estimate the magnitude and significance of the potential social impacts associated with organisations in the product system. The PRPs can be defined based on international standards, local legislation or industry best practices, norms and socio-economic context, experts' judgments, comparison between alternatives (Garrido et al. 2016). An example of a reference scale approach is the subcategory assessment method (SAM) (Ramirez et al. 2014), based on norms and on the socio-economic context, and it is based on a four-level scale (*A*, *B*, *C*, or *D*). An example of the application of the SAM method is provided in Sect. 7.5.

As far as the impact pathway (IP) approach is concerned, it is based upon the identification and quantification of social mechanisms, in a way similar to what is done in the environmental life cycle impact assessment with midpoint and endpoint indicators. The social mechanisms are represented by social impact categories, category indicators and characterisation models. Inventory results are therefore connected with impact categories (usually described as midpoint impact categories) and category endpoints (usually described by endpoint impact categories) (Neugebauer et al. 2017). Although some impact categories and their indicators have been developed, the IP approach is still at an early stage and lacks a systematic framework. Examples can be found in Weidema (2006), who proposed quality adjusted life years (QALYs) as a unit of impact measurement for human well-being analogous to disability adjusted life years (DALYs) being used as a measurement unit for damage to human health in environmental LCA. He used six damage categories for human life: health; life and

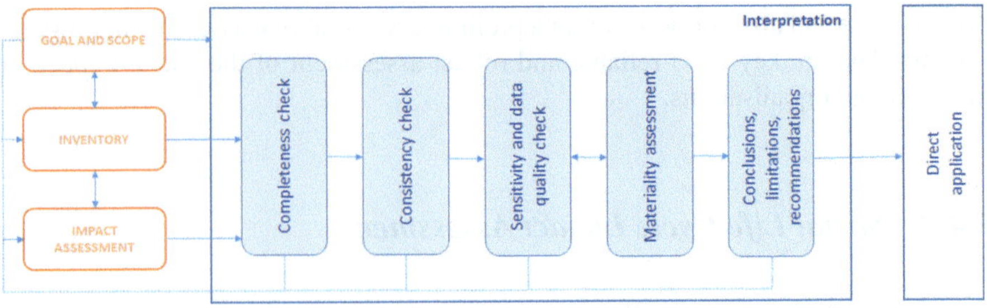

Fig. 7.3 Steps of the interpretation phase. *Source* Adapted from Hauschild et al. (2018) and Laurent et al. (2019) and included in the revised version of the social LCA guidelines (UNEP 2020)

longevity; autonomy; safety, security and tranquillity; equal opportunities and participation and influence. Feschet et al. (2012) assessed the social impact of a product using Preston pathway, an empirical relationship, between real life expectancy at birth and real per capita income.

An increased number of impact pathway is expected to occur in the next future, also thanks to the project promoted by the life cycle initiative that is addressing the development of a linkage between the top-down process that led to the creation of the SDGs and all the bottom-up knowledge, data and methodology in the life cycle sustainability assessment area.[3]

7.4.4 *Interpretation*

The interpretation is the final phase of the S-LCA study, in which all the previous phases are critically reviewed. When the iterative process is concluded, the results of the S-LCIA phase are checked and discussed in depth, and this discussion forms a basis for conclusions, recommendations and decision-making in accordance with the goal and scope definition.

In order to be interpreted, results have to be analysed at different levels: information and data might be aggregated and/or broken down at the level of life cycle phases, subcategories, stakeholders, or also at process level for extracting insights.

Like for LCA, the interpretation phase is built upon the requirements of ISO 14044 (2006) and consists of the following steps, illustrated in Fig. 7.3:

- Completeness check
- Consistency check
- Sensitivity and data quality check
- Materiality assessment

[3] For more information and updates, please refer to the project's Website https://www.lifecycleinitiative.org/activities/key-programme-areas/technical-policy-advice/linking-the-un-sustainable-development-goals-to-life-cycle-impact-pathway-frameworks/.

- Conclusions, limitations and recommendations.

The *completeness check* aims at reviewing each assessment phase to ensure that all the relevant issues, outlined in the goal and scope phase, have been satisfied in the inventory and impact assessment phase, i.e. that all pertinent data and information have been gathered and processed in relation to the relevant stakeholders, the results have met the objective of the study and that insights allow to draw conclusions from the life cycle evaluation.

The *consistency check* aims at ensuring that the methods applied in the inventory and impact assessment step and the data used are consistently applied throughout the study and are in accordance with the goal and scope of the study.

The *sensitivity check* aims at determining whether and to what extent the conclusions of the S-LCA study may be affected by the assumptions made during the previous steps. Assumptions may be related to data, value judgments, activity variable, calculation of the social performance and social impacts, aggregation and weighting.

When the different checks have been performed, the results should be further interpreted to determine the significance of the selected issues. This step of the interpretation aims at identifying significant social performances or impacts, risks, stakeholders' categories, life cycle phases of processes, in accordance with the goal and scope of the study. In the context of S-LCA, the significance is related to the concept of materiality.

When the results have been thoroughly analysed in relation to their completeness and consistency, and the material aspects of the study have been identified, conclusions can be drawn. This includes highlighting limitations and giving recommendations for improvement actions provided to the decision maker. Limitations might refer to the type and quality of the data used or to the referencing system adopted, the scoring system applied or the weighting criteria adopted, for example, aggregating the reference scale results into a subcategory result. It could be important to involve the stakeholders in this last step, extending the representativeness to those who might be affected by the decision of the study. This is the step where main questions raised during goal and scope find an answer.

Finally, S-LCA studies can be combined with/integrated to other evaluation methodologies, such as other life cycle methodologies (e.g. E-LCA, LCC), evaluation methods, ecodesign and multi-criteria methods. When such a combined study is carried out, either as a life cycle sustainability assessment or part of it, consistency must be ensured in system boundaries definition, function of the system, decision-making context and interpretation of the results.

7.5 S-LCA Case Study

In this section, an illustration of a case study of S-LCA is presented and described, following the structure of the S-LCA guidelines. The case study refers to "Cuore di

Fig. 7.4 Cuore di bue tomato, object of the case study

bue" tomato and shows how to implement in practice the reference scale assessment method "SAM", introduced in Sect. 7.4.3 (Fig. 7.4).

The objective of this case study is to identify and evaluate the social aspects of the life cycle of an Italian variety of tomato called "Cuore di Bue" produced by an Italian cooperative. This case study provided an opportunity to address issues of significant social interest for the cooperative and the product (e.g. human rights, community welfare, racial discrimination, child labour) and therefore to highlight the strengths and weaknesses of the company towards its own stakeholders.

7.5.1 Goal and Scope Definition

The object of the study is the tomato Cuore di bue, whose function is to contribute to satisfying a person's food needs for 1 kg of tomatoes.

Product function	Functional unit	Reference flow
To contribute to satisfying a person's food needs	The provision of a defined quantity of tomato, including its packaging	1 kg of tomatoes

The system boundaries are from cradle to gate and include the following phases: Production, processing/packaging and distribution. The use phase has been considered not from the technological point of view but in relation to the interaction with consumers, as main stakeholder.

The present study has focused on three stakeholders, identified based on their relevance: Workers, local community and consumers.

System boundaries definition

The definition of the system boundaries is necessary to identify and determine the process units that must be included in the study itself. It aims to define what are the

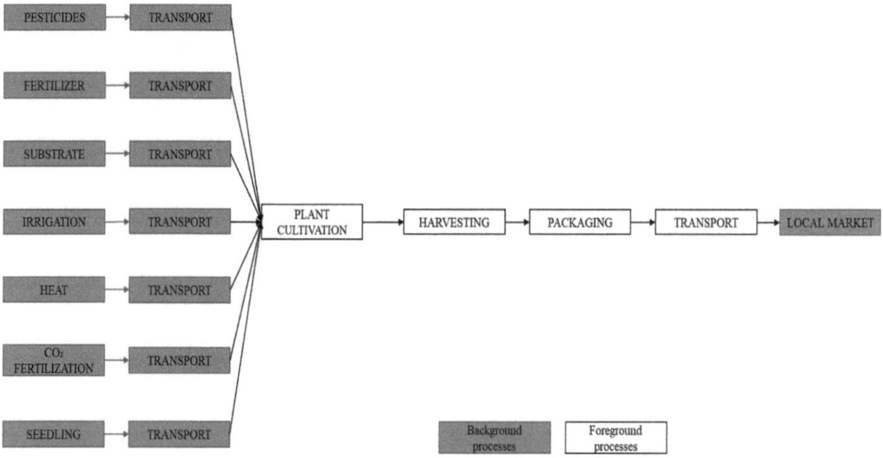

Fig. 7.5 System boundary of the tomato cuore di bue

fundamental processes for obtaining the tomato cuore di bue. To this end, they are identified and subsequently represented graphically, using a flow chart (Fig. 7.5).

The first segment of the chart involves "raw materials" necessary for the cultivation of the product. They consist of seedlings, coconut slabs (which represent the substrate on which the seedlings are placed) and plant protection products. The second process consists of the transport of the main materials, from the supplier to the company producing the tomato cuore di bue. Then, cultivation occurs, which is the stage during which the workers position the seedlings on the substrate, connect them to a centralised and computerised system, which supplies the nutrients and all the types of phytosanitary it needs. Then, the harvesting occurs, which takes place in different periods of the year, not only in open air but also in greenhouses.

When the tomatoes are grown and collected, they are then selected manually and packed in cardboard. The company also uses plastic boxes, but in this case, it is the customer who supplies the company with them and sends them back once they have been emptied.

As last step of the analysed system, the packed tomato is distributed to the users, either final consumers or retailers.

Cut-off criteria

Cut-off criteria are used to define what can be excluded from the study because it will not affect the outcomes, due to its low relevance. The criterion used in the study is represented by the working hours. In practice, this means that only those processes for which a defined amount of working hours are needed will be included in the study. For applying this criterion, it is necessary first to calculate the total number of working hours required for each process contributing to the creation and distribution of the "cuore di bue" tomato in relation to the reference flow (1 kg).

The identification of the working hours per process can be determined by estimates, simply multiplying the number of workers by the hours worked per week, considering the weeks worked in a year and then dividing this ratio by the total production.

This could be summarised by the following equation:

$$\mathrm{Wh} = \frac{W \cdot h \cdot n}{p} \tag{7.1}$$

where Wh is the working hours, W is the number of workers, h represents the hours per week, n the number of weeks/ year, and p the total production.

Working hours are multiplied by the required quantity of the functional unit, according to the formula:

$$\mathrm{WFU} = \mathrm{Wh} \cdot c \tag{7.2}$$

where WFU represents the amount of working hours per functional unit, and c is the amount of all materials necessary to produce 1 functional unit of Tomato Cuore di Bue.

After calculating the working hours necessary for each process, we obtain a summary table (Table 7.2).

Table 7.2 Summary of working hours per process unit

Process	Company	Transportation	Labour minutes
Phytosanitary distribution	Company B	Company B	1.7255
Plant production	Company C	Company C	1.465
Coconut slab production	Company D	Company D	4.653
Plantation, harvest and packing	Company A		98.988
Cardboard box production	Company E	Company E	120
Final delivery (transport)	Company F		702
	Company G		1476
	Company A		360
		Total	107.5293

The sum of the working hours necessary for the annual production of 1,965,000 kg of the "cuore di bue" tomato is 107,529.3 working hours. Consequently, 0.05473 h is necessary to produce 1 kg of tomatoes (functional unit).

We can now relate the hours of each process to the functional unit. The calculation of the percentages of hours (x) is obtained through the Eq. 7.3:

$$x = \frac{100 \cdot Y}{107.5293} \tag{7.3}$$

where Y corresponds to the hours relating to the single company.

The cut-off applied in the study excluded those processes that had a percentage of working hours less than 1%; thus, according to this, the processes that concern the production of boxes (120 h) and the final transport (360 h) are excluded from the study.

7.5.2 Life Cycle Inventory

To collect data from the involved stakeholders, different questionnaires were developed and distributed, each for the different groups:

- Questionnaire for the company administration (office workers);
- Questionnaire for employees directly involved in the tomato production (labourers).

The aim of interviewing different people is to allow triangulation, that is, to compare data amongst different sources of information to validate data of the inventory analysis, as summarised in Table 7.3.

To further validate the answers received, national laws and norms and the National Collective Worker Agreement (CCNL) in the agricultural and floriculture sector (Parti sociali 2010) have been considered and analysed.

Table 7.3 Triangulation data by stakeholder type

Stakeholder	Primary data	Triangulation
Workers	Business owner of the organisation	Workers of the organisation (72 interviewed)
Consumers	Marketing responsible of the organisation	Websites (as Istituto Nazionale di Statistica (ISTAT), www.agricolturanotizie.com) and local health authority (as Azienda Sanitaria Locale (ASL) in Italy)
Local community	Marketing and human resources responsible of the organisation	District responsible identified directly by the organisation

Table 7.4 Questionnaire for worker's stakeholder group

Subcategory	Number of questions	Results of the inventory questionnaire
Freedom of association	3 questions to the workers/11 questions to the organisation representative	No workers were members of a union but members of the cooperative company
Child labour	2 questions to the worker/7 questions to the organisation representative	Presence of policy against child labour
Fair salary	4 questions to the workers/10 questions to the organisation representative	The lowest salary is equal or higher than the minimum wage of the Italian agriculture sector
Working hours	6 questions to the worker/4 question to the organisation representative	Weekly number of average working hours is compliant with the law of the sector
Forced labour	3 questions to the worker/5 questions to the organisation representative	Presence of policy against forced labour
Equal opportunities/discrimination	9 questions to the worker/4 questions to the organisation representative	The organisation promotes equal opportunities for workers
Health and safety	12 questions to the workers/13 questions to the organisation representative	The organisation invests in and trains its employees with relation to accident prevention programs
Social benefit/social security	2 questions to the worker/3 questions to the organisation representative	The organisation provides more than two social benefits listed in the basic requirement

Source Petti et al. (2018)

A summary of the questionnaires administered, and the obtained results are reported in Tables 7.4, 7.5 and 7.6, for the different stakeholder groups.

7.5.3 Impact Assessment Through the Subcategory Assessment Method (SAM)

The impact assessment is the third phase of a S-LCA study. As mentioned in Sect. 7.4.3, different impact assessment methods exist; in this case study, the subcategory assessment method (SAM) has been applied, which is a reference scale assessment type of method. SAM structures the evaluation of the social performances according to four classes (*A*, *B*, *C* and *D*) based on the company's behaviour. Data need to be collected and compared to basic requirements (BRs), which are defined

Table 7.5 Questionnaire for local community stakeholder group

Subcategory	Number of questions	Results of the inventory questionnaire
Access to material resources	6 questions to the organisation/5 questions to the local community representative	Certificate Global Gap and Lotta Integrata
Access to immaterial resources	3 questions to the organisation/2 questions to the local community representative	No evidence of the promotion of community services (health/education/information sharing)
Delocalisation and migration	6 questions to the organisation/5 questions to the local community representative	No evidence of resettlement caused by the organisation
Cultural heritage	4 questions to the organisation/3 questions to the local community representative	Community and its subsistence is considered an activity of cultural heritage preservation
Safe and healthy living conditions	5 questions to the organisation/4 questions to the local community representative	Certificate Global Gap and Lotta Integrata
Respect of indigenous rights	3 questions to the organisation/2 questions to the local community representative	Communities and the regions were already occupied by similar activities, and there is no conflict with the local community
Community engagement	6 questions to the organisation/5 questions to the local community representative	The organisation actively participates in events of the local community ("sagras and banco alimentai")
Local employment	3 questions to the organisation/2 questions to the local community representative	It uses local employees
Secure living conditions	3 questions to the organisation/2 questions to the local community representative	The organisation does not reveal any conflicts or problems with the local community proven by the absence of judicial appeals to the organisation

Source Petti et al. (2018)

according to legislation or organisational practices and country context, resulting in different levels (Ramirez et al. 2014):

- Level A means that the organisation shows a proactive behaviour with respect to the basic requirement, since it promotes and satisfies the requirement also towards its suppliers or value chain;
- Level B highlights that the organisation respects the basic requirements on the basis of international and local standards;
- Classes C and D identify the aspects that do not satisfy the basic requirement.

Table 7.6 Questionnaire for consumer's stakeholder group

Subcategory	Number of questions	Results of the inventory questionnaire
Health and safety	6 questions	The company does not receive any complaints on health and safety issues from its consumers, but the company does not promote healthy and safety practises and policies with its own business partners
Feedback mechanism	3 questions	No feedback mechanism is available, but the consumers can still reach the company by telephone or by e-mail
Privacy	3 questions	No policy to guaranty consumers' privacy No protection of consumer's data supplied by Internet No policy or actions to protect the suppliers' privacy
Transparency	3 questions	The company does not communicate its social corporate responsibility but implement environmental and social tools for the impact assessment such as water footprint and LCA
End-of-life responsibility	2 questions	The company gives clear information for consumers on the end-of-life treatment of its product, but it does not promote policy and practises with its business

Source Petti et al. (2018)

Referring to the "cuore di bue" case study, the results of SAM impact assessment and evidences are illustrated in Table 7.7.

To each of the classes, a numerical scale has been associated ($A = 4$, $B = 3$, $C = 2$, $D = 1$), which transforms qualitative data into quantitative ones. In Figs. 7.6 and 7.7, an extract of a graphical summary of the assessment of different processes is reported.

7.5.4 Interpretation of Results

The details of the interpretation steps are not reported here, but rather the aspects raised as outcomes of the study are presented. In particular, the study pointed out the absence of proactive actions towards suppliers and other players in the value chain. For example, the company highlighted poor management of customer satisfaction systems; in this regard, the creation of a Website, for a direct contact with the customer, could represent an improvement and an excellent showcase for the product.

As an example, in Table 7.8, potential measures the company could adopt to improve its social impact are reported.

Table 7.7 Results of the application of the SAM method to the cuore di bue tomato case study

Stakeholder groups	Subcategory	Results of SAM	Evidence
Consumer	Health and safety product	B	The organisation invests and trains its employees in relation to accident prevention programs
	Feedback mechanism	C	There are no measures which enable the consumer to make complaints, such as providing a suggestion box on the help desk or a customer care section on the Website
	Consumer privacy	C	There is no formal policy on privacy within the organisation
	Transparency	C	The organisation has no formal report on social responsibility but demonstrates practises to its suppliers
Workers	Health and safety product	B	The organisation invests and trains its employees in relation to accident prevention programs
Local community	Safe and healthy living conditions	B	Certificate Global Gap and "Lotta Intograta"
	Access to immaterial resource	D	No evidence of the promotion of community services (health/education/information sharing)

Source Petti et al. (2018)

7.5.5 *Initiatives and Outlooks*

S-LCA is a relatively new discipline and an expanding field of research. Building upon the UNEP-SETAC Guidelines for S-LCA of products and the complementary methodological sheets, the field of S-LCA started establishing a framework building on the ISO 14040 and 14044 LCA standards. Through conferences, published journal articles, seminars and industry group publications, the methods are spreading, evolving and gaining in maturity.

In addition, several initiatives have been started, aimed at either further developing the methodology and making it more robust for the decision-making process and at promoting the use of S-LCA in the industrial context. A prominent initiative is the ***revision of the guidelines for social life cycle assessment*** (S-LCA) of products in the framework of the life cycle initiative, completed in 2020 in cooperation with the Social LC Alliance. S-LCA guidelines and the methodological sheets have played and will play a decisive role in the S-LCA implementation. The revision was needed to incorporate new methods, experiences and progresses of the recent years and consisted of two phases: a substantial technical revision, carried out with

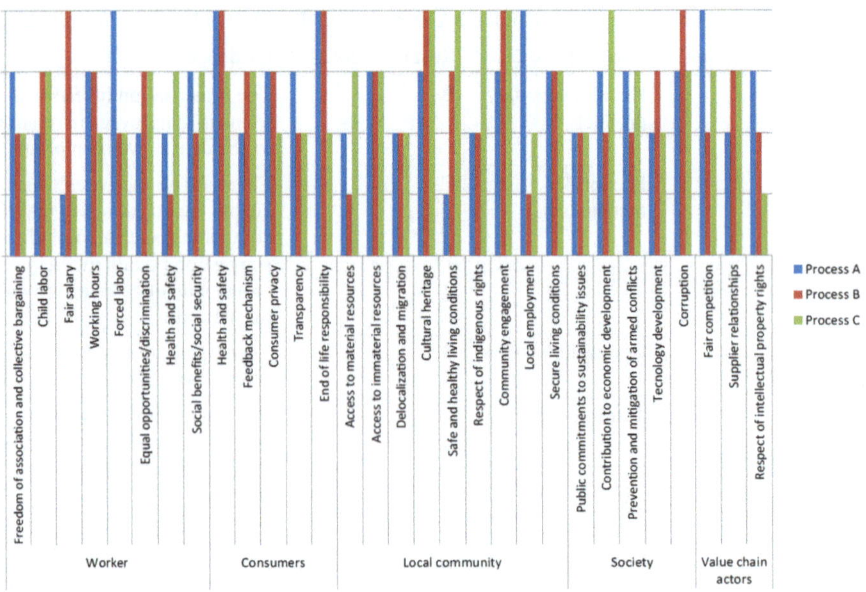

Fig. 7.6 Bar graph to summarise the product system social impact. *Source* Petti et al. (2018)

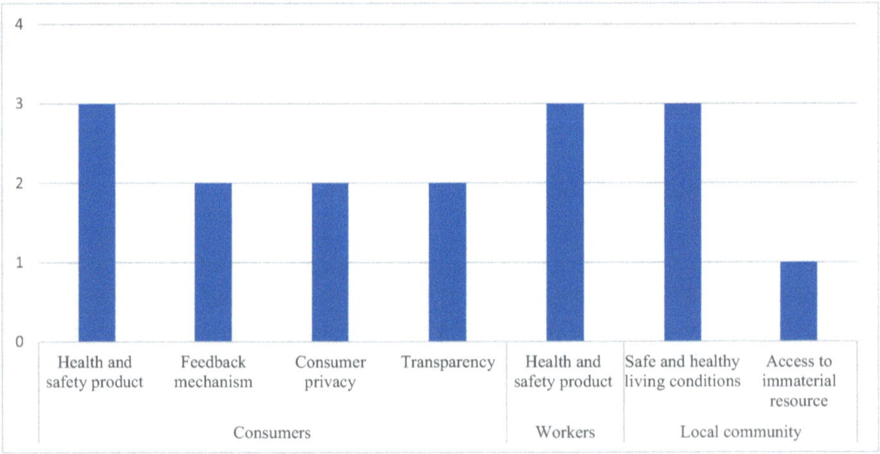

Fig. 7.7 Extract of the results based on Table 7.7. *Source* Petti et al. (2018)

the contribution of several experts at international level, and a road testing phase with companies and other organisations involved in the application of the updated guidelines on a range of products or organisations and industrial sectors.

In parallel, during the last years, a private business-led initiative coordinated by Prè Consultant started and is still ongoing: the ***Roundtable for Product Social***

Table 7.8 Summary of the SAM assessment results and relative action plan measures

Stakeholder groups	Subcategory	Results of SAM	Evidence
Consumer	Health and safety product	B	The organisation invests and trains its employees in relation to accident prevention programs
	Feedback mechanism	C	There are no measures which enable the consumer to make complaints, such as providing a suggestion box on the help desk or a customer care section on the Website
	Consumer privacy	C	There is no formal policy on privacy within the organisation
	Transparency	C	The organisation has no formal report on social responsibility but demonstrates practises to its suppliers
Workers	Health and safety product	B	The organisation invests and trains its employees in relation to accident prevention programs
Local community	Safe and healthy living conditions	B	Certificate Global Gap and "Lotta Integrata"
	Access to immaterial resource	D	No evidence of the promotion of community services (health/education/ information sharing)

Source Petti et al. (2018)

Metric, which proposes a practical method named product social impact assessment (PSIA) for the quantification of the social performances of products (Goedkoop et al. 2020). Members of the Roundtable are representatives from large companies who have identified and proposed a defined list of indicators and qualitative and quantitative methods for their quantification, with the ultimate goal to practically support organisations into assessing the potential social impacts of a product along its life cycle.

Other initiatives started with a focus on specific sectors; an example is the ***Social Life Cycle Metrics for Chemical Products***, a social metrics guidance prepared by key players of the chemical sector. The guidance has been officially released in June 2016, and latest updates have been published in 2018; it can be used to assess and report the social impact and value of products in a life cycle perspective.

In addition to scientific- and industry-led initiatives, the topic of S-LCA is also gaining attention at policy level. In particular, the DG Grow of the European Commission, in cooperation with UN Environment and the Life Cycle Initiative, on April 2019 organised the first social LCA metrics workshop with the aim of monitoring the status of social LCA and opening debate on future steps. This was an important step forward in the improvement and identification of S-LCA as methodology to evaluate social performances of products and organisations, as a key element for competitiveness.

S-LCA is an evolving field, and main developments are envisaged, both at the level of methodology and interpretation of results. We have identified the following main areas of developments, together with recommendations for moving forward S-LCA:

- Criteria and indicators for S-LCA;
- Impact assessment;
- Positive impacts;
- Communication of S-LCA results;
- Interdisciplinary approach for S-LCA.

Regarding the latter aspect, S-LCA stands out as an integration framework that can accommodate social information obtained with other methods and tools. It is thus important to strengthen the cooperation with other initiatives, to involve stakeholders directly into the analysis and to give relevance to the results. All these methodological and practical issues are particularly crucial when using S-LCA within the life cycle sustainability assessment (LCSA) framework. As the international context becomes increasingly demanding on social issues, S-LCA may be a good way for helping companies to be compliant with ONU's sustainable development goals. Thus, thanks to its complementarity with other reporting tools and standards, S-LCA can be used in combination with other techniques or tools to provide further information and help linking social impacts at the company level to the product's life cycle stages.

Exercises

(1) Why social LCA is important? How does it complement with environmental LCA?

(2) Explain what is stakeholder categories and subcategories? How they can be used in a social life cycle inventory? Feedback: Stakeholder categories can be used to select relevant social issues for evaluation in a product system, and they can be linked to different socially relevant themes, i.e., the subcategories. The subcategories are relevant to allow social inventory indicators selection and measures.

References

Ciroth, A., Eisfeldt, F.: PSILCA—a product social impact life cycle assessment database. Database version 1.0 Documentation Version 1.1 (2016)

Cochran, P.L.: The evolution of corporate social responsibility. Bus. Horiz. **50**, 449–454 (2007)

Feschet, P., Macombe, C., Garrabé, M.: Social impact assessment in LCA using the Preston pathway. Int. J. Life Cycle Assess. **18**, 490–503 (2012)

Fet, A.M.: Environmental management and corporate social responsibility. Clean Technol. Environ. Policy **8**, 217–218 (2006)

Garrido, S.R., Parent, J., Beaulieu, L., Revéret, J.P.: A literature review of type I SLCA—making the logic underlying methodological choices explicit. Int. J. Life Cycle Assess. https://doi.org/10.1007/s11367-0161067-z

Goedkoop, M.J., de Beer, I.M, Harmens, R., Peter, S., Dave, M., Alexandra, F., Anne, L.H., Diana, I., Diana, V., Ana, M., Elizabeth, M.-F., Carmen, A., Ipshita, R., Urs, S., Megann, H., Massimo, C., Thomas, A., Viot; J.-F., Alain, W.: Product social impact assessment handbook - 2020. Amersfoort, November 1st (2020)

Grießhammer, R., Benoît, C., Dreyer, LC., Flysjö, A., Manhart, A., Mazijn, B., Méthot, A.,Weidema, B.P.: Feasibility study: integration of social aspects into LCA, Discussion Paper from UNEP-SETAC Task Force Integration of Social Aspects in LCA meetings in Bologna (January 2005), Lille (May 2005), Brussels (November 2005) and Freiburg (May 2006)

Guinée, J.B., Heijungs, R., Huppes, J., Zamagni, A., Masoni, P., Buonamici, R., Ekvall, T., Rydberg, T.: Life cycle assessment: past, present, and future. Environ. Sci. Technol. **45**(1), 90–96 (2011)

Hauschild, M.Z., Dreyer, L.C., Jørgensen, A.: Assessing social impacts in a life cycle perspective—lessons learned. CIRP Ann. Manuf. Technol. **57**, 21–24 (2008)

Hauschild, M.Z., Bonu, A., Olsen, S.I., Chapter 12: Life cycle interpretation. In: Hauschild, M.Z., Rosenbaum, R.K., Olsen, S.I. (eds) (2018) Life Cycle Assessment - Theory and Practice. Springer, ISBN 978-3-319-56474-6 (2018)

ISO 14044: 2006 Environmental Management—Life Cycle Assessment—Requirements and Guidelines (2006)

ISO 14040:Environmental Management —Life Cycle Assessment—Principles and Framework (2006)

ISO 26000: Guidance on Social Responsibility (2010)

Jørgensen, A., Le Bocq, A., Nazarkina, L., Hauschild, M.: Methodologies for social life cycle assessment. Int. J. LCA **13**(2), 96–103 (2008)

Jørgensen, A., Finkbeiner, M., Jørgensen, M.S., Hauschild, M.Z.: Defining the baseline in social life cycle assessment. Int. J. Life Cycle Assess. **15**, 376–384 (2010)

Kloepffer, W.: Life cycle sustainability assessment of products. Int. J. LCA **13**(2), 89–95 (2008)

Klöpffer, W.: Life-cycle based methods for sustainable product development. In: Life-Cycle Approaches to Sustainable Consumption Workshop Proceedings, pp. 33–138 (2002)

Laurent, A., Weidema, B., Bare, J., Xun, L.J., De Souza, D.M., Pizzol, M., Sala, S., Schreiber, H., Thonemann, N., Verones, F.: Methodological review and detailed guidance for the life cycle interpretation phase. J. Ind. Eco. **24**(5), 986–1003 (2019). https://doi.org/10.1111/jiec.13012

Lehtonen, M.: The environmental—social interface of sustainable development: capabilities, social capital, institutions. Ecol. Econ. **49**, 199–214 (2004)

Parti Sociali: 30 luglio 2010 Contratto provinciale di lavoro degli operai agricoli e florovivaisti della provincia di Cuneo (2010)

Petti, L., Sanchez Ramirez, P.K., Traverso, M., Ugaya, C.M.L.: An Italian tomato "Cuore di Bue" case study: challenges and benefits using subcategory assessment method for social life cycle assessment. Int. J. Life Cycle Assess. **23**, 569–580 (2018). https://doi.org/10.1007/s11367-016-1175-9201621, 106–117

Macombe, C., Feschet, P., Garrabé, M., Loeillet, D., 2nd.: International seminar in social life cycle assessment—recent developments in assessing the social impacts of product life cycles. Int. J. LCA **18**(9), 940–943 (2011)

Macombe, C., Leskinen, P., Feschet, P., Antikainen, R.: Social life cycle assessment of biodiesel production at three levels: a literature review and development needs. J. Clean. Prod. **52**, 205–216 (2013)

Martinez-Blanco, J., Lehmann, A., Chang, Y.J., Finbeiner, M.: Social organisational LCA (SOLCA)—a new approach for implementing social LCA. Int. J. LCA **20**, 1586–1599 (2015)

Neugebauer, S., Emara, Y., Hellerström, C., Finkbeiner, M.: Calculation of fair wage potentials throughout products′ life cycle—introduction of a new midpoint impact category for social life cycle assessment. J. Clean. Prod. **143**, 1221–1232 (2017)

Pelletier, N., Ustaoglu, E., Benoit, C., et al.: Social sustainability in trade and development policy. Int. J. Life Cycle Assess. **23** (2018)

Ramirez, P.K.S., Petti, L., Haberland, N.T., et al.: Subcategory assessment method for social life cycle assessment. Part 1: methodological framework. Int. J. Life Cycle Assess. **19**, 1515–1523 (2014)

Sala, S., Farioli, F., Zamagni, A.: Progress in sustainability science: lessons learnt from current methodologies for sustainability assessment: Part 1. Int. J. LCA **18**, 1653–1672 (2013)

Scorelca: Social LCA, Sustainable Development, CSR: State of Research? What are the Methodological Needs? Final report 2017, ETUDE N° 2016–04 (2017)

UNEP/SETAC: Guidelines for Social Life Cycle Assessment of Products—A Social and Socio-Economic LCA Code of Practice Complementing Environmental LCA and Life Cycle Costing, Contributing to the Full Assessment of Goods and Services within the Context of Sustainable Development. Paris, pp.104 (2009)

UNEP/SETAC: The Methodological Sheets for Subcategories in Social Life Cycle Assessment (S-LCA). United Nations Environment Programme and SETAC (2013)

Warr, T., Hunkeler, D., Klöpffer, W., Pesonen, H.L., Ciroth, A., Brent, A.C., Pagan, R.: Environmental Life Cycle Costing: A Code of Practice. Pensacola, SETACproducts. United Nations Environment Programme, Paris. ISBN 978-1-880611-87-6 (2011)

Weidema, B.P.: The integration of economic and social aspects in life cycle impact assessment. Int. J. Life Cycle Assess. **11**, 89–96 (2006)

Zanchi, L., Delogu, M., Zamagni, A., Pierini, M.: Analysis of the main elements affecting social LCA applications: challenges for the automotive sector. Int. J. Life Cycle Assess. **23**, 519–535 (2018)

Chapter 8
Product Ecodesign

Daniela C. A. Pigosso

8.1 Emergence and Importance of Ecodesign

In the late 1990s, ecodesign emerged as a proactive approach for integrating environmental issues into the early phases of the product development process, targeted at enhacing the products' competitiveness at the same time as minimising the overall environmental impacts across the product life cycle phases (Bhamra et al.1999; Van Weenen 1995).

More recently, ecodesign has been used in a more systemic context (Ceschin and Gaziulusoy 2016), supported by the concepts such as (i) product–service system (in which the function is delivered through a service offering) covered in Chap. 9: Product–Service Systems (PSS) of this book (Pigosso and Mcaloone 2016; Sundin and Bras 2005; Birkeland 2005;) and (ii) Circular Economy, which aims to decouple value creation from the resource consumption (Dalhammar 2016; Mcaloone and Pigosso 2017; Kjær et al. 2018; Bocken et al. 2016). Furthermore, a number of initiatives have attempted to expand ecodesign from an environmental/economic point of view to embrace social sustainability aspects (Pigosso and McAloone 2015), leading to the concept of sustainable design (Spangenberg et al. 2010; Rodrigues et al. 2015; Manzini 2014; Ceschin and Gaziulusoy 2016). Specifically on social aspects and impacts, Chap. 7: Social LCA addresses a methodology that can support the consideration of social issues within ecodesign.

Ecodesign is often defined as actions taken in the product development aimed at minimizing environmental impacts throughout the product's life cycle without compromising other essential criteria such as performance, functionality, aesthetics, quality and cost. It integrates environmental issues in the design process relating what is technically possible with what is ecologically necessary and socially acceptable,

D. C. A. Pigosso (✉)
Mechanical Engineering Department, Technical University of Denmark (DTU), Nils Koppels Allé, Building 404, room 230, 2800 Kgs. Lyngby, Copenhagen, Denmark
e-mail: danpi@dtu.dk

J. A. de Oliveira et al. (eds.), *Life Cycle Engineering and Management of Products*,
https://doi.org/10.1007/978-3-030-78044-9_8

given the growing perception of the need to safeguard the environment in a sustainable development context (Johansson 2002; Van Weenen 1995; Pigosso et al. 2013).

The terminology for the concept of carrying out the product development process with the integration of environmental issues has changed over the past few decades. The original term, green design, has been replaced by ecological design, environmentally sensitive design or ecodesign (Brezet and Van Hemel 1997), design for the environment (Ehrenfeld and Hoffman 1993) and environmentally responsible design (Dermody and Hanmer-Lloyd 1995). It is interesting to note that the use of terminology also varies from continent to continent. While the term "Design for the Environment" is more used in the United States of America, the term ecodesign is more widely adopted on the European continent.

The ecodesign practice has become essential for companies that have recognized environmental responsibility to be vitally important for long-term success (Bakshi and Fiksel, 2003; Rodrigues et al. 2016).

In this context, some of the key drivers for ecodesign implementation are (Bakshi and Fiksel 2003; Jeswiet and Hauschild 2005; Byggeth and Hochschorner 2006; Brezet and Van Hemel 1997; Pigosso and Mcaloone 2017; Bocken et al. 2016; Bakker et al. 2014): (i)

- Compliance with an ever increasing set of product-reated environmental legistation (e.g., Directive 2002/96/EC on Waste Electrical and Electronic Equipment—WEEE);
- Reduction of costs due to a more efficient and effective resource utilisation;
- Increased demand for sustainable products, driven by enhanced societal awareness;
- New business opportunities and enhanced corporate image;
- Growing perception of products' environmental performance as a competitive edge;

In this chapter, the main fundamentals of ecodesign, its integration into the product development process and the main tools and practices are presented. The chapter is complemented with examples of products with better environmental performance and exercises to help fix the content presented.

8.2 Integration of Ecodesign into Product Development

Although products are fundamental to the quality of life and wealth, the growing consumption of products is, directly or indirectly, at the origin of most of the environmental impacts caused by society (Commission of the European Community 2001).

The environmental impacts are generated over the products' entire life cycle (Fig. 8.1), from the extraction of raw materials and manufacturing to use and final disposal (Nielsen and Wenzel 2002; Baumann et al. 2002).

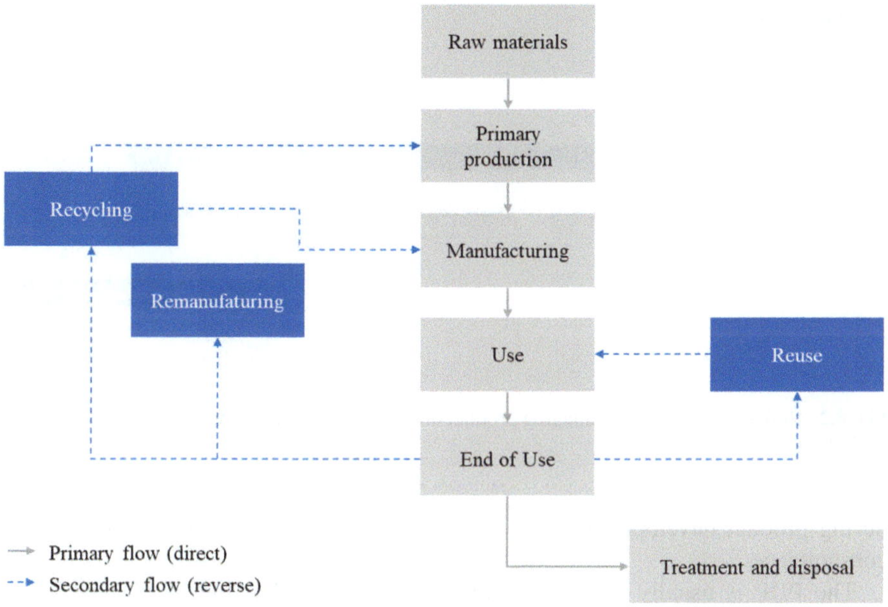

Fig. 8.1 Material life cycle of products

The product's material life cycle is shown in Fig. 8.1. The main (primary) flow consists of the phases of extraction of the raw material, processing by basic industries, manufacturing, use and end-of-use, ultimately resulting in the final treatment and disposal of the product. Each of these phases consumes inputs/resources (such as energy and raw material) and generates waste which must be properly treated and disposed or yet reused through end-of-life strategies (reuse, remanufacturing, reconditioning and recycling). More details about the product life cycle are presented in Chap. 1: Introduction to Life Cycle Engineering and Management (LCEM).

The environmental impacts caused by products throughout their material life cycle are the result of decisions and definitions taken in the early stages of the product development process (PDP) (Byggeth and Hochschorner 2006; Luttropp and Lagerstedt 2006; Vezzoli and Manzini 2008; Poole et al. 1999) - therefore the importance of integrating environmental considerations into the product development process. The product development process is a critical business process to increase companies' competitiveness mainly because it leads to a greater diversity of products which are capable of fulfilling the customers' needs. Developing products consists of a set of activities through which it is aimed—based on market needs and technological possibilities and restrictions and considering the company's competitive and product strategies—to arrive at a product's development specificities and its production process, so that the manufacturing is able to produce it and accompany it after its launch (Rozenfeld et al. 2006). If sustainability requirements are taken into account

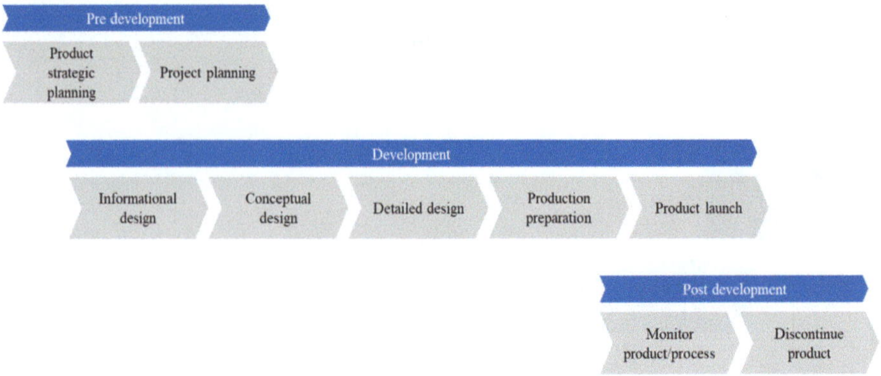

Fig. 8.2 Reference model for managing product development process (adapted from Rozenfeld et. al (2016))

during product development, more sustainable manufacturing will be the natural consequence.

The PDP is usually represented by a reference model (Fig. 8.2), such as the one proposed by Rozenfeld et. al. (2016). The reference model is structured in three macro-phases (pre-development, development and post-development). Pre-development contains two phases (strategic product planning and development planning). The development consists of five phases (informational design, conceptual design, detailed design, production preparation and product launch). Finally, monitoring product/process and discontinuing product are the two phases that make up post-development.

The reference model describes activities, expected results, responsibilities, available resources, support tools and information needed or generated in the process and consists of a collection of best practices in product development. Reference models are used in the product development process to establish a common language for all professionals in different areas of knowledge involved in the project, helping in communication and integration between them (Rozenfeld et al. 2006).

The greatest opportunities for environmental improvements on a product are in the early stages of its development process, in which the degrees of freedom in establishing the product's characteristics and the potential for environmental improvements are great (Fig. 8.3). Estimates indicate that 60 to 80% of a product's total environmental impact is established in these phases. As the product's characteristics and details are determined, the degrees of freedom gradually decrease. In the final stages of the process, product knowledge is great, but the possibilities for changing the design are small due to the large number of decisions that have already been taken during the process. At this point, the opportunities for environmental improvement are restricted to production, logistics, recycling, etc. (McAloone and Bey 2009a, b).

The good news is that most of the factors that enable the successful integration of ecodesign into product development process are the same factors that are recognized as essential to the success of product development itself. In this way, a company that

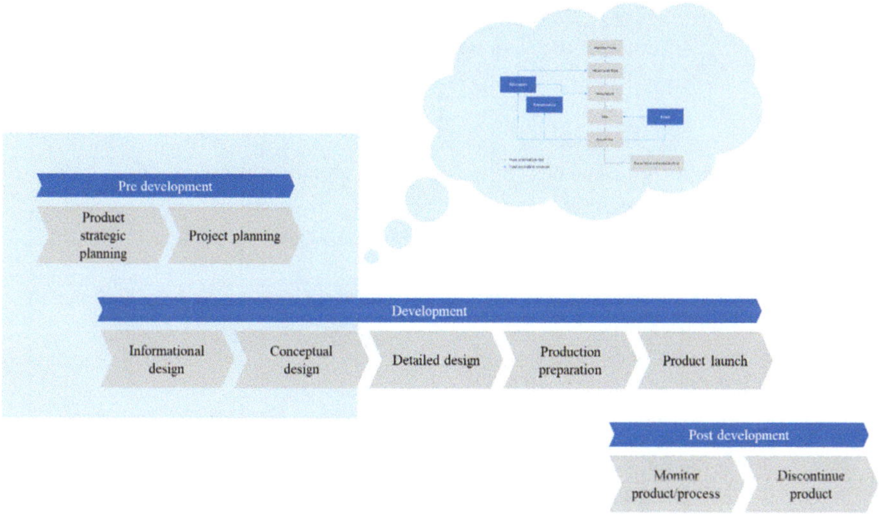

Fig. 8.3 Product development and the material life cycle of products

manages product development properly and in a structured way is more likely to succeed when integrating ecodesign into product development (Johansson 2002).

8.3 Tools and Good Practices for Ecodesign

Although companies are increasingly aware of the need to carry out the ecodesign integration into the product development process management, several companies still face challenges in identifying which ecodesign practices, methods and tools must be implemented.

In the last decades, several ecodesign methods and tools (any systematic mean to deal with environmental issues during the product development process) have been developed for the assessment of environmental impacts, highlighting potential problems and conflicts and facilitating the choice between different aspects through the comparison between ecodesign strategies (Pigosso et al. 2014a, b).

Operational practices are directly related to the material life cycle (Fig. 8.1), and their application varies according to the product characteristics and life cycle phases/aspects that have the greatest potential environmental impact (Borgianni et al. 2019). They are composed of ecodesign strategies, which provide guidance based on a number of design options[1] (Pigosso et al. 2014a, b). The design options are associated with each specific strategy and present design alternatives so that the specific strategy is met. Ecodesign methods and tools can be associated with specific strategies and design options, helping their implementation.

[1] The specific ecodesign strategies correspond to details of the generic strategies.

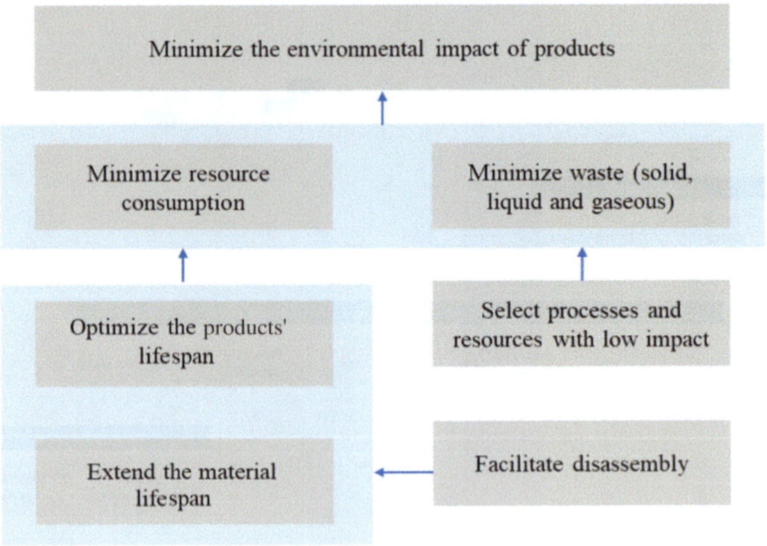

Fig. 8.4 Functional hierarchy between ecodesign strategies

According to Vezzoli and Manzini (2008), the strategies involved in integrating environmental requirements into the product (goods and services) development process are as follows:

- Minimize resource consumption (including material and energy);
- Minimize energy consumption;
- Select processes and resources with low environmental impact;
- Optimize the product's lifespan;
- Extend the material's lifespan;
- Facilitatedisassembly.

The functional hierarchy of these strategies is shown in Fig. 8.4. Facilitate product disassembly through "design for disassembly" is, for example, an effective strategy both for extending the material's lifespan and for optimizing the product's lifespan also implying in minimizing the consumption of resources and in the selection of processes and resources with low environmental impact (Vezzoli and Manzini 2008).

The guidelines for each of the ecodesign strategies are detailed below (Vezzoli and Manzini 2008):

- **Strategy: Minimize resource consumption**
  - Minimize material content;
  - Minimize losses during the production process;
  - Minimize packaging, keeping the product intact during transportation and storage;
  - Encourage efficient consumption systems of consumables;

  - Encourage the use of systems with flexible material consumption;
  - Minimize material consumption during the product development process.

- Strategy: Minimize energy consumption
  - Minimize energy consumption during the extraction and processing of raw materials and production;
  - Minimize energy consumption during transportation and storage;
  - Select systems with efficient operation stages for optimized energy consumption;
  - Undertake dynamic energy consumption;
  - Minimize energy consumption during the product development process.

- Strategy: Select processes and resources with low environmental impact
  - Select non-hazardous and non-toxic materials;
  - Select non-hazardous and non-toxic energy sources;
  - Select renewable or biocompatible materials;
  - Select renewable and biocompatible energy sources.

- Strategy: Optimize product's lifespan
  - Develop products with the appropriate lifespan;
  - Facilitate updates and adaptations;
  - Facilitate maintenance;
  - Facilitate repair;
  - Facilitate reuse;
  - Facilitate remanufacturing.

- Strategy: Extend the material's lifespan
  - Adopt the cascade approach;
  - Select materials with the most efficient recycling technologies;
  - Facilitate the removal and transportation at the product's end of life;
  - Identify different material;
  - Minimize the number of different and incompatible material;
  - Facilitate cleaning;
  - Facilitate composting;
  - Facilitate incineration.

- Strategy: Facilitate disassembly
  - Reduce and facilitating disassembly and separation operations;
  - Design modular parts and components;
  - Ensure easy accessibility of the junction elements;
  - Use reversible junction systems;
  - Use permanent junction systems that can be easily opened;
  - Co-design special technologies and features for crushing separation.

The prioritization of ecodesign strategies to be implemented can be done through product life cycle assessments, which can be quantitative (using the life cycle assessment [LCA] technique discussed earlier in this book) or semi-quantitative (using quantitative matrices for impact assessment) (Pigosso et al. 2011). In this chapter, the Design for Environment matrix (DfE matrix) (Eagan et al. n.d.) is presented as a semi-quantitative tool for identifying product hotspots (that is, life cycle phases and environmental aspects with the greatest potential for improvement through the application of ecodesign strategies) (Borgianni et al. 2019).

The DfE matrix can be used to quantify the potential environmental impacts of a product by assessing the environmental aspects involved throughout the stages of its life cycle. The results obtained by applying the method can be used to compare the product under development with an existing product (useful for redesigning existing products) or to compare design alternatives for a new product under development, complementing the economic, customer value and manufacturability parameters, which must also be assessed.

The environmental aspects defined by the method are related to the consumption of materials and energy and to the generation of liquid effluents, solid waste and gaseous emissions and make up the matrix's columns. The life cycle phases considered by the method include pre-manufacture, manufacture, distribution and packaging, use and maintenance and end of life and correspond to the matrix's lines (Chart 8.1).

Each cell in the matrix (defined by the cross between environmental aspects and phases of the product life cycle) is filled out through a scoring system guided by questions pre-defined by the method, presented below in a guideline format.

The total score obtained by filling in all the cells in the matrix indicates, in a semi-quantitative way, the phases of the product life cycle under analysis in which the greatest environmental impacts occur and which must, therefore, be optimized by product developers in order to improve the product's environmental performance. The total score of the matrix is a relative measure of the product's attributes and complements the economic, customer value and manufacturability parameters. Reference

Chart 8.1 Design for environment matrix (DfE Matrix) (Yarwood; Eagan)

Life cycle phase	Environmental aspect					
	Materials	Energy consumption	Solid waste	Liquid effluent	Gaseous emissions	**Total**
Pre-manufacture						
Manufacture						
Packaging and distribution						
Use and maintenance						
End of life						
Total						

data provide additional information to answer the questions for filling the matrix, which can be adapted according to the company's needs (Eagan et al. n.d.).

A.1: Pre-manufacture X Materials

- What percentage of suppliers has a formal environmental management system (EMS) in progress?

  0% or unknown = 0 points
  1 to 5% = 2 points
  6 to 25% = 3 points
  25 to 50% = 4 points
  > 50% = 5 points.

A.2: Pre-manufacture X Energy Consumption

- What percentage of suppliers has formal energy conservation practices in progress?

  0% or unknown = 0 points
  1 to 5% = 2 points
  6 to 25% = 3 points
  25 to 50% = 4 points
  > 50% = 5 points

A.3: Pre-manufacture X Solid Waste

- What percentage of suppliers has ISO 9000 or ISO 14000 in progress or regularly publish company environmental reports?

  0% or unknown = 0 points
  1 to 5% = 2 points
  6 to 25% = 3 points
  25 to 50% = 4 points
  > 50% = 5 points

A.4: Pre-manufacture X Liquid Effluents

- What percentage of suppliers has a water conservation program?

  0% or unknown = 0 points
  1 to 5% = 2 points
  6 to 25% = 3 points
  25 to 50% = 4 points
  > 50% = 5 points

A.5: Pre-manufacture X Gaseous Emissions

- What percentage of suppliers has a formal program to minimize gas emissions in progress?

0% or unknown = 0 points
1 to 5% = 2 points
6 to 25% = 3 points
25 to 50% = 4 points
> 50% = 5 points

B.1: Manufacture X Materials

For this product or component:

- Is the largest possible amount of recyclable materials used in the product? (yes = 1, no = 0)
- Is the use of hazardous materials avoided or minimized? (yes = 2, no = 0)
- Is the amount of material used in the product minimized? (yes = 1, no = 0)
- Is the number of different types of materials used in the product minimized? (yes = 1, no = 0)

B.2: Manufacture X Energy Consumption

For this product or component:

- Is the intensive use of energy in manufacturing processes minimized? (yes = 2, no = 0)
- Is cogeneration, heat exchange or other techniques used to use the wasted energy? (yes = 2, no = 0)
- Is transportation between manufacturing and product assembly locations minimized? (yes = 1, no = 0)

B.3: Manufacture X Solid Waste

For this product or component:

- Is material loss minimized and reuse during manufacturing optimized to the maximum? (yes = 1, no = 0)
- Are suppliers encouraged to minimize the quantities and types of package for their products? (yes = 1, no = 0)
- Are opportunities for reusing and reducing package waste maximized when components/products are transported between facilities? (yes = 1, no = 0)
- Is the intentional introduction of lead, cadmium, mercury and hexavalent chromium avoided? (yes = 2, no = 0)

B.4: Manufacture X Liquid Effluents

For the manufacturing of this product or component:

- Are alternatives to the use of solvents and toxic oils investigated? (yes = 2, no = 0)
- Are opportunities for capturing and reusing liquid byproducts generated during the manufacturing process maximized? (yes = 1, no = 0)

- Is the emission of pollutants in water avoided or minimized? (yes = 2, no = 0)

B.5: Manufacture X Gaseous Emissions

For the manufacturing of this product or component:

- Is the generation of gases which cause global warming and the destruction of the ozone layer avoided? (yes = 2, no = 0)
- Is the generation of hazardous air pollutants avoided during the manufacturing process? (yes = 2, no = 0)
- Is the use of solvents, paints and adhesives with high evaporation rates of volatile organic compounds eliminated or minimized? (yes = 1, no = 0)

C.1: Packaging and Distribution X Materials

For this product or component:

- Is the use of reusable packaging for distribution between the company's facilities explored? (yes = 1, no = 0)
- Is the use of reusable packaging for distribution between the company and its suppliers explored? (yes = 1, no = 0)
- Are recyclable materials used in packaging for product transportation and delivery? (yes = 1, no = 0)
- Are recycled materials used in packaging for product transportation and delivery? (yes = 1, no = 0)
- Is the number of different types of materials used in packaging minimized? (yes = 1, no = 0)

C.2: Packaging and Distribution X Energy Consumption

For this product or component:

- Are reusable packaging materials used, with the least volume and weight possible, while maintaining transportation functions? (yes = 5, no = 0)

C.3: Packaging and Distribution X Solid Waste

For this product or component:

- Is it possible to easily separate the packaging materials enabling recycling and reuse? (yes = 1, no = 0)
- Is most of the packaging used recycled? (yes = 2, no = 0)
- Are the different types of packaging materials marked for easy identification? (yes = 2, no = 0).

C.4: Packaging and Distribution X Liquid Effluents

For this product or component:

- Is maximum prevention taken on the leakage of hazardous liquids during transportation? (yes = 5, no = 0).

C.5: Packaging and Distribution X Gas Emissions

For this product or component:

- Is the use of chlorinated polymers or plastics that can produce dangerous gas emissions avoided if incinerated at low temperatures in packaging for transportation and consumption? (yes = 3, no = 0)
- Is the use of brominated flammability retardants that can produce toxic gas emissions if incinerated at low temperatures in packaging? (yes = 2, no = 0)

D.1: Product Use and Maintenance X Materials

For that product or component:

- Is it facilitated the product's disassembling for updating, repair or reuse? (yes = 1, no = 0)
- Are product parts or components available for repair? (yes = 1, no = 0)
- Are potential barriers to recycling, such as the use of additives, metal treatments on plastic, the application of paint on plastic or the use of material of unknown composition, avoided? (yes = 2, no = 0)
- Is the plastic used clearly identified by type of polymer? (yes = 1, no = 0).

D.2: Product Use and Maintenance X Energy Consumption

For this product or component:

- Is energy consumption minimized when using the product? (yes = 2, no = 0)
- Are options for adjusting energy consumption based on the intensity of product or component activity offered? (yes = 3, no = 0)

D.3: Product Use and Maintenance X Solid Waste

For this product or component:

- Is the use of disposable components such as batteries and cartridges avoided? (yes = 2, no = 0)
- Are junction elements, such as screws and pressure fasteners, used with the same type of heads? (yes = 1, no = 0)
- Is the use of adhesives and welding avoided to facilitate disassembly, reuse and recycling? (yes = 1, no = 0)
- Is the repair and/or update of components' facilitated? (yes = 1, no = 0).

D.4: Product Use and Maintenance X Liquid Effluents

For this product or component:

- Is the release of water polluting substances avoided during the product use ? (yes = 5, no = 0)

D.5: Product Use and Maintenance X Gaseous Emissions

For this product or component:

- Is the emission of air pollutants avoided during use and maintenance of the product? (yes = 2, no = 0)
- Is the emission of gases which cause global warming and the destruction of the ozone layer avoided during the use and maintenance of the product? (yes = 3, no = 0)

E.1: End of Life X Materials

For this product or component:

- Is the reuse and recycling of materials facilitated? (yes = 1, no = 0)
- Is the identification and separation of materials by type facilitated? (yes = 1, no = 0)
- Is the use of materials that need to be disposed as hazardous waste avoided? (yes = 1, no = 0)
- Is the intentional introduction of lead, cadmium, mercury and hexavalent chromium in product materials avoided? (yes = 2, no = 0)

E.2: End of Life X Energy Consumption

For this product or component:

- Are plastic parts and fibers that can be safely used for energy generation, such as incineration, used? (yes = 2, no = 0)
- Is the use of hazardous materials that need to be transported as hazardous waste to industrial landfills avoided? (yes = 3, no = 0)

E.3: End of Life X Solid Waste

For this product or component:

- Is the existence of internal or external infrastructure to recover/recycle solid waste ensured by the company? (yes = 2, no = 0)
- Is the connection between different materials that may hamper their separation avoided? (yes = 3, no = 0)

E.4: End of Life X Liquid Effluents

For this product or component:

- Is it possible to recover problematic hazardous liquids during disassembly? (yes = 5, no = 0)

E.5: End of Life X Gaseous Emissions

For this product or component:

- Is the release of substances that cause ozone depletion and/or global warming avoided during the final disposal of the product or component? (yes = 2, no = 0)
- Is the recycling of gases contained in the product performed during disassembly so that they are not lost? (yes = 1, no = 0)
- Is the release of air polluting substances avoided during the final disposal of the product or component? (yes = 2, no = 0).

If the question is not applicable to the analyzed product, it is not necessary to answer it, and the cell must remain blank. The DfE Matrix (Eagan et al. nd) can be used to carry out benchmarking studies between different products (e.g., different product concepts and/or products competing in the market) and to identify potential focal areas for the product redesign (those with lower scores).

For readers who want to go deeper into the theme, another interesting guide describes seven steps for implementing ecodesign (Mcaloone and Bey 2009a), also available in Portuguese (McAloone and Bey 2009b). The steps are as follows: (1) Describe the context for using the product; (2) Create an overview of environmental impacts; (3) Create the environmental profile and identify the root causes; (4) Outline the network of interested parties; (5) Quantify environmental impacts; (6) Create concepts with better environmental performance; (7) Develop an environmental strategy.

8.4 Ecodesign Case Studies

Several examples of products with better environmental performance can be found in books such as the one published by (Vezzoli and Manzini 2008). The book presents examples applying various ecodesign strategies (Fig. 8.4) and is a recommended reading for all readers of this chapter.

In addition, it is interesting to note that several companies already have a well-structured process for implementing ecodesign to the product development process such as Philips, Steelcase, Adidas, Patagonia and Natura. In general, we recommend readers to research the sustainability reports of companies in order to see how the best ecodesign practices have been implemented.

In this chapter, the focus will be on presenting how the DfE Matrix can be applied to a specific example with the analysis of results and possible actions in the product development process, so that new concepts with better environmental performance can be developed.

Imagine that you are part of the design team of the fictional company Printec, which develops multifunctional printers (including copying, printing and scanning). The team is currently working on the redesign of a multifunctional with the aim at becoming a market leader in terms of environmental performance in view of the great increase in the customers' awareness and new market opportunities, especially in Europe.

The team defines that one of the first steps in the information design phase is to identify the current environmental profile of the printer that is being comercialized in the market in order to identify which are the main impacts throughout the life cycle. The team identifies several possible tools available but decides to apply the DfE Matrix considering the amount of existing data and also the time taken to carry out this initial task.

A 4-hour workshop is held at Printec. Stakeholders in this new project are invited to participate in the assessment session of the existing printer (including product engineers, designers, supplies, material specialists, production engineers, marketing experts, etc.). In addition, people involved in after sales and support for the current printer are invited, given their knowledge of the use and end-of-life phases. The process is facilitated by the project manager, and the final result of the calculated matrix is shown in Chart 8.2.

The values obtained in the matrix are based on the scoring system guided by the questions proposed by the DfE matrix. The questions have been answered based on the information made available by PrinTec. For example, in the case of Pre-manufacture X Materials, the result obtained is 3 points (meaning that 6 to 25% of the company's suppliers have EMS). Another example: Distribution and packaging X Liquid effluents: the final score is 5 points because the company takes maximum precaution regarding the leakage of hazardous liquids during transportation.

The analysis of the DfE Matrix results at PrinTec provided the following insights for the company:

- Pre-manufacture: The phase scores only 5 points, mostly due to the lack of available data at PrinTec regarding environmental practices applied within the value chain and suppliers. A greater involvement with suppliers to obtain more accurate data is required. Improvement opportunity: a team has been defined to obtain

Chart 8.2 Example of applying the DfE Matrix at PrinTec

Life cycle phase	Environmental aspect					
	Materials	Energy consumption	Solid waste	Liquid effluent	Gaseous emissions	Total
Pre-manufacture	3	0	*2*	0	0	5
Manufacture	2	5	*3*	5	5	20
Packaging and distribution	3	5	*0*	5	5	18
Use and maintenance	*0*	*0*	*1*	*5*	*5*	*11*
End of Life	3	3	*2*	0	5	13
Total	11	13	*8*	15	20	***67***

information from suppliers to obtain more accurate data. On the basis of the results an awareness process in the value chain might be necessary.

- Manufacture: The participation of production engineers in the workshop has been essential for the assessment of this phase. Improvement opportunities identified include: B.1: use of recyclable materials, minimizing the consumption of materials and different types of materials used in the product and B.3: reusing and reducing packaging waste, potentially collaborating with suppliers.
- Packaging and distribution: The workshop participants concluded that the focus on product development tends to be very low in relation to packaging and distribution, and the main considerations are related to the volume of packaging to reduce transportation costs. Improvement opportunities identified include: C.1: use of recycled materials in packaging preferably with a few different types of materials; C.3: initiatives to increase the recyclability of packaging such as separation and identification of different types of materials.
- Use and Maintenance: According to the workshop participants, the use phase of the printer has been greatly influenced by design decisions made when the current printer has been developed, which brings several opportunities for improvement on the development of the new printer, including: D.1: design for repair, disassembly and recycling. D.2: minimization of energy consumption during use; D.3: minimization of consumables and design to facilitate maintenance/repair, disassembly, reuse and recycling.
- End of Life: Similarly to the previous phase, most of the decisions made during product development have influenced the score obtained by the current printer in this phase. Several improvement opportunities have been identified for the new printer, including E.1: design for reuse of components and recycling of materials; E.2: use of materials that can be safely incinerated (important specifically for the European market); E.3: choice and type of junction between different materials; E.4: new technologies for removing ink from cartridges.

The analysis of the results obtained allows to conclude that the life cycle phase with the greatest potential for improvement is the use and maintenance phase, with a current score of only 11 points (out of 25 possible). Similarly, the "Solid Waste" aspect has the greatest potential for improvement due to low scores mainly in the packaging and distribution, use and maintenance and end-of-life phases (due to the lack of information in the pre-manufacture phase, the team decided to disregard it for this specific project). The design team then decided to focus its design efforts on these two phases, also including energy consumption throughout the printer's lifespan.

The team selected environmental performance indicators for each of the prioritized areas, defining specific goals for improving environmental performance (e.g., 100% use of recycled materials in packaging). The definition of these environmental requirements must then be considered together with the other technical and economic performance requirements of the product, in the conceptual design phase. Thus, it is worth emphasizing the need to align ecodesign indicators and requirements with the organization's performance measurement systems (more information can be found in Chap. 11: Environmental Management Systems and Environmental Performance Measurement).

If the reader is interested in deepening the best practices, tools and methods for implementing ecodesign, there is already a vast literature on the subject that can be searched (Pigosso et al. 2014a, b, 2015).

8.5 Final Considerations

Although companies are increasingly aware of the need to integrate ecodesign into the product development process, it is still unclear which ecodesign practices must be implemented by companies. The main challenges for the implementation of ecodesign in companies are related to five main areas (Dekoninck et al. 2016):

- Strategy: calculating the business case, ensuring top management commitment and developing a long-term strategy for the implementation of ecodesign.
- Tools: identifiying the right tools for implementation, and dealing with limitation of existing tools (e.g., to obtain data to quantify life cycle assessment);
- Collaboration: creating awareness, allocating responsibilities internally in the company and ensuring transparency in the value chain;
- Management: dealing with internal and external resistance; managing customer requirements and implementing new business models;
- Knowledge: nurturing knowledge and experience internally in the company; obtaining new types of data not directly available in the company.

In the coming years, trends related to the implementation of ecodesign are related to the circular economy (see Chap. 10: Corporate Sustainability—defining business strategies and models based on circular economy) and the development of product–service systems (see Chap. 9: Product–Service Systems (PSS)), in addition to incorporating social innovation issues (see Chap. 7: Social LCA).

8.6 Proposed Exercises

Calculation exercise:

(1) Calculate the environmental profile using the DfE Matrix for three products with different environmental profiles:

   I. Personal computers
   II. Passenger plane for commercial flights
   III. PET packaging for drinks

Based on the results obtained, identify the main opportunities for improving the environmental performance for each of the products (in terms of life cycle phases and environmental aspects) and reflect on the differences in results obtained according to the type of product and its environmental profile.

Case study analysis exercise:

(2) Assess three different products regarding the ecodesign strategies that have been implemented in the products:

I. Non-durable consumer goods: Natura's SOU line packaging
II. Durable consumer goods: Brastemp's white line Inverse refrigerator
III. Capital good: BMW i3 electric car.

Reflection questions:

(3) What is the market demand for products with better environmental performance?
(4) Products with better environmental performance are always more expensive. Comment on the statement showing examples that confirm or falsify that statement.
(5) How can more efficient products (e.g., with greater energy efficiency) encourage greater consumption?

FEEDBACK

Personal computer (fictional example).

Life cycle phase	Environmental aspect					
	Materials	Energy consumption	Solid waste	Liquid effluent	Gaseous emissions	Total
Pre-manufacture	4	4	4	*4*	4	20
Manufacture	3	2	2	*3*	5	15
Packaging and distribution	*1*	*0*	*3*	*5*	*2*	*11*
Use and maintenance	3	5	3	*0*	2	13
End of life	3	5	5	*0*	3	16
Total	14	16	17	*12*	16	***75***

Reflections on the results:

- Pre-manufacture: In this phase, a score of 20 points was obtained due to the company's concern with the origin of the materials used in its product. In all aspects assessed, it was possible to observe the engagement of suppliers to be between 6 and 25%. However, it is perceived the need to advance further in this process, so that at least 50% of the company's suppliers have EMS, to guarantee a better environmental performance of the final product.

- Manufacture: The score in this phase was 15 points. The fact that the computer is an electronic product with more than 60 toxic substances makes it difficult to reduce the use of these types of substances. In addition, it was not possible to obtain information on the use of recyclable material. In terms of energy consumption, the company does not have programs to recover energy. Regarding solid waste, suppliers are not encouraged to reuse or reduce their packaging. As proposals for improvement, the need to invest in the development of new technologies to reduce the use of toxic substances and the identification of parts of recyclable product is highlighted. In addition, the company should establish partnerships with research institutions to explore the development of energy efficient technologies.
- Packaging and distribution: The company obtained only 11 points in the "packaging and distribution" phase. The main areas for improvement are related to increasing the use of reusable packaging inside the company and within the value chain. In addition, the company should invest in the development of new products with less weight in order to optimize transportation. Modular design should also be explored, so that the computers can be more easily disassembled, repaired and recycled. Finally, the company should explore the establishment of partnerships with other companies in order to avoid the incineration of products and the generation of emissions.
- Product use and Maintenance: the product has not been designed to be disassembled and updated. If the computer fails within the warranty period, the company chooses to deliver a new product to the customer. This has increased the generation of solid waste. Together with the product development team, a new version of the computer is being developed so that the product can be easily repaired, updated and reused.
- End of Life: The computer's new version has improvements in its composition with modularized parts that allow reducing disassembly and recycling times. All product's components are identified with a code so that the company can monitor its life cycle. In addition, the company will have a new front on remanufactured products that can be purchased at a lower price and with a guarantee equal to that of the new product.

Passenger plane for commercial flight (fictional example)

Life cycle phase	Environmental aspect					
	Materials	Energy consumption	Solid waste	Liquid effluent	Gas emissions	Total
Pre-manufacture	5	3	5	*3*	4	20
Manufacture	4	3	4	*5*	5	21
Packaging and distribution	5	5	4	*5*	2	21
Use and maintenance	*3*	*5*	*4*	*0*	*5*	*17*
End of life	5	5	2	*0*	3	15

(continued)

(continued)

Life cycle phase	Environmental aspect					
	Materials	Energy consumption	Solid waste	Liquid effluent	Gas emissions	Total
Total	22	21	19	*13*	19	***94***

Reflections on the results:

- Pre-manufacture: In this phase, the assessed company obtained a total of 20 points. However, it is important to highlight that the company does not have consolidated information from all suppliers, which leads to a lack of knowledge due to information related to energy consumption and effluents generated. A possible proposal for improvement is to apply questionnaires to all suppliers where they have questions related to these two points.
- Manufacture: The score obtained in this phase was 21 points. The company's main challenges are related to the development of technologies to reduce energy consumption for the product's manufacture. It is important to highlight that, although it is not an easy task, it would be possible to create alternatives with companies that offer more sustainable technologies, without compromising security requirements. In addition, it is important that the company improves engagement so that its suppliers increase the rates to reuse and reduce their packaging, such as implementing reverse logistics.
- Packaging and Distribution: With 21 points, the company must still work to increase the recycling of the packaging used and develop more alternatives that allow to improve the separation of materials, thus increasing recycling and the quality of recycled materials.
- Product use and Maintenance: With 17 points in the assessment, the current model of the company's commercial plane has been developed with the aim at extending its life cycle from 25 to 40 years. For this, remanufacturing has been adopted as an end-of-life strategy to facilitate the recovery and updating of components (e.g., the landing gear). Components that can no longer be recovered will be recycled both for internal use in the company (for the manufacture of new components) and for external use through the recycled material's sale.
- End of Life: Although the company tries to recover the product at the end of its lifespan, it still presents difficulties due to infrastructure and labor issues. Currently, the company is working to hire certified companies that can recover the discarded product. In addition, the company is developing a new model of commercial plane that facilitates the recovery of parts.

PET packaging for drinks (fictional example)

Life cycle phase	Environmental aspect					
	Materials	Energy consumption	Solid waste	Liquid effluent	Gas emissions	Total
Pre-manufacture	2	*0*	*2*	*0*	*0*	*4*
Manufacture	5	1	4	2	*3*	15
Packaging and distribution	2	0	2	5	*2*	11
Use and maintenance	5	5	5	0	*0*	15
End of life	5	5	5	5	*3*	23
Total	19	11	18	12	*8*	***68***

Reflections on the results:

- Pre-manufacture: In this phase, the assessed company obtained a total of four points due to the lack of integration of programs related to the environment and the low level of engagement with suppliers. As a proposal for improvements, the company could create a prior assessment to verify whether suppliers have EMS, energy conservation programs, sustainability reports, etc. In addition, the company can create partnerships with universities so that work that allows the use of tools to improve the origin of materials used in the manufacture of its products can be developed. Finally, the company can create a system of sustainability indicators to measure its progress annually.
- Manufacture: The score in this phase was 15 points. The justification is based on the company's lack of resources to develop energy reduction programs and alternatives so that the recyclable material is processed more efficiently. Currently, the company has problems with water pollutants and has no process for treating these effluents. As an improvement proposal, we cite the possibility of the company to rent more efficient equipment inside the product–system service approach to be able to reduce energy consumption in manufacturing.
- Packaging and Distribution: The company does not have an environmental management program, which means that it does not have initiatives to help improving recycling. As a proposal for improvements, the company and its team can create new programs that allow the recycling of packaging internally and in collaboration with its suppliers.
- Product use and Maintenance: The company and its team started to develop partnerships with other packaging companies to set up reverse logistics programs that allow them to recover their waste and their recycling processes to be done only by companies that have better technologies in order to avoid the generation of gases.
- End of Life: With a total of 23 points, an alternative to improve product recovery in the end-of-life stage is the development of incentives for consumers to return the product, increasing recycling rates and minimizing the inappropriate disposal of packaging.

With this exercise, it was possible to identify how environmental issues can be integrated into product development in addition to understanding their importance given the direct and indirect influence on all phases of the product's life cycle. The environmental profile of each product varies greatly according to the product and the company that develops that product. The examples are fictional and aim at exemplifying how the matrix's results can be interpreted to identify opportunities for improvement. The score obtained by the DfE Matrix must not be used for external communication of product performance but rather as an internal tool to compare different products/models under development, for benchmarking studies and to identify opportunities for improvement.

Case study analysis exercises:

(6) Assess three different products regarding the ecodesign strategies that have been implemented in the products:

I. Non-durable consumer goods: Natura's SOU line packaging

In Natura's SOU line, ecodesign has been implemented with the product dematerialization specifically in the material used for the packaging, which brought 70% less plastic volume and formulas with fewer ingredients.

Strategies implemented:

(a) Minimize material consumption (minimize material content; minimize packaging, keep the product intact during transportation and storage)
(b) Increase the material's lifespan (select materials with the most efficient recycling technologies; minimize the number of different and incompatible materials)

(7) Durable consumer goods: Brastemp's white line inverse refrigerator. Brastemp's white line inverse refrigerator offers products with greater energy efficiency. It consumes approximately 25% less energy when compared to economical refrigerators because they have an intelligent compressor that measures periods of greater or lesser of the refrigerator's consumption.

Strategies implemented:

(a) Minimize energy consumption (select systems with energy-efficient operation stages; undertake dynamic energy consumption)

(8) Capital asset: BMW i3 electric car: developed for sustainable mobility, it has an inspiring and fully electric design. In addition, the car has several sustainable and recycled materials, which enables a reduction of weight at the same time as extending the overall life time of the car.

Strategies implemented:

(a) Minimize energy consumption (select systems with energy-efficient operation stages)

(b) Minimize toxic emissions (select non-toxic or non-hazardous energy sources)
(c) Use renewable or biocompatible resources (select renewable or biocompatible materials; select renewable and biocompatible energy sources)

Reflection exercises:

(9) What is the market demand for products with better environmental performance?

Consumers are increasingly aware and concerned with environmental issues and the impacts caused by the products they consume.

(10) Products with better environmental performance are always more expensive. Comment on the statement showing examples that confirm or falsify that statement.

Products with better environmental performance are not always more expensive and can actually be cheaper than "traditional" products. As the increase in environmental performance is directly related to the increase in efficiency throughout the product's lifespan, significant savings can be achieved. Imagine, for example, the case of the refrigerator that consumes less energy because it is more efficient. This means that, throughout the entire use of the refrigerator (10–15 years), the user will save at least 25% on the electricity bill. Even if the product is more expensive to purchase (which is not always the case), the cost over the entire life cycle will be much less than a less efficient refrigerator.

(11) Can more efficient products (e.g., with greater energy efficiency) encourage greater consumption?

Usually, consumers tend to unconsciously use the product more if they realize that the costs associated with its use are not high (that is, in the case of more efficient products). Imagine, for example, that a consumer who used an old model of car has just switched to a newer model, which is 50% more efficient. Considering the distance traveled to be the same, that consumer will spend 50% less on fuel. With the money saved, the consumer decides to travel to the beach on a holiday, for example, which leads to greater fuel consumption (and product use) due to its increased efficiency.

Proposal of script for the application of DfE Matrix

Company:		Assessed product:		
Phase	Question		Points	Comments and opportunities for improvement
A.1: Pre-manufacture X Materials	What percentage of suppliers has a formal environmental management system (EMS) in progress?	0% or unknown = 0 points		
		1 to 5% = 2 points		
		6 to 25% = 3 points		
		25 to 50% = 4 points		
		> 50% = 5 points		
A.2: Pre-manufacture X Energy Consumption	What percentage of suppliers has formal energy conservation practices in progress?	0% or unknown = 0 points		
		1 to 5% = 2 points		
		6 to 25% = 3 points		
		25 to 50% = 4 points		
		> 50% = 5 points		
A.3: Pre-manufacture X Solid Waste	What percentage of suppliers has ISO 9000 or ISO 14,000 in progress or regularly publish company environmental reports?	0% or unknown = 0 points		
		1 to 5% = 2 points		
		6 to 25% = 3 points		
		25 to 50% = 4 points		
		> 50% = 5 points		
A.4: Pre-manufacture X Liquid Effluents	What percentage of suppliers has a water conservation program?	0% or unknown = 0 points		
		1 to 5% = 2 points		
		6 to 25% = 3 points		
		25 to 50% = 4 points		
		> 50% = 5 points		

(continued)

(continued)

Company:		Assessed product:		
Phase	Question		Points	Comments and opportunities for improvement
A.5: Pre-manufacture X Gas Emissions	What percentage of suppliers has a formal program to minimize gas emissions in progress?	0% or unknown = 0 points		
		1 to 5% = 2 points		
		6 to 25% = 3 points		
		25 to 50% = 4 points		
		> 50% = 5 points		
B.1: Manufacture X Materials	For this product or component:	Is the largest possible amount of recyclable materials used in the product? (yes = 1, no = 0)		
		Is the use of hazardous materials avoided or minimized? (yes = 2, no = 0)		
		Is the amount of material used in the product minimized? (yes = 1, no = 0)		
		Is the number of different types of materials used in the product minimized? (yes = 1, no = 0)		
B.2: Manufacture X Energy Consumption	For this product or component:	Is the intensive use of energy in manufacturing processes minimized? (yes = 2, no = 0)		
		Is cogeneration, heat exchange or other techniques used to use the wasted energy? (yes = 2, no = 0)		

(continued)

(continued)

Company:		Assessed product:		
Phase	Question		Points	Comments and opportunities for improvement
		Is transportation between manufacturing and product assembly locations minimized? (yes = 1, no = 0)		
B.3: Manufacture X Solid Waste	For this product or component:	Is material loss minimized and reuse during manufacturing optimized to the maximum? (yes = 1, no = 0)		
		Are suppliers encouraged to minimize the quantities and types of packaging for their products? (yes = 1, no = 0)		
		Are opportunities for reusing and reducing packaging waste maximized when components/products are transported between facilities? (yes = 1, no = 0)		
		Is the intentional introduction of lead, cadmium, mercury and hexavalent chromium avoided? (yes = 2, no = 0)		
B.4: Manufacture X Liquid Effluents	For manufacturing this product or component:	Are alternatives to the use of solvents and toxic oils investigated? (yes = 2, no = 0)		

(continued)

(continued)

Company:		Assessed product:		
Phase	Question		Points	Comments and opportunities for improvement
		Are opportunities for capturing and reusing liquid byproducts generated during the manufacturing process maximized? (yes = 1, no = 0)		
		Is the emission of water pollutants avoided or minimized? (yes = 2, no = 0)		
B.5: Manufacture X Gas Emissions	For manufacturing this product or component:	Is the generation of gases that cause global warming and the destruction of the ozone layer avoided? (yes = 2, no = 0)		
		Is the generation of hazardous air pollutants avoided during the manufacturing process? (yes = 2, no = 0)		
		Is the use of solvents, paints and adhesives with high evaporation rates of volatile organic compounds eliminated or minimized? (yes = 1, no = 0)		
C.1: Packaging and Distribution and X Materials	For this product or component:	Is the use of reusable packaging for distribution between the company's facilities explored? (yes = 1, no = 0)		

(continued)

(continued)

Company:		Assessed product:		
Phase	Question		Points	Comments and opportunities for improvement
		Is the use of reusable packaging for distribution between the company and its suppliers explored? (yes = 1, no = 0)		
		Are recyclable materials used in packaging for transportation and product delivery? (yes = 1, no = 0)		
		Are recycled materials used in packaging for transportation and product delivery? (yes = 1, no = 0)		
		Is the number of different types of materials used in packaging minimized? (yes = 1, no = 0)		
C.2: Packaging and Distribution X Energy Consumption	For this product or component:	Are reusable packaging materials used, with the least volume and weight possible, while maintaining transportation functions? (yes = 5, no = 0)		
C.3: Packaging and Distribution X Solid Waste	For this product or component:	Is it possible to easily separate the packaging materials enabling recycling and reuse? (yes = 1, no = 0)		
		Is most of the packaging used recycled? (yes = 2, no = 0)		

(continued)

(continued)

Company:		Assessed product:		
Phase	Question		Points	Comments and opportunities for improvement
		Are the different types of packaging materials marked for easy identification? (yes = 2, no = 0)		
C.4: Packaging and Distribution X Liquid Effluents	For this product or component:	Is maximum prevention taken on the leakage of hazardous liquids during transportation? (yes = 5, no = 0)		
C.5: Packaging and Distribution X Gas Emissions	For this product or component:	Is the use of chlorinated polymers or plastics that can produce dangerous gas emissions avoided if incinerated at low temperatures in packaging for transportation and consumption? (yes = 3, no = 0)		
		Is it avoided the use of brominated flammability retardants that can produce toxic gaseous emissions if incinerated at low temperatures in the packaging? (yes = 2, no = 0)		
D.1: Product Use and Maintenance X Materials	For this product or component:	Is it facilitated the product's disassembling for updating, repair or reuse? (yes = 1, no = 0)		
		Are product parts or components available for repair? (yes = 1, no = 0)		

(continued)

(continued)

Company:		Assessed product:		
Phase	Question		Points	Comments and opportunities for improvement
		Are potential barriers to recycling, such as the use of additives, metal treatments on plastic, the application of paint on plastic or the use of materials of unknown composition, avoided? (yes = 2, no = 0)		
		Are the plastics used clearly identified by type of resin? (yes = 1, no = 0)		
D.2: Product Use and Maintenance X Energy Consumption	For this product or component:	Is energy consumption minimized when using the product? (yes = 2, no = 0)		
		Are you offered options for adjusting energy consumption based on the intensity of product or component activity? (yes = 3, no = 0)		
D.3: Product Use and Maintenance X Solid Waste	For this product or component:	Is the use of disposable components such as batteries and cartridges avoided? (yes = 2, no = 0)		
		Are junction elements, such as screws and pressure fasteners, used with the same type of heads? (yes = 1, no = 0)		

(continued)

(continued)

Company:		Assessed product:		
Phase	Question		Points	Comments and opportunities for improvement
		Is the use of adhesives and solders avoided to facilitate disassembly, reuse and recycling? (yes = 1, no = 0)		
		Is it facilitated the components' repairing and/or updating, preferably their total replacement? (yes = 1, no = 0)		
D.4: Product Use and Maintenance X Liquid Effluents	For this product or component:	Is the release of water polluting substances avoided when using the products? (yes = 5, no = 0)		
D.5: Product Use and Maintenance X Gas Emissions	For this product or component:	Is the emission of air pollutants avoided during the use and maintenance of the product? (yes = 2, no = 0)		
		Is the emission of gases that cause global warming and the destruction of the ozone layer avoided during the use and maintenance of the product? (yes = 3, no = 0)		
E.1: End of Life X Materials	For this product or component:	Is the reuse and recycling of materials facilitated? (yes = 1, no = 0)		

(continued)

(continued)

Company:		Assessed product:		
Phase	Question		Points	Comments and opportunities for improvement
		Is the identification and separation of materials by type facilitated? (yes = 1, no = 0)		
		Is the use of materials that need to be disposed as hazardous waste avoided? (yes = 1, no = 0)		
		Is the intentional introduction of lead, cadmium, mercury and hexavalent chromium in product materials avoided? (yes = 2, no = 0)		
E.2: End of Life X Energy Consumption	For this product or component:	Are plastic parts and fibers that can be safely used for energy generation, such as incineration, used? (yes = 2, no = 0)		
		Is the use of hazardous materials that need to be transported as hazardous waste to industrial landfills avoided? (yes = 3, no = 0)		
E.3: End of Life X Solid Waste	For this product of component:	Is the existence of internal or external infrastructure for the company to recover/recycle solid waste ensured? (yes = 2, no = 0)		

(continued)

(continued)

Company:		Assessed product:		
Phase	Question		Points	Comments and opportunities for improvement
		Is the connection between different materials that may difficult their separation avoided? (yes = 3, no = 0)		
E.4: End of Life X Liquid Effluents	For this product or component:	Is it possible to recover problematic hazardous liquids during disassembly? (yes = 5, no = 0)		
E.5: End of Life X Gas Emissions	For this product or component:	Is the release of substances that cause ozone depletion and / or global warming during the final disposal of the product or components avoided? (yes = 2, no = 0)		
		Is the recycling of gases contained in the product performed during disassembly so that they are not lost? (yes = 1, no = 0)		
		Is the release of air polluting substances avoided during the final disposal of the product or component? (yes = 2, no = 0)		

References

Bakker, C., Wang, F., Huisman, J., Den Hollander, M.: Products that go round: exploring product life extension through design. J. Clean. Prod. (2014). https://doi.org/10.1016/j.jclepro.2014.01.028

Bakshi, B.R, Fiksel, J.: The quest for sustainability: challenges for process systems engineering. AIChE J. **49**(6), 1350–1358 (2003). https://doi.org/10.1002/aic.690490602/abstract

Baumann, H., Boons, F., Bragd, A.: Mapping the green product development field: engineering, policy and business perspectives. J. Clean. Prod. **10**(5), 409–425 (2002). https://doi.org/10.1016/S0959-6526(02)00015-X

Bhamra, T.A., Evans, S., McAloone, T.C.: Integrating environmental decisions into the product development process. I. The early stages. In: EcoDesign'99: First, no. September: 1–5. http://ieeexplore.ieee.org/xpls/abs_all.jsp?arnumber=747633

Birkeland, J.: Design for ecosystem services a new paradigm for ecodesign. Building **2005**(September), 1–8 (2005)

Bocken, N.M.P., de Pauw, I., Bakker, C., van der Grinten, B.: Product design and business model strategies for a circular economy. J. Ind. Prod. Eng. (2016). https://doi.org/10.1080/21681015.2016.1172124

Borgianni, Y., Maccioni, L., Pigosso, D.: Environmental Lifecycle Hotspots and the Implementation of Eco-Design Principles: Does Consistency Pay Off?, pp. 165–176 (2019). https://doi.org/10.1007/978-981-13-9271-9_16

Brezet, J.C., Van Hemel, C.: Ecodesign: A Promising Approach to Sustainable Production and Consumption (1997)

Byggeth, S., Hochschorner, E.: Handling trade-offs in ecodesign tools for sustainable product development and procurement. J. Clean. Prod. **14**(15–16), 1420–1430 (2006). https://doi.org/10.1016/j.jclepro.2005.03.024

Ceschin, F., Gaziulusoy, I.: Evolution of design for sustainability: from product design to design for system innovations and transitions. Des. Stud. **47**, 118–163 (2016). https://doi.org/10.1016/j.destud.2016.09.002

Communities, Commission of the European: Green Paper on Integrated Product Policy. Environment: Office for Official Publications of the European Communities, Brussels, Belgium (2001). http://scholar.google.com/scholar?hl=en&btnG=Search&q=intitle:GREEN+PAPER+ON+INTEGRATED+PRODUCT+POLICY#0.

Dalhammar, C.: Industry attitudes towards ecodesign standards for improved resource efficiency. J. Clean. Prod. **123**, 155–166 (2016). https://doi.org/10.1016/j.jclepro.2015.12.035

Dekoninck, E.A., Domingo, L., O'Hare, J.A., Pigosso, D.C.A., Reyes, T., Troussier, N.: Defining the challenges for ecodesign implementation in companies: development and consolidation of a framework. J. Clean. Prod. **135**, 410–425 (2016). https://doi.org/10.1016/j.jclepro.2016.06.045

Dermody, J., Hanmer-lloyd, S.: Greening new product development: the pathway to corporate environmental excellence. Greener Manage. Int. **11**, 73–88 (1995)

Eagan, P.: Development of a Streamlined, Life-Cycle, Design for the Environment (DFE) Tool for Manufacturing Process Modification : A Boeing Defense & Space Group Case Study. Matrix (n.d.)

Ehrenfeld, J.R., Hoffman, A.J.: Becoming a green company: the importance of culture in the greening process. In: Greening of Industry Conference. Boston (1993)

Jeswiet, J., Hauschild, M.: EcoDesign and future environmental impacts. Mater. Des. **26**(7), 629–634 (2005). https://doi.org/10.1016/j.matdes.2004.08.016

Johansson, G.: Success factors for integration of ecodesign in product development: a review of state of the art. Environ. Manag. Health **13**(1), 98–107 (2002). https://doi.org/10.1108/09566160210417868

Kjaer, L.L., Pigosso, D.C.A., McAloone, T.C., Birkved, M.: Guidelines for evaluating the environmental performance of product/service-systems through life cycle assessment. J. Clean. Prod. **190**, 666–678 (2018). https://doi.org/10.1016/j.jclepro.2018.04.108

Luttropp, C., Lagerstedt, J.: EcoDesign and the ten golden rules: generic advice for merging environmental aspects into product development. J. Clean. Prod. **14**, 1396–1408 (2006). https://doi.org/10.1016/j.jclepro.2005.11.022

Manzini, E.: Making things happen: social innovation and design. Des. Issues **30**(1), 57–66 (2014)

McAloone, T.C., Bey, N.: Melhoria Ambiental Por Meio Do Desenvolvimento de Produtos - Um Guia (2009a)

McAloone, T.C., Bey, N.: Environmental Improvement through Product Development—A Guide (2009b)

Mcaloone, T.C., Pigosso, D.C.A.: From ecodesign to sustainable product/service-systems: a journey through research contributions over recent decades. In: Stark, R., Seliger, G., Bonvoisin, J. (eds.) Sustainable Manufacturing. Sustainable Production, Life Cycle Engineering and Management, pp. 99–111. Springer (2017). https://doi.org/10.1007/978-3-319-48514-0_7

Nielsen, P.H., Wenzel, H.: Integration of environmental aspects in product development: a stepwise procedure based on quantitative life cycle assessment. J. Clean. Prod. **10**, 247–257 (2002)

Pigosso, D., McAloone, T.: How can design science contribute to a circular economy? Proc. Int. Conf. Eng. Des., ICED **5**, 299–307 (2017)

Pigosso, D.C.A., McAloone, T.C., Rozenfeld, H.: Characterization of the state-of-the-art and identification of main trends for ecodesign tools and methods: classifying three decades of research and implementation. J. the Indian Inst. Sci. **95**(4) (2015)

Pigosso, D.C.A, McAloone, T.C.: Best practices for the integration of social sustainability into product development and related processes. In: 20th International Conference 'State of the Art' Sustainable Innovation & Design (2015)

Pigosso, D.C.A., Mcaloone, T.C.: Maturity-based approach for the development of environmentally sustainable product/service-systems. CIRP J. Manuf. Sci. Technol. **15**, 33–41 (2016). https://doi.org/10.1016/j.cirpj.2016.04.003

Pigosso, D., McAloone, T.C., Rozenfeld, H.: Systematization of best practices for ecodesign implementation. In: Proceedings of International Design Conference, DESIGN. Vol. 2014–Janua (2014a)

Pigosso, D., McAloone, T.C., Rozenfeld, H. Systematization of best practices for ecodesign implementation. In: International Design Conference—DESIGN 2014, pp. 1651–1662. Dubrovnik, Croatia (2014b)

Pigosso, D., Rozenfeld, H., McAloone, T.C.: Ecodesign maturity model: a management framework to support ecodesign implementation into manufacturing companies. J. Clean. Prod. **59**, 160–173 (2013). https://doi.org/10.1016/j.jclepro.2013.06.040

Pigosso, D.C.A., Souza, S.R., Ometto, A.R., Rozenfeld, H.: Life cycle assessment (LCA): Discussion on full-scale and simplified assessments to support the product development process. In: 3rd International Workshop—Advances in Cleaner Production. São Paulo (2011)

Poole, S., Simon, M., Sweatman, A., Bhamra, T.A., Evans, S., McAloone, T.C.: Integrating environmental decisions into the product development process: Part 2 The later stages. In: Proceedings. EcoDesign'99: First International Symposium On Environmentally Conscious Design and Inverse Manufacturing, pp. 334–337. IEEE (1999). http://ieeexplore.ieee.org/xpls/abs_all.jsp?arnumber=747633

Rodrigues, V.P., Pigosso, D.C.A., McAloone, T.C.: KPIs for Measuring the Sustainability Performance of Ecodesign Implementation into Product Development and Related Processes: A Systematic Literature Review (2015)

Rodrigues, V.P., Pigosso, D.C.A., McAloone, T.C.: Framework for Measuring the Sustainability Performance of Ecodesign Implementation (2016). In . http://www.sustain.dtu.dk/.

Rozenfeld, H., Forcellini, F.A., Amaral, D.C., Toledo, S.L.D., Silva, J.C.D., Alliprandini, D.H., Scalice, R.K.: Gestão Do Desenvolvimento de Produtos: Uma Referência Para a Melhoria Do Processo, 1st edn. Saraiva, São Paulo (2006)

Spangenberg, J.H., Fuad-Luke, A., Blincoe, K.: Design for sustainability (DfS): The interface of sustainable production and consumption. J. Clean. Prod. **18**(15), 1485–1493 (2010). https://doi.org/10.1016/j.jclepro.2010.06.002

Sundin, E., Bras, B.: Making functional sales environmentally and economically beneficial through product remanufacturing. Production **13**, 913–925 (2005). https://doi.org/10.1016/j.jclepro.2004.04.006
Vezzoli, C., Manzini, E.: Design for Environmental Sustainability, 1st edn. Springer, London (2008)
Van Weenen, J.C.: Towards sustainable product development. J. Clean. Prod. **3**(1–2), 95–100 (1995). http://www.sciencedirect.com/science/article/pii/095965269500062J

Chapter 9
Product-Service Systems (PSS)

Henrique Rozenfeld, Maiara Rosa, Sânia da Costa Fernandes, Marina de Pádua Pieroni, Carolina Queiroz Souza, Érica Gonçalves Rezende, and Cristina Targas Gurian

9.1 Motivation and Context

Imagine that you are, for example, a traditional supplier of capital assets such as forklifts. Other companies will purchase the forklifts and use them on their shop floor to move certain materials. This relationship is classified as Business-to-Business (B2B), as your customer is a company that uses your product to offer solutions to other customers.

The typical characteristics of a business of this nature are:

- Your revenue varies according to the time of the year, as your market may be seasonal;
- You want to sell more and more forklifts. Otherwise, your revenue would run out. Thus, you supply high-quality forklifts with scheduled obsolescence to ensure that your customers will buy more forklifts in the future;
- Many times, technical assistance is offered by third parties, as you are not concerned with this aspect of your business;
- You may have never thought about what environmental impacts your forklift truck might have during its life cycle (i.e., production, use and end-of-life).

Now think about the implications of those characteristics for your customer, who buys the forklift. He needs to invest initial capital, provide maintenance (or request technical assistance) and, after a while, change the forklift.

Keep this context in mind, but let us change the focus of our analysis. A forklift is depreciated over time, and it generates costs regarding maintenance. However, there is a significant period where the forklift is available but not in operation. Then, the

H. Rozenfeld (✉) · M. Rosa · S. da C. Fernandes · M. de P. Pieroni · C. Q. Souza · É. G. Rezende · C. T. Gurian
Department of Production Engineering, University of São Paulo—USP, Av. Trab. São Carlense, 400—Arnold Schimidt Park, São Carlos 13566-590, Brazil
e-mail: roz@usp.br

J. A. de Oliveira et al. (eds.), *Life Cycle Engineering and Management of Products*,
https://doi.org/10.1007/978-3-030-78044-9_9

costs are diluted in a shorter operation time than possible, making the product less advantageous for the customer.

Let us now look at this case from the perspective of environmental impacts. Think that, from time to time, your customer will have to discard the old forklift and purchase a new one due to scheduled obsolescence. This will lead to a continuous generation of waste from forklifts whose lifespan could be extended. And what about the parts that are replaced throughout the life cycle every time we do maintenance? Usually, the parts are discarded, and, in a few cases, the material is recycled. However, only a few components replaced during maintenance actually wore out. The company could reuse what has already been produced to manufacture a new product, reducing manufacturing costs and the business model's environmental impact.

Some of those reflections are related to significant changes in the way of doing business. Have you thought about why some industrial segments are undergoing a business revolution, such as the music and video industries, for example? Streaming technology and people's connectivity enabled opportunities not even imagined years ago.

There is also a revolution in the transport and hotel markets. You must know that the largest company providing individual mobility does not have any vehicles in its assets. As well as the largest people hosting network does not have any accommodation properties.

Given all these examples, we may question ourselves about:

- leaving the traditional business model, which is based only on selling products;
- focusing more and more on the customer but, at the same time, considering other stakeholders who could impact the business;
- utilizing the potential of new technologies; and
- ensuring new solutions to be economically, socially, and environmentally sustainable.

The answer that fits all these questions is: leaving the paradigm of being a product supplier and becoming a service provider that uses the product to deliver what the customer/user wants or needs. A system composed of this product and its associated services is called a Product-Service System (PSS).

To become a PSS provider, a company that works in the product-oriented paradigm must go through a transformation process known as servitization. In this process, you must mainly change your mindset in order to reinvent the way you do business, which often requires innovations in the products themselves and in the way you relate to customers and other stakeholders. In addition, as we all share the same planet and live in society, this paradigm shift must have fewer negative impacts on the environment and society. We must mind the dimensions of sustainability so that, for example, the social impact does not increase with solutions in which labor relations become unregulated without protecting the minimum dignity of people.

9.2 Goals

Within the context presented, this chapter aims at presenting the basic concepts about PSS, as well as some related concepts and success cases in PSS, so that you can:

- know the potentials of adopting this approach and the challenges that impact this adoption;
- relate PSS to other approaches focusing on Life Cycle Engineering and Management (LCEM), which is the central theme of this book;
- know how to develop a PSS in practice.

9.3 Definition of PSS

The term product-service system has been formally introduced by Goedkoop et al. (1999) as "a marketable set of products and services capable of meeting, together, the needs of a user" (Goedkoop et al. 1999). Products and services are combined in a system, which is a collection of tangible and intangible elements organized around interactions through a set of common goals to be achieved (Cavalieri and Pezzotta 2012; Goedkoop et al. 1999). The services are provided through products developed in a context in accordance with the practices in use by customers (Tan 2010).

There is no standard definition for PSS yet, which depends on the focus of analysis in the research field (Mougaard 2015). However, the existing definitions may cover different aspects, including the concept of system, customer needs, tangibility and intangibility, networks and infrastructure, social aspects and partnerships, and environment impact (Annarelli et al. 2016). The term PSS is used by the research communities of information systems, business management, operations management, marketing, service sciences, engineering, and design (Boehm and Thomas 2013; Lightfoot et al. 2013). However, several terms have been used in the literature to refer to PSS, as illustrated in Fig. 9.1. The font size of the terms in this figure represents the frequency with which each term appeared in the literature review.

In this chapter, we will use the following definition: **PSS is an integrated product and service offering, which delivers value to customers through the value creation components of an innovative business model that considers the needs of stakeholders. The value creation components of the business model are the processes, resources, people, and partners in the value chain. Resources include the technology and infrastructure applied to create value**.

9.4 Types of PSS

A PSS can be classified into typologies in order to allow that, through the descriptions of possible PSS variations, researchers and organizations can predict PSS behaviors

Fig. 9.1 Terms related to PSS. *Source* Authors

and make adequate decisions (Park et al. 2012). The PSS classification can also be useful for defining the business model (Reim et al. 2015). The most widespread typology in the literature is proposed by Tukker (2004). This tipology is presented below.

Tukker's (2004) classification defines three categories, which are illustrated in Fig. 9.2:

- Product-oriented PSS;
- Use-oriented PSS;
- Result-oriented PSS.

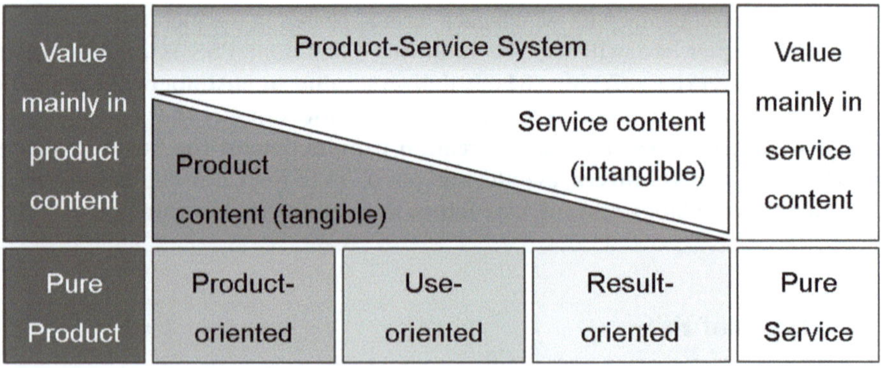

Fig. 9.2 Types of PSS. *Source* Adapted from Tukker (2004)

In a product-oriented PSS, the supplier traditionally sells the product, offering and charging for additional services that guarantee the product's functionality, usability, and durability. A typical example is cars. We buy a car and, therefore, we own the product. However, during the car's life cycle, we still pay for maintenance and repair services offered by the car supplier, besides taxes regarding the vehicle ownership.

In a use-oriented PSS, the provider owns the product and does not transfer its ownership to the customer after the transaction. The provider must maximize the product use phase by developing robust products to extend their life cycle, reducing maintenance and remanufacturing costs since they are under the provider's responsibility. The commercial transaction can be carried out through:

- Product leasing: the customer pays a regular fee for using the product over a given period. This is the case of the automotive industry in the B2B relationship. A company in need to use automobiles does not purchase them. The company pays for using them, and, at the end of a period, it may purchase the product for a residual price. A variation of this type of PSS is when the customer returns the product to the provider at the end of the use phase. This is the case of "Brastemp pure water," which presented in the "PSS cases" section of this chapter;
- Renting/sharing: similar to leasing, the customer pays a fee for using the product in a given period. The difference is that, in this case, the use of the product can be shared by more customers. This is the case of equipment rentals, such as cranes in the construction industry or forklifts for factories. The most popular examples of this type of PSS are the rental/sharing of bicycles, scooters, motorcycles, and automobiles in large cities. In the "PSS cases" section of this chapter, you can consult the case of "Beepbeep";
- Product pooling: it is a particular case of rental business model in which different customers use the product simultaneously, dividing the payment among themselves. This is the case of the Uber pool, where, through the app, more than one person can use the same vehicle, sharing the costs for transportation.

In a result-oriented PSS, the provider sells a result or performance from the product use. The provider owns the product and can customize different services associated with the value offer, which can be subdivided into:

- Activity management: The client outsources some activities to the provider, who takes responsibility for delivering the results of that activity. One example is the "JBT Foodtech" case presented in this chapter's "PSS cases" section. It occurs when the provider is responsible for operating the equipment and achieving the customer's desired results.
- Payment per service unit: The customer pays for a result that the product provides through its use. This is also the case for "JBT Foodtech" presented in this chapter's "PSS cases" section. But here, the customer is the one who operates the equipment. Payment is based on the number of oranges processed;
- Functional result: The provider negotiates with the client to deliver a functional result that adds value. The provider charges a fee for a functional unit and is

completely free to decide how she/he will deliver this result. This is the case of companies that provide, for example, a temperature in a given room or environment. The provider fully controls the air conditioning heating or cooling system and ensures a given temperature to be kept continuously. Another example is when the provider offers a good harvest to a farmer, being responsible for applying pesticides, monitoring the plantation, etc. In the "PSS cases" section of this chapter, we present the case of "Pay-per-lux - Philips," in which the customer pays for lumens that she/he received in the lighting of his facilities.

Despite the popularity of Tukker's typology (2004), its application depends on the individual interpretation of each scenario, which results in ambiguity and confusion. This is mainly due to the typology being purely descriptive and based on PSS examples without defining more specific criteria for comparison. There are other ways to classify a PSS, but since Tukker's typology (2004) is the most known and used, this chapter limits its scope to this typology. For other typologies, see the publications of Roy (2000), Van Halen et al. (2005), and Meier et al. (2010).

We have already presented the basic definition and types of PSS. Although many proposed PSS definitions in the literature state that PSS is related to sustainability, a PSS is not necessarily sustainable unless conceived to be so (Barquet et al. 2016a). Note that we do not cite sustainability in the basic definition we have presented. Therefore, we will now explore the definitions that link PSS to sustainability and define what a sustainable PSS is.

9.5 PSS and Environmental, Social and Economic Sustainability

The most cited sustainable development definition was published in a report produced by the World Commission on Environment and Development (WCED) in 1987, with the title "Our common future" (United Nations 1987). It defines sustainable development as "development that meets the current needs without compromising the ability of future generations to meet their own needs" (United Nations 1987). This concept of sustainable development has evolved to define the three pillars of sustainability known by the acronym TPL (Tripple Bottom Line), which define three dimensions of performance measurement: environmental, social, and financial (Slaper and Hall 2011). These three pillars are equivalent to the three sustainability perspectives: environmental, social, and economic (Vezzoli et al. 2018).

From the environmental perspective, the concern is to reduce the greenhouse effect impacts, ozone layer depletion, eutrophication, acidification, fog smoke, toxic emissions, and other types of pollution. The **social perspective** considers equity, social cohesion, and equitable distribution of resources based on the "principle that everyone has the right to the same access to global natural resources." One of the central issues in this perspective is the eradication of poverty due to poor distribution and armed conflicts that occur in the world (Vezzoli et al. 2018). The **economic**

perspective is broader than the financial one defined in TPL, which aimed only at achieving profit in its original version. The principle of this perspective is that the production and consumption development model must be guided by the environmental conditions and the search for social equity, i.e., this perspective is associated with the other two perspectives of sustainability. Economic value goes beyond financial value and competitiveness.

In his provocatively entitled book "Cannibals with forks: The Triple Bottom Line of 21st Century Business", Elkington explores paradigm shifts in seven dimensions[1] for a sustainable future (Elkington 1997). In the "value" dimension, he states that new values are globally emerging in the direction of sustainability. However, several managers still believe that businesses must pursue the creation of economic value without considering social or ethical values. In this context, PSS development has the potential to balance these aspects of sustainable development.

Initially, the concept of PSS emerged directly related to sustainability. According to Mont (2002), the adoption of PSS must minimize the production and consumption environmental impact. The author defines the PSS as "a system of products, services, relationships and support infrastructure that is developed to be competitive, satisfy the needs of consumers and have a lower environmental impact compared to traditional business model." The PSS approach may reduce environmental impact mainly because the product belongs to the service provider in some cases, who is interested in making the product last longer (Manzini et al. 2001).

However, few companies are committed to sustainability when adopting the PSS approach. Although sustainability is at the origin of the term PSS and its potential to reduce environmental and social impacts, several applications are currently not sustainable, and many studies do not consider sustainability (Tukker 2015). The PSS approach does not guarantee environmental and social improvements if it is not specifically designed for this purpose (Tukker 2015; Tukker and Tischner 2006). In other words, PSS is not a panacea for sustainability issues, and offering a PSS is not always better than selling products (Tukker 2015). Achieving sustainability depends on a profound transformation of the business model towards this goal.

If there is no paradigm shift, the PSS will not be sustainable. For example, the lifespan of shared products is shorter than in traditional cases, as users do not take good care of the product (Tukker 2015). Another example is the bike-sharing solutions in China. In some cities in China, bike-sharing solutions multiplied due to a government policy to reduce car use. However, many of those providers had insufficient financial returns. This deregulation created cemeteries for thousands of discarded bicycles, leading to a major environmental impact.

Today, we differentiate PSS from sustainable PSS (Barquet 2016b; Bacchetti et al. 2016), also known as S-PSS. In 2002, Manzini and Vezzoli stated in a study published by UNEP (United Nation Environment Program) that PSS does not necessarily lead to sustainable solutions. Still, it has the potential to support sustainability (Manzini and Vezzoli 2002).

[1] The seven dimensions that will undergo paradigm shifts in the future are: market, values, transparency, technology in life cycle, partnership (value chain), time, corporate governance.

Therefore, this chapter defines **sustainable PSS as a PSS[2] that aims to obtain environmental and social benefits and the economic benefits to the PSS provider and all stakeholders** (Manzini and Vezzoli, 2002; Bacchetti et al. 2016; Vezzoli et al. 2018).

In this chapter, it is essential to mention the circular economy approach, which is increasingly discussed in many forums and is increasingly being adopted by companies. We will discuss the circular economy concept by comparing it with sustainability and analyzing its relation to the PSS.

Circular Economy is a regenerative system in which resources, waste, emissions, and energy loss are minimized by slowing down, closing, and narrowing the cycles of materials and energy. This can be achieved through robust product designs with long life, maintenance, repair, reuse, remanufacturing, reconditioning, and recycling (Geissdoerfer et al. 2017).

The circular economy aims at keeping products, components, and materials at their peak of value and usefulness for longer. It is possible to say that the most important point among the principles and concepts in the theme of Circular Economy is the existence of a systemic view of activities that is no longer focused on the process but the system as a whole (holistic view). This integration will link the economic gain with sustainability and bring non-specific solutions, but connected in the whole process. These can no longer be planned in isolation, and must now add value to all parts of the chain in new ways of connections and business (Blomsma and Brennan 2017). This way, the circular economy can be treated as a strategy to achieve sustainability. Circular business models add to the sustainable business models the characteristics related to the material and energy cycles mentioned in the definition of circular economy (Geissdoerfer et al. 2018b).

According to the mentioned definition of circular economy, we can define a circular PSS as a PSS inserted in a regenerative system in which resources, waste, emissions, and energy loss are minimized by slowing down, closing, and narrowing the material and energy cycles.

However, a company can adopt the principles of "closing the cycles" of Circular Economy and, even though, not achieve sustainability if, for example, its employees work in poor working conditions (Geissdoerfer et al. 2018a). Closing the loops strategies have already explicitly been part of several sustainable development approaches. Many other circular economy strategies also already existed implicitly in the concept of sustainability, such as slowing down, narrowing the cycle, and minimizing resources, waste, emissions, and energy loss. However, sustainability still is more comprehensive and consolidated than the circular economy since it considers TPL, among other concepts. Thus, in this chapter, we will only deal with the concept of sustainable PSS.

The considerations over the material and energy cycles show the importance of dealing with the PSS life cycle concept. Therefore, in the next section, we will discuss the relationships of PSS and Life Cycle Engineering and Management.

[2] Consider the definition of PSS on the previous section. This definition is complementary.

9.6 PSS and Life Cycle Engineering and Management

A sustainable PSS must consider the aspects discussed in the previous section and, of course, the life cycle. A PSS life cycle has two perspectives: information and material (Fig. 9.3).

The information perspective considers the acquisition, creation, control, updating, storage, and disposal of information at all stages of the PSS material life cycle and involving all processes. Even before the material cycle begins, the first phases of the information cycle involve managing the innovation, developing, and implementing the PSS. The material cycle is repetitive and occurs after the development of the PSS is completed and its production is released (Pieroni 2017). The material cycle begins with the extraction of raw materials, production, distribution, and use of PSS and ends with the application of end-of-life strategies (Pieroni 2017). The two perspectives can be classified into three major groups (Fig. 9.3):

- Beginning of Life (BOL)
- Middle of Life (MOL)
- End of Life (EOL).

In use-oriented and result-oriented PSS, according to Tukker's (2004) classification, the provider is the owner of the physical product. Therefore, to increase the use of the asset, the provider will attempt to incorporate more robust products. Robust products may be achieved by carrying out projects oriented to remanufacturing. It

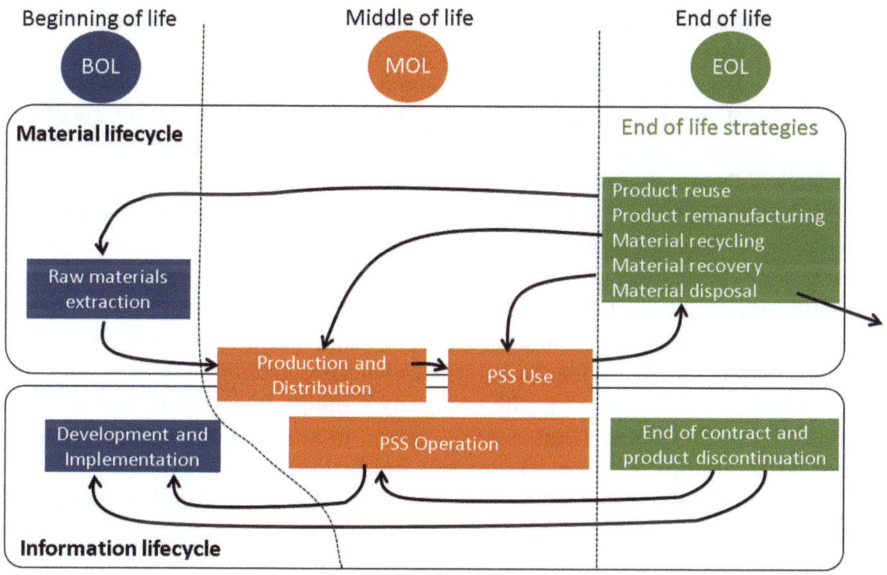

Fig. 9.3 PSS life cycle phases according to two PSS perspectives: material and information life cycle. *Source* Adapted from Pieroni (2017)

makes it easier to bring the product back through reverse logistics (since the provider owns the product) and remanufacturing, enabling the company to reuse the product for other customers. For more information on this, see Chap. 12: Green Supply Chain Management in this book.

Now that we know the definition of PSS and sustainable PSS, as well as the phases of their life cycle, we will present the benefits and barriers of this new approach.

9.7 Benefits and Barriers

Initially, we will present the benefits that a supplier or provider has when adopting the PSS approach and the benefits for customers or users, the environment, and, finally, for the government and society. Then, we will present the barriers that need to be overcome for the implementation of the PSS.

One benefit is **market growth** (Manzini et al. 2001). With PSS, the customer does not need to purchase a product (in the case of use-oriented and result-oriented PSS), obtaining the complete solution of services and products by a much smaller monthly disbursement (Tukker 2004). The sale is no longer seasonal, and the provider receives **recurring revenue**. This generates **greater predictability and long-term cash flow stability** (Alonso-Rasgado et al. 2004; Zancul et al. 2016).

The PSS solutions require greater interaction between customers and provider, generating frequent **feedback** from customers, what creates **long-term relationships** between those involved, **making it easier to obtain information for improving products and services**. This closer relationship can impact **customer loyalty, ensuring the sale of consumables**.

Other benefits of the increased information flow between customer and producer are **opportunities for incremental innovation** as obtaining **frequent feedback** makes it easier to solve customer problems. In addition, **technical assistance can be optimized** as there is better communication with the product user and control over the product through traceability and monitoring. Thus, there is **greater knowledge of the product in the application environment** (Wang et al. 2011) and a greater **guarantee of correct operation of the product** (Tan et al. 2010).

Finally, one can have a **better brand value, improving the organization's image** for customers and other stakeholders (Wagner et al. 2013). For cases of sustainable PSS, the provider must meet the agreed service level and avoid/reduce environmental and social impacts. In this case, **the company's image becomes efficient** (quick response) and **sustainable** (Baines et al. 2007). In cases of sustainable PSS, the provider must **comply with environmental legislation**.

In cases of result-oriented PSS, as in the "pay-per-lux - Philips" case, the PSS provider manages a function provided for the customer. It creates an intangible value linked to the offer, making it a **differential oriented to the customers' needs**.

The main benefit for the customer is **higher satisfaction** due to perceiving more value through PSS solutions (Tukker 2004). This point may also be linked to the **elimination of responsibilities attached to product ownership**, i.e., depending on the type of PSS, the customer is no longer responsible for the repair, maintenance, and disposal of the product (Barquet et al. 2013). If the customer does not purchase the product and pays only for its use, function, or result, he **does not need to pay a high purchase price**. Thus, he can take advantage of the product and its associated services **since there is no need for high investment**. Instead, the customer can make **smaller, predictable payments according to the usage**.

In the Service Level of Agreement (SLA), there may be a clause about maintaining the product in good condition. This kind of contractual clause brings **less risk of stopping** the operation of the product since it may guarantee a certain level of availability of the product and preventive maintenance.

Another benefit would be **access to several services**, payment methods, and options in the market due to the different types of PSS in the market, also related to the different levels of customization of the solution offered to the final consumer. The PSS differential is the possibility of its **customization to meet the specific needs** of every client (Annarelli et al. 2016).

The main environmental benefit is **reducing the use of natural resources** (Cook et al. 2006; Annarelli et al. 2016). This benefit is perceived when the provider adopts a more sustainable approach (Reim et al. 2015), i.e., when the provider designs products with extended durability (more robust) and recyclability. Doing so ensures the possibility of product reuse, repair, reconditioning, and remanufacturing, according to some end-of-life strategies that will be addressed in Chap. 12: Green Supply Chain Management (Cook et al. 2006; Annarelli et al. 2016). Consequently, there is an **increase in the product's life cycle** since the design does not aim at programmed obsolescence (Cook et al. 2006).

Finally, this type of PSS **"dematerializes" products and uses fewer resources**, as customers perceive the value of services and alter the system's life cycle planning (Cook et al. 2006), allowing for a high **environmental gain**.

However, one of the biggest challenges is changing the mentality necessary to make this business model viable both from the perspective of the provider and the customer, who is used to owning the product (observe the barriers to the adoption of PSS). All stakeholders involved (providers and customers mainly) must have a life cycle-oriented mindset.

Shifting to PSS generates **minor problems and costs for providers**, which are related to assuming the responsibility for purchasing, using (i.e., providing the function), replacing, and maintaining products. However, taking this responsibility ultimately benefits society. Shifting to PSS also generates **more jobs in the service sector** (Mont 2002) as the chain of service providers around the PSS requires more people deliver value.

Extending the product life cycle requires **efficient management of solid waste** (Mont and Lindhqvist 2003), resulting in environmental benefits. The society also benefits from a **more environmentally sustainable economy**, for to achieve benefits related to other sustainability perspectives, there must be incentives (governmental and cultural) such as the option to stop using a car and start using a bicycle to go to work every day. In this example, the creation of bicycle path infrastructure is a factor that encourages the use of bicycles. The government needs to promote sustainable lifestyles and consumption patterns through **public policies** (Manzini et al. 2001) in order to obtain benefits for society, making it more sustainable. An example is stimulating the use of electric cars and discouraging individual cars at certain times (rotation) and locations (traffic tax in cities' downtowns). Furthermore, the users' mindset should be a stimulus itself for using more sustainable solutions.

After presenting the main benefits associated with the use of PSS, we will present the main barriers.

The main barrier is **product-oriented culture** (Tan 2010; Barquet 2015) and **consumption attached to product ownership** (Baines et al. 2007). This barrier is easily seen in the context of emotional attachment with objects and luxury products. For example, owning a Rolls-Royce car is a sign of status and power. Potential buyers of a Rolls-Royce would hardly shift from owning a car towards a use-oriented service. Furthermore, there might exist internal **strategic conflicts** from choosing between selling more products and increasing the product's lifespan along with providing services. In addition, the complexity of a PSS may lead to legal issues since the law may have gaps regarding the characteristics of this kind of solution (Hojnik 2016). Finally, **there are no or few incentives in the governmental sphere to create environmentally better solutions** (Mont 2002).

Other barriers related to research and work in the PSS area are: the **fragmentation of researches** in PSS (Baines et al. 2017), which are not focused on practical application (Martinez et al. 2010). PSS studies are numerous, always presenting "new" approaches but **without consolidating practical PSS development and implementation methodologies**. Part of these difficulties stems from the **lack of consensus on PSS definitions** and proven business models. We also note that many PSS implementations are not yet delivering the expected financial return.[3]

In Fig. 9.1 of this chapter, when we present the definition of PSS and the related terms, we realize that the term servitization stands out right after PSS. This term has been created by Vandermerwe and Rada (1988) even before the formalization of the term PSS. Today it is a consensus that this term represents the process of transforming a company so that it becomes a PSS provider. Servitization, therefore, involves the innovation of the business model. For this reason, the next section discusses the relation between these concepts.

9.8 PSS, Servitization, and Business Model Innovation

Servitization is the process that creates additional value by transforming a traditional business model into one that offers PSS (Martinez et al. 2010). A business model defines the mechanisms for creating, delivering, and capturing value by the company (Osterwalder and Pigneur 2010). Servitization implies changing an existing business model to a new one (Baines et al. 2017; Martín-Peña et al. 2017), i.e., servitization is

[3] Search the internet for the car2go case, which is no longer applied in several cities due to lack of financial return.

directly related to the business model innovation. Such a concept involves creating a completely new business model or redesigning an existing business model in already established organizations (Bocken et al. 2014).

The PSS development can also lead to the business model innovation. The difference between servitization and PSS development is in the scope of these approaches. If the company does not have the mindset and culture of offering services, the business model transformation starts with a servitization process by aimed at creating its first PSS offering. Within the context of Life Cycle Engineering, the resulting PSS should be sustainable. For this, during the PSS development, ecodesign strategies, tools, and methods should be used (which are presented in Chap. 8: Product Ecodesign). In addition, the impact of the PSS should be evaluated throughout its life, i.e., it is necessary to predict the impacts of the PSS and try to minimize them during the development phase. One way to assess the PSS in operation is using the LCA technique - Life Cycle Assessment (see Chap. 3: Life Cycle Assessment (LCA)—Definition of Goals and Scope).

Any transformation based on the business model innovation should be placed in the context of change management (see Fig. 9.4), as it deals with the most determining factor for the success of any transformation and, therefore, of servitization: people.

In short, servitization can be considered a business model innovation. Like any transformation, servitization should be placed in the context of change management. Servitization covers the PSS development and is focused on changing the companies' mindset and the culture of a product orientation to become a PSS provider. After the paradigm shift is already incorporated into the organization's culture, the creation of new PSS occurs through the process of developing a sustainable PSS. The development process should incorporate the ecodesign and the life cycle assessment approaches.

After knowing the servitization and PSS theories, a question remains: how to follow the path of servitization? To answer this question, we present below a

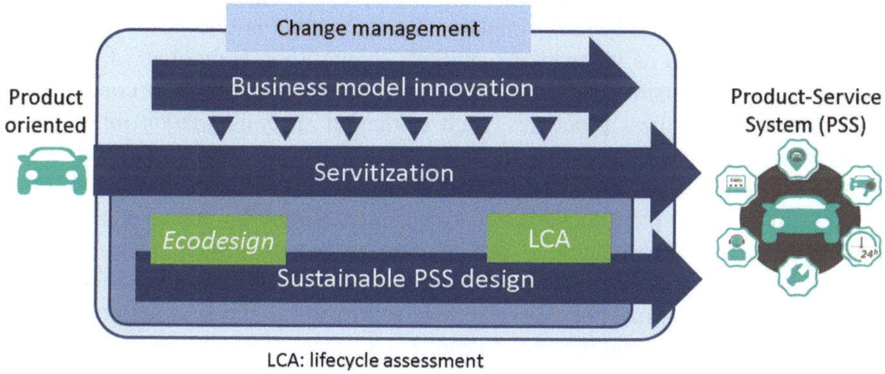

Fig. 9.4 Conceptual view of the relationship between change management, business model innovation, servitization, and PSS development. *Source* Authors

servitization methodology, which follows the concept discussed in this section of integrating the approaches illustrated in Fig. 9.4.

9.9 Servitization Methodology

The servitization methodology aims to help companies that wish to transform themselves from product suppliers into PSS providers. It can also be used for the development of a new PSS within the organization. As the name implies, it is a methodology, i.e., a collection of methods. The methods of this methodology are organized by activities that are grouped according to a common objective. The methodology activities can be combined in different ways depending on the characteristics of the servitization project to be developed. This fact makes the methodology flexible. The selected activities are associated with the most appropriate method(s), that may also be selected.

In this section, the main characteristics and groups of activities that enable the implementation of the servitization methodology will be presented. For more details, you can access the methodology's online content.[4]

In Fig. 9.5, it can be seen that the activity groups are placed in two planes. In a plan, there are the change management and project management activities because these activities permeate the entire process of servitization and relate to all activities.

Change management creates the conditions to prepare and motivate people who make up the organization (as well as the partner organizations) in relation to servitization. Change management activities are integrated with project management activities. Each servitization can be considered a project with well-established beginning and end. Thus, it is necessary that this project is managed properly.

The activity groups from the other plan in Fig. 9.5 are described below.

The business analysis assesses the organization's current business external and internal environments to identify trends and map the company's current situation. Also, this activity should be performed to understand other important aspects related to the business, such as value chain, market, competitors, technology, legislation, and stakeholders. Business analysis should preferably be carried out continuously through parallel business processes, such as market analysis, market intelligence, technological prospecting, technological surveillance, among others. It provides subsidies for strategic planning and servitization itself. A good practice is to synthesize the information from these analyzes into a *roadmap*, known as a *technology roadmap*, to interrelate technology, product, and market on a time scale.

In the context of this methodology, strategic planning involves the establishment of actions and organizational goals in relation to servitization. It is in strategic planning that you decide to go through servitization.

---

[4] The servitization methodology is available online through the link http://www.pdp.org.br/servmtd/ and is one of the methodologies derived from the innovation flexible methodology (http://www.flexmethod4innovation.com.br/).

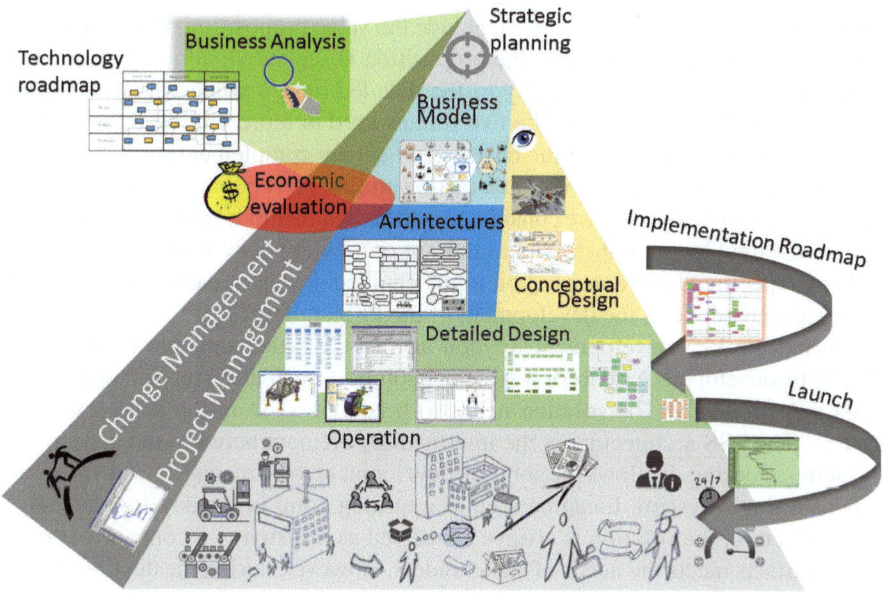

Fig. 9.5 Hierarchical perspective of the servitization methodology. *Source* Authors

The value proposition involves the definition of a market segment, the identification of who the stakeholders are, the understanding of their needs, desires, pains, problems, market and technological opportunities, as well as the understanding of legislation barriers. This activity group also involves the use of creativity techniques to create PSS value propositions based on the information collected previously. The value proposition brings the characteristics of the solution (product and service offering) that can potentially deliver value to stakeholders. More than that, the value proposition should explicit the potential benefits that can be perceived by the stakeholders. However, the entire solution needs to be designed to allow systems, services, among other dimensions of the solution, to be structured (architecture) and detailed (detailing). This conception of the solution is performed through conceptual design. This design is detailing carried out in parallel to the value proposition, when necessary. Conceptual design represents products and services with a level of abstraction that does not fit the business model, as it goes beyond the sticky notes. For the design of services, systems maps, personas, empathy maps, journeys maps etc. can be used. Some representations overlap with the PSS architecture. When we adopt the current physical product and just want to add services to the product offering, the conceptual design of the product does not occur because it already exists. However, it is necessary to analyze whether the characteristics of the current product are suitable for servitization.

The activities of defining the main elements of the business model that create, deliver and capture value are based on the previously established value proposition. Tipically, the business model representations (such as the Canvas proposed by

Osterwalder and Pigneur 2010) used include the value proposition. In the case of the servitization methodology, the value proposition is separated, allowing us to move from the value proposition directly to the detailed design when we are not going to change the business model. The two frames used to represent the value proposition and the other elements of the business model are complementary. The business model may have multiple levels of abstraction and the most popular level used is only superficial. In other words, we consider that the business model representation needs a more detailed level of abstraction (and not only with sticky notes). The PPS can be better specified before moving on to the detailed design of each of the elements. This is usually done in the development of the PSS architecture.

During the PPS development, several artifacts, whether tangible or intangible, should be developed. As part of a systemic solution, such artifacts cannot be developed independently. The solution needs to be observed from the perspective of a system. The PSS architecture is the intermediary element between the concept and the detailed desing. It allows the solution development to be managed in an integrated way as a single system. It can be seen as a turning point in the development. Even in architecture, the solution is always conceived as a system – a concept where the set of artifacts meets the needs of stakeholders. However, during the detailed design, different professionals will be needed to develop each PSS artifact. For example, service designers will be able to design PSS services in detail while programmers will be asked to develop the code for the necessary PSS software, and engineers will design the hardware that makes up PSS. In this detail, it is necessary to ensure that the artifacts will not be developed completely independently, but respecting the interfaces and interactions between the artifacts that will guarantee that the solution will follow the proposed concept (Rozenfeld et al. 2018). The PSS architecture represents the elements of the solution in an interconnected way, associating their functions with them. If the solution architecture is very complex (which is usually the case with solutions like PSS), it can be divided into layers, such as product, service, infrastructure, among other possibilities, according to the solution under development.

The economic viability analysis aims to assess the economic and financial characteristics of the business models generated, resulting in indicators used for decision making. The basic input for the economic viability study is the initial information specified in the business model. That information allows to estimate revenues and costs. It evolves iteratively as the solution is detailed. Economic viability allows us to select alternatives, optimize or simply discontinue the PSS development. In the case of PSS, one of the great financial advantages is obtaining recurring revenues. However, if the value of the asset is very high, making its implementation as PSS unfeasible, a PSS installation fee should be charged to offset the value of the fixed asset.

A PSS solution can require a lot of adaptations and changes in an organization. Those changes are not always feasible in the short term. Thus, it is recommended to split the solution implementation for it to be evolutionary, implementing a simplified version (such as an MVP: Minimum Viable Product) as soon as possible so that it can be tested. The implementation roadmap represents the main results that must be achieved during servitization, demonstrating how the solution will evolve from MVP to the final solution.

Detailed design activities involve the detailed development of all the elements that make up the PSS. At this time, the development of the elements can be separately assigned to professionals specialized in each type of technology: software, hardware, processes, services, infrastructure, among others. These professionals will detail and test the elements while respecting the integration requirements derived from the architecture. This is important to allow that the final solution to funciton as a system. Obviously, existing artifacts do not need to go through this process. For example, in a case of servitization based on a pre-existing product, the parts of the product do not need to be detailed again (except for possible modifications). The detailed design will provide all the information necessary for the company to implement the PSS, including production processes, service provision, among others.

The launching activities aim at starting the value chain operation, which involves the final acquisitions, installation of machinery, people training, establishment of the necessary partnerships not yet been defined in previous steps, approval of processes, issuing of the necessary certifications, among others.

The operating activities involve those required during the PSS's middle of life (MOL) to ensure that the solution is properly offered. It includes continuous improvement of the PSS based on stakeholder feedback and the evolution of the solution according to the roadmap. At the end of life (EOL) of the PSS, strategies are triggered, such as product reuse and remanufacturing, material recycling, recovery, and disposal.

BOL, MOL, and EOL have economic, environmental, and social impacts associated with them. As in a product's life cycle, it is interesting to predict, measure them and work to reduce negative impacts. Clearly, the PSS has other dimensions that need to be considered in addition to the product, such as services, infrastructure, etc. as previously mentioned. Thus, there are techniques adapted to assess the solutions' sustainability. One of them is the guide for assessing the environmental performance of PSS proposed at the Technical University of Denmark (Kjaer et al. 2017). However, there are several other techniques to assess the sustainability of a PSS. Usually, those techniques are based on the LCA and focus mainly on the economic and environmental dimensions (Doualle et al. 2015). The main difference between performing LCA for a pure product and for a PSS is that, in the case of a PSS, the support processes and systems need to be included in the analysis in addition to the analysis of the product itself. We should consider that the support services, processes, and systems have material flows and, consequently, the potential to generate environmental impact (Kjaer et al. 2017). A point of attention for applying the LCA to the PSS is the proper definition of the scope to be analyzed and the level of detail desired, since the complexity of a PSS combined with an excessive level of detail can make the analysis unfeasible (Kjaer et al. 2017).

Each of the groups of activities for servitization has challenges to be overcome. It is difficult to say which activity is the most challenging, as it depends on the know-how that the company has, its previous experience, and the PSS which is being developed. A company that knows its customers very well, keeping them in constant contact to understand their needs and get feedback on solutions, will find the design of value proposition easy. On the other hand, a company approaching an entire new market segment can spend a significant amount of resources and time to accomplish the same acticity. Likewise, if the PSS is based on a higly complex technical product (such as an airplane turbine, for example) and with highly complex services (such as predictive maintenance based on remote sensing), the defining of the architecture and the detailed design will spend much more time and resources compared to a PSS with low complexity products and services (such as a monthly supply of office pens based on a predefined demand). However, one factor that is constantly reinforced by companies providing PSS is the need to establish service levels that are achievable (is it possible to guarantee that a machine will be running in 99% of its working time?). Also, it is necessary that the service levels are compatible with the risks to be assumed and with the expected profits. Establishing a good service contract has a high impact on the success or failure of a PSS. It is a task that requires a good understanding of all factors of the business model and an investment of people's time, and legal, tax, and technical knowledges.

9.10 Cases of PSS

This section present real cases of PSS in the Brazilian and global markets. The cases evolve over time, which can add new concepts. We, therefore, recommend that you always keep up to date with new cases and new versions of the presented cases. Considering the cases presented, the greatest advantages for the PSS provider are the recurrent revenues, customer loyalty, and the possibility of reaching a market segment that previously did not have the financial capacity to acquire the offerings.

9.10.1 Pure water, Brastemp

Brastemp, a Brazilian manufacturer of household appliances, offers a subscription service for water purifiers. Subscriptions vary depending on the type of purifier and its features. It includes online services in a relationship program for subscribers.

The great advantage is that the customers does not have to worry about maintenance of the equipment and the acquisition of a new one.

9.10.2 Beepbeep

Beepbeep is a Brazilian startup, located in São Paulo, that offers mobility through 100% electric cars (currently only Renault Zoe cars). Through a mobile app, the user can find and reserve a vehicle. The vehicles should be found and returned to the company's stations, which are distributed throughout the city. Payment for the service consists of an initial fee plus an additional fee per minute of use.

People who cannot afford to buy vehicles get a personalized modality for special cases, combining it with the use of public transport for their daily activities.

9.10.3 JBT Foodtech

JBT Foodtech is a subsidiary of JBT Corporation (American multinational) and specializes in manufacturing equipment for the food processing industry. The product, called Citrus Extrator, is the core of the citrus processing system and is offered as a service. The business model focuses on providing integrated solutions, which also include fruit storage, receipt and handling of fruits, citrus pulp processing, waste processing, among others. During the off-season, the equipment undergoes preventive maintenance. At its end of life, the equipment is returned to JBT, which replaces it with a "new one" to continue to deliver the citrus processing function to the customer. The provider collects the equipment and remanufactures it, reusing 80% of the material.

With this solution, customers (orange juice producers) can only worry about their core businesses. They also do not need to buy the equipment. The solution causes less environmental impact, as a large part of the material is reused.

9.10.4 Meyer Solutions in Printing Technologies

Meyer is a Brazilian company that operates in the B2B market and offers innovative and continuous solutions in print management. It offers the printing service, being responsible for the delivery and installation of equipment, maintenance with specialized technical support, delivery of parts and consumables, and removal of equipment. Instead of selling the printing equipment, the company provides customizable product rentals. This enables the best use of resources and operational security for customers (companies that need image processing, such as medical clinics).

By paying the usage fee, the customer has access to equipment that is always up to date and they do not need to worry about end of life of products. The provider earns loyalty when purchasing consumables.

9.10.5 Pay-Per-Lux—Philips

Philips has a business model whose value proposition is to keep the lighting of the place. Offering the product as a service, Philips is responsible for the entire lighting service, from the solution design, equipment installation, maintenance, and updates during the service offering. Customers pay a fee for contracting the service and for the lighting consumed, and no longer for the purchase of light bulbs. At the end of the contract, the light bulbs can be returned to the production process. Raw materials are reused, optimizing recycling and reducing waste. The subway system in Washington (United States) was one of the first to adopt this model. This PSS is called by Philips as "light as a service".

This solution reduces the operating cost for the customer, who does not have to worry about having an associated team to purchase the light bulbs and electrical material. This solution minimizes the environmental impact as the provider uses optimized processes and products.

9.10.6 Vigga™

VIGGA™ is a Danish company operating in the textile industry. This company has clothing brands for babys, children and pregnant women. Instead of selling the clothes, the company provides the possibility of renting organic clothes for children and pregnant women in the right size. In this way, parents no longer need to buy different-sized of clothes for their growing children, avoiding the waste of resources and the generation of waste. The company is responsible for collecting used clothes and making a new use available for other children and pregnant women. For this, the company performs quality verification and washing of clothes. This circular solution saves mothers' time, money, and resources.

The customers have access to a greater number of options than they would have if they purchased the product. The product's life and usage are also extended to more than one customer. This tends to cause less environmental impact.

9.11 Final Considerations

The practical application of PSS does not follow the academic evolution. Although there are many initiatives by companies that wish to provide or are already providing PSS offers, there is still a predominance of companies that are structured under the logic of the traditional business model. The biggest barriers lie in the difficulty in financial justification and on the mindset focused on the product ownership paradigm, both from the perspective of the provider and the customer. PSS is a trend. At the same time, it is observed that it is not a "panacea" that will take companies to a new level of competitiveness.

Change management can assist in the incorporation of new skills, capabilities, and behaviors for individuals within the organization towards an integrated development, either internally or among other companies in the value chain. In addition, it is necessary to adopt a new approach to communication and engagement with customers, who are used to owning the products and not paying for the services.

We might highlight the importance of defining the value proposition, which encompasses the product and service offering and its potencial value. We should consider the co-creation of value between the provider and the customer as well as the entire articulation of the value chain. The value proposition needs to make clear the potential benefits for stakeholders and, in particular, for customers. The focus on value is reinforced in the early stages of developing a PSS. However, we should keep in mind that the real value of a PSS is perceived by stakeholders and not only by customers, especially in the design of a sustainable PSS, where law requirements, environmental factors, and communities should be considered.

In the detailed design, after defining the PSS architecture, it is possible to work separately on each of the technologies and elements of the PSS, such as on the hardware, software, on the electronic and mechanical components, specification of services, etc. But until reaching this stage, the PSS representation should integrate products, services and the necessary infrastructure. Logically, the actors and their responsibilities are part of this holistic view of PSS design.

The term "system" of the PSS should also be emphasized. It comprises the entire infrastructure and the value chain. As in many cases the co-creation of value between stakeholders occurs, the PSS design covers the specification of all organizations involved in the development process as well as the configuration of the integration between them. This requires new mindset for those who develop the offerings and a greater integration between the stakeholders in the value chain. We have to put ourselves in the "other's place". Empathy methods are essential in this process. In this context, service and design consultancy, calcutation of operational costs and ownership cost for customers, can be a great competitive advantage for those who offer PSS. The provider would also need to "design" the partners and customers' business models in order to distribute responsibilities and gains. After all, integrated development requires commitment and sharing of risks and benefits in the new ecosystem.

What about sustainability? This is an essential issue in developed countries and is starting to become the order of the day in developing countries. Although sustainable development is a need in the face of climate change and social impacts, the practice of sustainability in developing countries is still more due to international pressure, legislation, and society than to the mindset of the people involved in the servitization process. Once again, the importance of mindset evolution towards the development of a PSS that considers the environmental, social, and economic aspects is highlighted. The servitization/PSS approach is considered one of the strategies of the circular economy and of Life Cycle Engineering and Management.

There are several sustainable PSS initiatives, but this practice is not yet widespread. Even because PSS may cause a rebound effect from users who would not be interested in a lower environmental or social impact of a PSS offering, but who would be focused on the financial benefits and facilities provided. The rebound effect means that users of a PSS will be able to have more resources available, which can be used in activities that cause greater environmental impact. This can make new products purchased (and therefore with all related impacts) to worsen environmental and social impacts. From the provider's point of view, in most cases, the financial perspective is still the basic condition for adopting the PSS.

An important aspect is to consider regional differences. The same servitization activities cannot be developed in developed countries, which generally have a consolidated manufacturing industry, and in developing countries, whose business capabilities may still be under development. Another factor is the difference between tax laws. We know of some PSS providers that do not use the term PSS because today they rent the equipment as a form of revenue. They say that if they called the business as PSS, they would risk paying taxes on services, which does not happen with rental model. These factors show the importance of offering flexible methodology, which can be adjusted for each case of servitization.

Being able to prototype the PSS in the early stages of development, especially services, is something that needs to be further investigated in order to support the PSS design. This fact is related to the intangibility of services. There are approaches that support product prototyping as a tangible asset. But efforts must be made to develop techniques that support the prototyping of services when integrated with products.

The challenge of achieving the economic viability of the PSS within a new, broader, and holistic logic should go through the definition of new assessment methods. Traditional methods of economic viability are not enough to support the decision of developing or not a PSS. More than that, in addition to the definition of the PSS economic value, it is also necessary to invest in methods which assess the intangible values that the PSS can provide to stakeholders. This assessement should be converted into quantifiable units that show how the benefits can impact the return for the business.

The digitization of manufacturing and products allows the monitoring of products throughout their life cycle, with an emphasis on the use/operation phase, which brings new challenges to the PSS design. Within the 4th Industrial Revolution, known as Industry 4.0, there are many concepts and technology offerings that can be used to develop and operationalize the PSS in a more interactive and intelligent way. During the development of a PSS, we should know the technological solutions that already exist in the market and how they can be used to add value to the stakeholders of a PSS. The use of data analysis (analytics) to know the behavioral pattern of customers and extract guidelines to guide new developments is an opportunity that is still little explored within the servitization context.

Incorporating new approaches proven in practice to PSS design is important to achieve synergy between the best management practices. Especially, incorporating this approach into life cycle management is primordial to achieve the expected benefits by causing the least possible environmental and social impacts, while ensuring financial health for the organizations involved along the value chain.

9.12 Proposed Exercises

Beta printers case: Beta is a Brazilian company that develops, manufactures, and sells different printers, including Beta Print H54. Beta Print H54 is a digital printer widely used by marketing agencies and photo stores. This printer accepts different types and sizes of paper. In addition to the machine, the company also offers maintenance and technical assistance services. Last year, Beta experienced an exponential drop in the sales numbers, and they keep dropping that has been repeating in the last months. The company' has carried out market research to understand what has been happening and received feedback from its customers:

- "The machine is expensive and has more features than we need"—freelance photographer;
- "Maintenance takes too long, and we lose customers while the machine is off"—photo-printing shop;
- "Maintenance is expensive"—marketing agency;
- "We love the machine. If it were not so expensive, we could have several machines and increase our sales. Sometimes, we are not able to serve all customers who come to us."
- "We already had a printer, but we had to sell it. Its maintenance was very costly. Currently, we outsource the printing service, which often lacks quality"—photographer.

Coincidentally, the new business manager has just participated in a product development conference on the theme "Product-service systems" and learned more about servitization. He gathered his team and decided to study the possibility of using Beta Print H54 as a basis to create a PSS. You are in charge of:

Exercise 1. Create a stakeholders map: Who would be the stakeholders in the servitization case of Beta Print H54? Think of everyone who somehow participates or influences the solution: customers, potential partners, regulatory agencies, users, etc. Use the model below and distribute the stakeholders according to their degree of importance and impact on the solution.

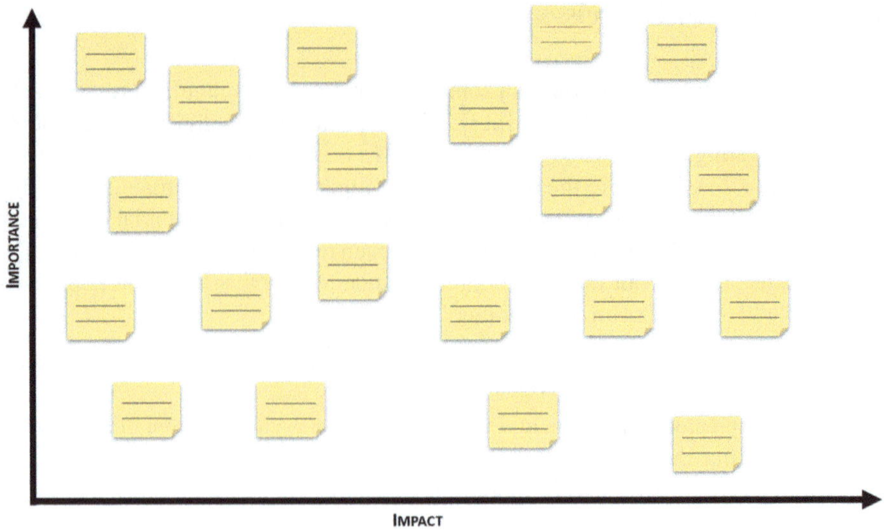

Feedback: Examples of possible stakeholders:

- Advertising agencies;
- Photographers;
- Photo studio
- Rival companies;
- Customers (photo buyers);
- Development partners

Exercise 2. Choose a stakeholder from the previous exercise and list her/his main problems, needs, and wishes related to photo printing. Then, imagine a solution that can solve those problems and meet those needs and wishes by integrating products and services.

Answer: e.g., Photographers.

- *problems: no money to buy the printing machine / pays too much for outsourcing the service;*
- *need: printing high-quality photo albums;*
- *wish: owning a photo printer;*

Solution: Service - Offering a photo-printing service for the photographer as a use-oriented PSS. The photographer pays a value proportional to the number of photos printed. You might need to add a membership fee or formulate a fixed monthly minimum contract. Maintenance services are under the provider's responsibility (BETA), who must guarantee high-quality printings for the photographer. Product - The printing machine should undergo modifications. It should become modular and customizable to fit each photographer's needs regarding paper type and size. In addition, sensors and IoT technologies should be installed to control usage, allowing the provider to charge the customer properly.

Exercise 3. Based on the solution that you proposed in Exercise 2, describe the entire life cycle of the solution, considering sustainable alternatives.

Answer.

BOL: Sustainable development / robust product; Create mechanisms to increase usage and availability, and reduce maintenance needs; Develop IOT technologies.

MOL: Offer training for using the product; maintenance.

EOL: Take the product back from the customer after the end of the contract and redirect it to another customer. At the product's end of life, remanufacture it or look for alternatives for better disposal (for example, reuse in another sector, reuse of parts, etc.*).*

Exercise 4. Reflect which actions should be done during the solution development to ensure that the solution is environmentally sustainable throughout its life cycle.

Answer:

The proposed solution should include robust products, which use the minimum possible amount of raw materials. A sustainable solution should also use inks free of potentially hazardous solvent, print photos on recyclable paper, among others. The company must provide services that improve usability to avoid or reduce waste. Also, it is essential to reflect on the end of life, planning remanufacturing, reusing parts, recycling, etc.

Read the case description of the washing service for *Tex* industrial cleaning cloths to answer the three exercises below.

Tex—Industrial Cloth Management is a company that offers cleaning cloths for industries, printing plants, and repair shops. The service includes providing cleaning cloths, washing them after use, and replacing them with new ones whenever necessary. After using the cleaning cloths, the customers should deposit them in specific containers for dirty cloths. After certain periods (specified in the contract), *Tex* removes the dirty cleaning cloths, washes them, and returns them to the customers. Each cleaning cloth performs this cycle 50 times. Despite the low price of cleaning cloths on the market, their disposal costs have increased meaningfully, making *Tex*'s service an attractive alternative. *Tex* uses the most adequate solvents during the washing process, and the water is reused several times along the washing and drying stages. In addition, the oil in the wastewater is used to generate energy at the *Tex* plant. See the service cycle offered by *Tex* in Fig. 9.6.

Exercise 6. Make some commentaries about the business model that structures Tex's PSS. Classify this PSS and describe the case characteristics that justify your choice. Point out the benefits that this solution provides for the customers.

Fig. 9.6 Tex - industrial cloth management

Answer

The service offers greater value to the customer. In Tukker's (2004) *classification, this is a use-oriented PSS. The aspects that justify this answer are that the service provider provides clean cloths for industrial cleaning and is paid for the cleaning and disposal process of used cleaning cloths. The customer does not have to worry about washing or disposing of the cloths, besides always having clean cloths available for use.*

Exercise 7. Analyze the Tex case regarding the advantages of offering PSS over the product sale. Analyze the case from the social, environmental, and economic perspectives.

Answer

Environmental, economic, and social benefits.

Adequate solvents are used in the washing process, while water and energy are reused several times in the washing and drying stages. The residual oils from the washing water disposal are used to generate energy for the Tex plant.

In the case of selling cleaning cloths as a pure product, there would probably not be a concern about wastewater disposal in the washing process. Also, according to the description of the case, disposing of the cleaning cloths would be costly.

If the plant produces its own energy with the residual oils, it needs less or does not need external energy. The residual water is treated in the plant and can be discharged into the common sewer. The cloth washing system is circular.

Exercise 8. Describe the main benefits and barriers (for the provider) of implementing a PSS regarding the two previous cases.

Answer

Case	*Benefits*	*Barriers*
Beta Print H54	*Increase in the number of customers* *Recurring revenue* *Development of robust products* *Maintenance reduction* *Customers loyalty and trust*	*High investment in assets* *Big customers may want to keep the product ownership*
Tex - cleaning cloths industrial washing	*Reduced environmental impact* *Recurring revenue* *Energy production*	*High investment in the washing process infrastructure* *Buying cloths as pure products is cheap* *It requires strict environmental regulations for the disposal of dirty cloths*

Exercise 9. Product-service systems are great alternatives for reducing consumption and are often seen as sustainable alternatives. However, this premise is not always true. See the fictional case below in which resource consumption becomes higher than a product-selling-based business model.

A telephone company sells communication services by building packages composed of a given amount of telephone calls, internet connection, and SMS texts. Those packages include a new smartphone being offered to the customers. This service is renewed every year, and a new cell phone is provided along, i.e., the customer subscribes to the communication service package and can change their smartphone for a new one yearly.

Describe the problem and suggest solutions based on the whole life cycle of the solution.

Answer

Problem: Changing the device every year increases consumption. If commercialization were in the purchase format, it would not occur since the cost to buy a new device is high.

Solution: Changing the smartphone at no cost when it stops working or charging a fee for the annual exchange — discouraging annual exchange. Another possibility is to keep this smartphone exchange model but reducing the monthly fees for customers with smartphones being used for extended periods (over 1-year-long).

Exercise 10. Based on your daily life, think about a problem that you have that could be solved with a PSS solution. Propose a solution, think about its value proposition, its business model, and its life cycle.

The activity can be carried out in groups. .

References

Alonso-Rasgado, T., Thompson, G., Elfström, B.: The design of functional (total care) products. J. Eng. Des. **15**(6), 515–540 (2004)

Annarelli, A., Battistella, C., Nonino, F.: Product service system: A conceptual framework from a systematic review. J. Clean. Prod. **139**, 1011–1032 (2016)

Bacchetti, E., Vezzoli, C., Landoni, P.: Sustainable Product-Service System (S. PSS) applied to Distributed Renewable Energy (DRE) in low and middle-income contexts: a case studies analysis. Procedia CIRP **47**, 442–447 (2016)

Baines, T., et al.: State-of-the-art in product-service systems. Proc. Inst. Mech. Eng. b: J. Eng. Manuf. **221**(10), 1543–1552 (2007)

Baines, T.S., et al.: Servitization: revisiting the state-of-the-art and research priorities. Int. J. Oper. Prod. Manag. **37**(2), 256–278 (2017)

Barquet, A.P.B., et al.: Employing the business model concept to support the adoption of product-service systems (PSS). Ind. Mark. Manage. **42**(5), 693–704 (2013)

Barquet, A.P., et al.: Sustainable product service systems–from concept creation to the detailing of a business model for a bicycle sharing system in Berlin. Procedia CIRP **40**, 524–529 (2016a)

Barquet, A.P., Seidel, J., Seliger, G., Kohl, H.: Sustainability factors for PSS business models. Procedia CIRP **47**, 436–441 (2016b)

Blomsma, F., Brennan, G.: The emergence of circular economy: a new framing around prolonging resource productivity. J. Ind. Eco. **21**(3), 603–614 (2017)

Bocken, et al.: A literature and practice review to identify Sustainable Business Model Element Archetypes. J. Clean. Prod. **65**, 42–56 (2014)

Boehm, M., Thomas, O.: Looking beyond the rim of one's teacup: a multidisciplinary literature review of Product-Service Systems in Information Systems, Business Management, and Engineering & Design. J. Clean. Prod. **51**, 245–260 (2013)

Cavalieri, S., Pezzotta, G.: Product-service systems engineering: state of the art and research challenges. Comput. Ind. **63**(4), 278–288 (2012)

Cook, M., Bhamra, T.A., Lemon, M.: The transfer and application of Product Service Systems: from academia to UK manufacturing firms. J. Clea. Prod. **14**(17), 1455–1465 (2006)

De Pieroni, M. P.: Proposal of a Business Process Architecture (BPA) Development Method for supporting the transition of manufacturing companies into Product-Service System (PSS) providers. University of São Paulo (2017)

De Zancul, E.S., et al.: Business process support for IoT based product-service systems (PSS). Bus. Process. Manag. J. **22**(2), 305–323 (2016)

Doualle, B., Medini, K., Boucher, X., Laforest, V.: Investigating sustainability assessment methods of product-service system. Procedia CIRP **30**, 161–166 (2015)

Elkington, J.: Cannibals with Forks: The Triple Bottom Line of Sustainability. Capstone Publishing, Gabriola Island (1997)

Geissdoerfer, M., et al.: The circular economy—a new sustainability paradigm? J. Clean. Prod. **143**, 757–768 (2017)

Geissdoerfer, M., Vladimirova, D., Evans, S.: Sustainable business model innovation: a review. J. Clean. Prod. **198**, 401–416 (2018a)

Geissdoerfer, M., et al.: Business models and supply chains for the circular economy. J. Clean. Prod. **190**, 712–721 (2018b)

Goedkoop, M. J., et al.: Product service systems, ecological and economic basics. The Hague, NE: Ministry of Housing, Spatial Planning and the Environment Communications Directorate (1999)

Hojnik, J.: The servitization of industry: EU law implications and challenges. Common Market Law Rev. **53**(6), 1575–1623 (2016)

Kjaer, L.L., Pigosso, D.C., McAloone, T.: A guide for evaluating the environmental performance of product/service-systems. Technical university of Denmark,.Department of mechanical engineering (2017)

Lightfoot, H., Baines, T., Smart, P.: The servitization of manufacturing: a systematic literature review of interdependent trends. Int. J. Oper. Prod. Manag. **33**(11/12), 1408–1434 (2013)

Manzini, E., Vezzoli, C.A.: Product-Service Systems and Sustainability: Opportunities for Sustainable Solutions. UNEP-United Nations Environment Programme (2002)

Manzini, E., Vezzoli, C., Clark, G.: Product-service systems: using an existing concept as a new approach to sustainability. J. Des. Res. **1**(2), 27–40 (2001)

Martinez, V., et al.: Challenges in transforming manufacturing organisations into product-service providers. J. Manuf. Technol. Manag. **21**(4), 449–469 (2010)

Martín-Peña, M.L., Pinillos, M.J., Reyes, L.E.: The intellectual basis of servitization: a bibliometric analysis. J. Eng. Technol. Manage. JET-M **43**, 83–97 (2017)

Meier, H., Roy, R., Seliger, G.: Industrial Product-service systems—IPS2. CIRP Ann. Manuf. Technol. **59**, 607–627 (2010)

Mont, O.: Clarifying the concept of product—service system. J. Clean. Prod. **10**, 237–245 (2002)

Mont, O., Lindhqvist, T.: The role of public policy in advancement of product service systems. J. Clean. Prod. **11**(8), 905–914 (2003)

Mougaard K (2015) A framework for conceptualisation of PSS solutions: on network-based development models. Thesis, Technical University of Denamark (2015)

Osterwalder, A., Pigneur, Y.: Business Model Generation: A Handbook for Visionaries, Game Changers, and Challengers. Wiley (2010)

United Nations: Report of the World Commission on Environment and Development: Our Common Future (1987)

Park, Y., Geum, Y., Lee, H.: Toward integration of products and services: taxonomy and typology. J. Eng. Technol. Manage. JET-M **29**(4), 528–545 (2012)

Reim, W., Parida, V., Örtqvist, D.: Product-Service Systems (PSS) business models and tactics—–a systematic literature review. J. Clean. Prod. **97**, 61–75 (2015)

Roy, R.: Sustainable Product-Service Systems. Futures—Sustainable Futures, vol. 44, n. Sustainable Futures, pp. 289–299 (2000)

Rozenfeld, H., Rosa, M., Fernandes, S.C.: Servitization methodology: PSS design, change management, or business model innovation? In: 23o Seminário Internacional de Alta Tecnologia, 2018, Piracicaba. Desenvolvimento de produtos inteligentes: desafios e novos requisitos. Piracicaba: UNIMEP, vol. 1, pp. 50–70 (2018)

Slaper, T.F., Hall, T.J.: The triple bottom line: What is it and how does it work. Indiana Bus. Rev. **86**(1), 4–8 (2011)

Tan, A.R.: Service-oriented product development strategies. Technical University of Denmark, Thesis (2010)

Tan, A.R., et al.: Strategies for designing and developing services for manufacturing firms. CIRP J. Manuf. Sci. Technol. **3**(2), 90–97 (2010)

Tukker, A.: Product services for a resource-efficient and circular economy - a review. J. Clean. Product. **97**, 76–91 (2015). https://doi.org/10.1016/j.jclepro.2013.11.049

Tukker, A., Tischner, U.: Product-services as a research field: past, present and future. Reflections from a decade of research. J. Clean. Product. **14**(17), 1552–1556 (2006). https://doi.org/10.1016/j.jclepro.2006.01.022

Tukker, A.: Eight types of product–service system: eight ways to sustainability? Experiences from SusProNet. Bus. Strateg. Environ. **13**, 246–260 (2004)

Van Halen, C., Vezzoli, C., Wimmer, R.: Methodology for Product Service System Innovation: How to Develop Clean, Clever and Competitive Strategies in Companies (2005)

Vandermerwe, S., Rada, J.: Servitization of business: adding value by adding services. Eur. Manag. J. **6**(4), 314–324 (1988)

Vezzoli, C., et al.: Sistema produto+ serviço sustentável: fundamentos (2018)

Wagner, L., Baureis, D., Warschat, J.: Developing product-service systems with innofunc®. Int. J. Ind. Eng. Manag. **4**, 1–9 (2013)

Wang, P.P., et al.: Status review and research strategies on product-service systems. Int. J. Prod. Res. **49**(22), 6863–6883 (2011)

Chapter 10
Corporate Sustainability—Defining Business Strategies and Models from Circular Economy

Fabio Neves Puglieri and Diego Rodrigues Iritani

10.1 Introduction

Like all areas of knowledge, environmental management has undergone a process of evolution in recent decades. It is even possible to say that this process of birth of environmental concerns on the part of business until what we have today as the state-of-the-art in environmental management is relatively young when compared to other areas.

Making a brief history, several authors, including Hoffman (2000) and Barbieri (2016), point out that at a global level, until the beginning of the 1970s, organizations adopted a null posture in relation to the environmental issue. This meant that there were no concerns on the part of organizations regarding their environmental aspects and impacts.

This scenario began to change, even in the 1970s, initially boosted by changes in the political and academic environment, that is, discussions and decisions made by government rulers, researchers, and other members of society. Some of the main results of this were the creation of the United Nations Environment Program (UNEP) and the Meadows report, both in 1972. As a reflex, several countries have made different commitments regarding economic development and the environment. The consequence of these agreements, especially those signed at the United Nations Conference on the Human Environment in Stockholm, in 1972, was the appearance of the first environmental legislations and, in order to comply with the commitments,

F. N. Puglieri (✉)
Department of Production Engineering, Federal University of Technology - Paraná—UTFPR, Rua Dr. Washington Subtil Chueire, 330—Jardim Carvalho, Ponta Grossa, PR 84017-220, Brazil
e-mail: puglieri@utfpr.edu.br

D. R. Iritani
Upcycle, Rua Maria Lucinda, 141, apto 12—Vila Zanardi, Guarulhos 07090-160, Brazil

J. A. de Oliveira et al. (eds.), *Life Cycle Engineering and Management of Products*,
https://doi.org/10.1007/978-3-030-78044-9_10

countries started to set limits on emissions in the form of laws, with which have come to be complied especially by companies.

In order to comply with environmental legislation, companies have adopted the first environmental management practices called reactive or end-of-pipe practices. As their names suggest, those practices are based on companies reacting to the first environmental laws aiming at acting on the environmental impacts they caused, such as air and water pollution by the release of gases and effluents, respectively. The method of action has been based on the use of filters in chimneys to retain part of the particulate material that has been released into the atmosphere or even effluent treatment and dilution stations to ease the severity to the environment. As a result, the environmental benefit has been often small, while the company's operating costs have increased due to the need to treat polluting agents.

However, since the 1980s, in addition to the government with legislation becoming more rigorous, other actors have begun to press for the companies' good environmental performance, such as Non-Governmental Organizations (NGOs), communities, and customers. This has occurred, among other things, due to the greater diffusion of environmental issues in society and its consequences on people's lives.

Just as an illustrative example, it was in 1986 that the Chernobyl disaster occurred, which had a very negative impact on the media at that time, especially on society about environmental impacts related to nuclear energy. Adding to this the high costs of treatment of end-of-pipe practices adopted in the previous decade, some companies have decided to start the search for less costly and more environmentally efficient alternative practices. At that moment, they verified that a good alternative would be to improve activities that potentially cause environmental impacts, that is, in the environmental aspects of their processes, products, and services.

This has become known as preventive positioning with some examples of practices to be cited, the 3Rs (Reduction, Reuse, and Recycling) aligned with environmental management systems and also with other environmental programs, such as Pollution Prevention and Cleaner Production which can be found in Chap. 2: Cleaner Production (CP). The results have been improvements in production processes, product modifications, introduction of new manufacturing technologies, and organizational development, which provided not only the environmental improvement of operations, reduced operating costs, and greater productive efficiency, but also a better relationship with stakeholders.

More recently, since the 1990s, some companies have managed to go further and adopt practices in their businesses that allow the creation of value and competitive advantages,[1] configuring a new moment in the history of environmental management in the company, which is the strategic posture. As an example, it is possible to mention eco-innovation in products, processes, and business models.

The change in business posture when incorporating these issues into strategic decisions, aiming at creating long-term value and competitive advantages, also represented a change in relation to the business actors who were involved until then

[1] Advantage obtained by the company in relation to competitors, differentiating it. It is a result of a competitive strategy and is characterized by being difficult to imitate.

with environmental sustainability. What until then were attributions of engineers and managers whose activities involved, directly or indirectly, the organization's environmental performance, has come to be considered a fundamental element of the top management's decisions, that is, directors', presidents', and majority stockholders'. In other words, the business's environmental dimension (in addition to the social dimension, not discussed here) also started to guide decisions at the corporate level. These decision-making and corporate management approaches will be developed in this chapter called Corporate Sustainability.[2]

Corporate Sustainability Strategies are closely related to the Life Cycle Thinking (LCT) concept previously introduced in Chap. 1: Introduction to Life Cycle Engineering and Management (LCEM). This concept involves expanding the business's perception that its environmental impacts are not limited to occur only within the company's physical limits. That is, it is also responsibility of decision makers to consider the search for raw materials and components with better environmental performance, to minimize environmental impacts in the use and transportation phase, and to think about solutions related to the end-of-life of products. Environmental sustainability strategies based on life cycle thinking can generate sustainable business models, which are based on eco-innovation and circular economy. Eco-innovation deals with the development and application of a business model which arises from a business strategy and which incorporates sustainability in all its operations. Because it is based on life cycle thinking, it needs to involve all partners in the value chain, changing products, processes, market approaches, and the company's own organizational structure, increasing its performance and competitiveness. Circular economy is a concept that refers to the creation of a restorative and regenerative economy, offering opportunities for creativity, innovation, and thus, such as eco-innovation, fostering new business models.

The strategic definition for environmental sustainability occurs through a process known as strategic planning for sustainability. At this stage, it is common to find several tools that can be used to assist the decision maker. The results of this strategic plan feed another process, which is the development of business models.

However, even before starting the strategic planning for sustainability and the definition of business models, it is important to emphasize that it is difficult for a company to obtain competitive advantages if there is no change in the top management's mindset. Thus, considering all these three elements (change of mindset, strategic planning for sustainability, and definition of sustainable business models), the methodology for developing sustainability at the corporate level to be presented in this chapter is structured as in Fig. 10.1.

In other words, based on the life cycle thinking, the importance of promoting mindset changes in organizations' decision makers will first be discussed. Then, it is presented how to carry out a strategic planning for sustainability from a framework and supporting tools, with competitive strategies that generate value for the business as outputs. Finally, these sustainable strategies can lead to sustainable business models, which will be presented and discussed later in the chapter.

[2] The term sustainability in this chapter does not include the social dimension.

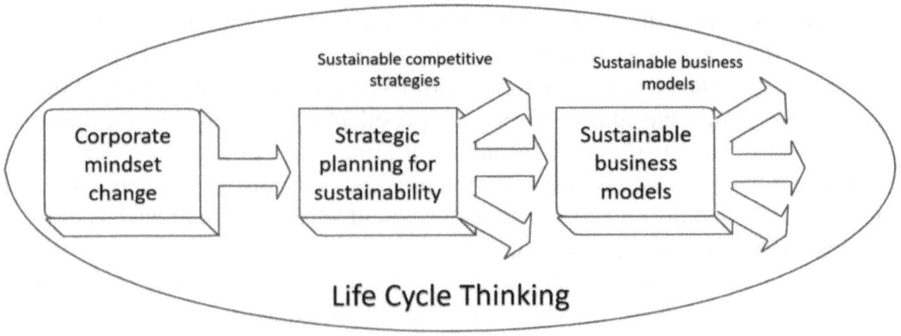

Fig. 10.1 Methodology for the development of corporate sustainability. *Source* Authors

Before starting the development of Fig. 10.1, it is important to note that currently many public and private organizations are adopting the 17 Sustainable Development Goals (SDGs), a global agenda (Agenda 2030) addressing the challenges of sustainable development, such as a corporate strategic positioning. In the case of the content developed in this chapter, as previously mentioned, social issues will not be addressed. However, many of these goals can be directly and indirectly met by circular business strategies and models such as industry, innovation, and infrastructure (SDG 9), responsible consumption and production (SDG 12), action against global climate change (SDG 13), and partnerships and means of implementation (SDG 17).

In other words, the methodology proposed here may allow the development of strategies and business models that operationalize several of the United Nations' Sustainable Development Goals. In addition, it presents a comprehensive perspective when considering opportunities throughout the organization's life cycle with a focus on generating competitive value for the business.

10.2 Corporate Mindset Change

Mindset can be understood as a person's attitude, habit, or particular way of thinking. In other words, strategic decision makers have a mindset which can influence the long-term strategies that will be defined for the business. These strategies, in turn, may have repercussions on the future of the business (for success or failure).

For more than fifty years, business strategies had been oriented to return value to shareholders,[3] being sources of growth for many organizations. However, the business environment has changed, and currently these practices are not seen as viable in the long term, as the commitment of organizations goes beyond their shareholders

[3] Those who can make a profit from the success of a business, such as a shareholder, for example.

and includes other stakeholders such as society and the environment itself. In other words, a mindset change becomes fundamental for doing business these days.

But changing the mindset is not a simple thing. As Grayson et al. (2008) affirm, it is a journey that involves a radical change in the top management's way of thinking regarding value creation. And this is where an interesting point is inserted: many decision makers do not see such a change and miss business opportunities. In other words, there is a need to see business opportunities and this must be done by changing the corporate mindset to sustainability.

For Kessel et al. (2017), a mindset for sustainability refers to a way of thinking in which the broad understanding of the ecosystem results in actions for the greater good of the whole. It is understandable and possible to affirm then that the business's focus becomes the stakeholders, expanding the vision of the organization beyond its physical boundaries and considering that the responsibility of its products/services and activities becomes the whole life cycle.

Some recent researches show a change in that mindset. According to the Massachusetts Institute of Technology (MIT), between 2010 and 2016 there was an increase of almost 400% in the perception of directors and executives on the degree of importance of corporate sustainability for the business. Other studies (UNRUH et al. 2016; Kiron et al. 2017) also reinforce this new behavior by showing that sustainability initiatives are not disconnected from growth and that are allowing several companies to generate long-term competitive advantages.

Summing up, it is an important input for sustainable strategic planning to change the mindset of decision makers regarding sustainability. Many profitable and competitive new business opportunities depend on the idealization and implementation of innovative business models, the success of which depends on strategic planning based on a life cycle thinking, which will possibly fail if there is no perception and commitment from the company's top management in face of sustainability.

10.3 Strategic Planning for Sustainability

According to the MIT's sustainability report with Boston Consulting Group (BCG) (2013), the companies that are leading the corporate sustainability movement have in common a fundamental element, which is a sustainability strategy. The same report points out that 90% of the companies interviewed consider a sustainability strategy important to remain competitive. Therefore, there is a need to plan these strategies.

Strategic planning can be defined, in general, as an administrative technique that establishes the direction of success for the entire organization, guiding its members in the desired future direction, taking advantage of opportunities, and avoiding threats. For this, the strategic planning literature presents a variety of models going from formalized frameworks to less structured models.

Some strategic planning models with a focus on sustainability, although few, are also presented in the literature. However, only one will be presented in this chapter along with some supporting tools. We then suggest to readers interested in knowing

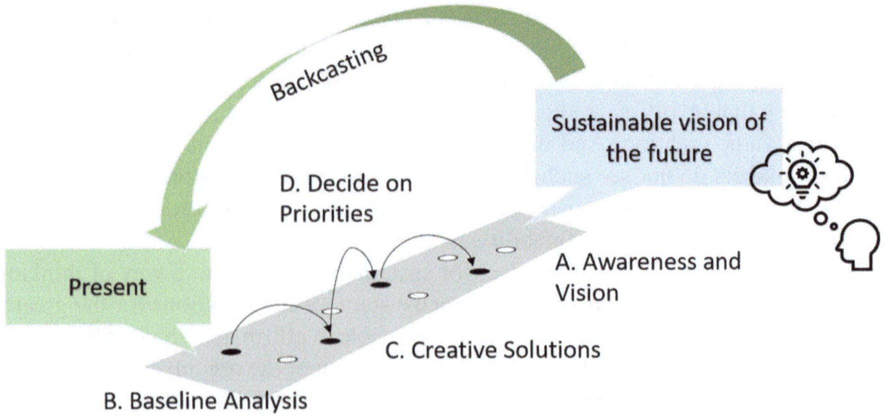

Fig. 10.2 ABCD framework. *Source* Adapted from The Natural Step (2019)

more about sustainable strategic planning models to look for more details in the research developed by Puglieri (2015), which is in the bibliographic references of this chapter, to find more details.

The strategic planning model for sustainability chosen to be presented here is the ABCD framework proposed by The Natural Step, which is illustrated in Fig. 10.2.

Firstly, it is important to highlight a concept that is the basis for The Natural Step's ABCD framework, which is Backcasting. Backcasting consists of defining a future vision of sustainability, both for society and for the company. According to Holmberg and Robèrt (2000), Backcasting is a method in which desired future conditions are visualized and steps are then defined to achieve those conditions. In other words, Backcasting is a strategic way of planning and it can be applied in the predicting of legal changes or in the market, finding business opportunities, and avoiding risks (Holmberg and Robèrt 2000).

Just to quickly exercise how Backcasting works, imagine yourself for a moment as a company's strategic decision maker. Now project yourself into the future, like 10 or 15 years, and try to visualize what the market, society, politics, and legislation will look like in that time frame. Ask yourself how sustainability will influence that future and try to imagine how your company could be succeeding in this environment and obtaining competitive advantages. Done! That's what Backcasting is all about.

Turning now to the framework, it starts at point A called Awareness and Visioning, which consists of aligning the company's thinking around an understanding of what sustainability is and questioning in which context of sustainability that the company is inserted. It is also here that a sustainable future vision for the organization is projected, based on the Backcasting which has already been explained.

Then it begins step B called Baseline Mapping. It is at this point that the company's activities are analyzed in relation to sustainability. In other words, products, services, energy, capital, and human resources are assessed from a life cycle perspective, allowing to identify critical issues, their implications for the business, and opportunities for change.

Step C is called Creative Solutions, which aims at developing potential strategies for the issues identified in the previous step and oriented towards the sustainable future vision defined in step A.

Finally, it is in step D, called Decide on Priorities, that strategic decision makers must prioritize strategies that maximize environmental, social, and economic returns. There is a recommendation for the company to make the choice, initially, for those strategies that are easier to implement and that bring a faster return on investment.

Although the framework makes no mention, we can cite several tools that can assist decision makers in this process. Three of these tools are PESTEL analysis, SWOT analysis, and Life Cycle Stakeholders, which can be used especially in step B of the framework.

The PESTEL analysis, which comes from the words Political, Economic, Social, Technological, Environmental, and Legal, is a tool that helps to identify current and future issues related to the organization's macro environment. When used for future forecasts, the PESTEL analysis can point out possible impact factors on the success of the organization's strategies.

In the political context, tariffs, restrictions, taxes, and even conflicts between nations must be observed due to sustainability. Some examples that can be mentioned are government incentives or political instability in regions that hold important inputs for the business.

On the economic side, opportunities for economic growth such as Gross Domestic Product (GDP) and inflation must be considered. In society, trends in social behavior changes must be analyzed. For example, generation Z and millennials care more about sustainability-related issues than older generations and this can be seen in the growing disinterest, in some countries, in the purchase of new cars, which has been replaced by business models focused on sharing and servitization, which will be discussed later. This can lead to major changes in the vehicle market in the future, changing the automakers' strategies and business models (see Ford's case-exercise later in this chapter).

In technologies, technological changes that may be more sustainable are analyzed, such as the impact of Industry 4.0, big data, internet of things, artificial intelligence, etc., directing new strategies and more environmentally sustainable business models.

In the environmental area, a possible scarcity of raw materials, the greater market demand for environmental labeling and the extinction of animal and plant species are examples of situations that may change the business environment in the future. Finally, new legislation may emerge prohibiting products and activities permitted today.

The SWOT analysis (Fig. 10.3) is another tool that is much widespread and that can help decision makers in the development of sustainable strategies.

SWOT analysis allows the identification of strengths, weaknesses, opportunities, and threats. When defining its sustainability strategy, the company must always maximize its internal competencies (strengths) that can allow the implementation of

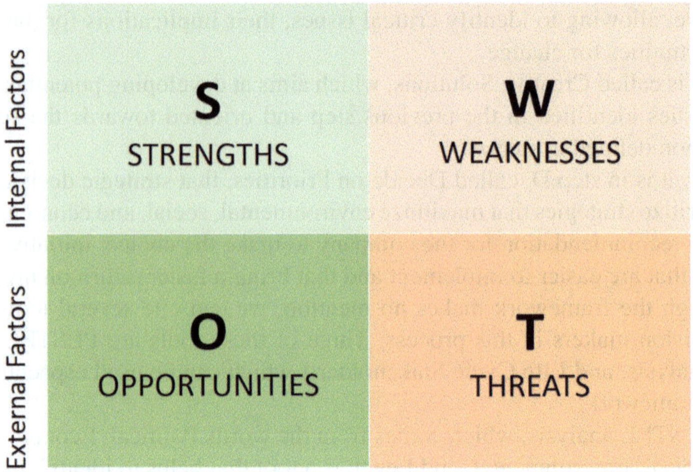

Fig. 10.3 SWOT analysis. *Source* Authors

sustainable competitive strategies while exploring possible external opportunities. At the same time, the organization must avoid its external threats and weaknesses.

Another strategic tool that can be indicated is Life Cycle Stakeholders (LCS) (UNEP 2017). It is a tool that allows the company to define a strategy based on the needs and expectations of the stakeholders that are throughout its life cycle. The general structure of the LCS can be seen in Fig. 10.4.

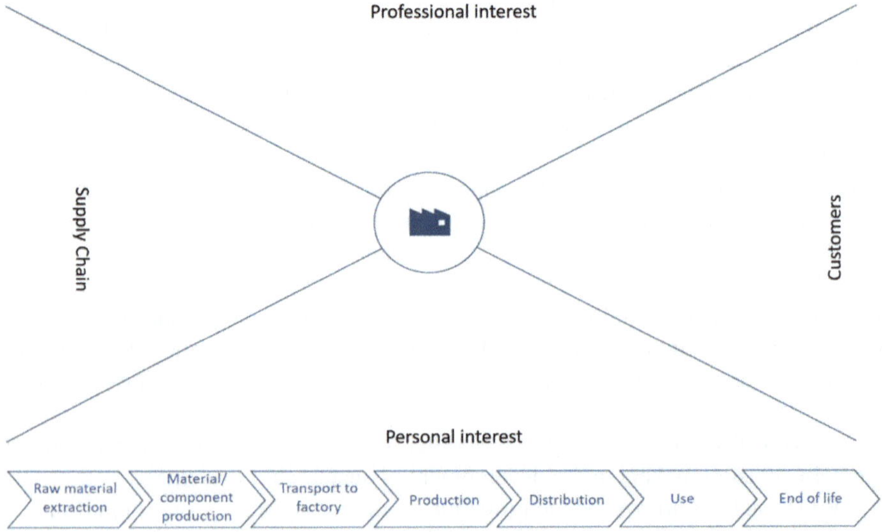

Fig. 10.4 Life cycle stakeholders. *Source* Adapted from UNEP (2017)

Some examples of stakeholders that can be classified throughout the life cycle are:

- Supply chain: raw material suppliers, component suppliers, logistics providers, and input suppliers;
- Consumers: logistics providers, customers, consumers, recyclers, salespeople, distributors, and retailers;
- Stakeholders with professional interest: government, stakeholders, employees, banks, insurance companies, trade associations, and universities;
- Stakeholders with personal interest: NGOs, community, and residents close to the company.

From the PESTEL and SWOT analyzes and the application of the LCS, the company's strategic decision makers will be able to gather important information, both current and future, regarding its life cycle. This information is basic input to guide, based on a strategic planning model (such as the ABCD framework), possible long-term competitive strategies for the business.

According to Porter (2016), professor at Harvard University, it is possible to identify three types of long-term generic competitive strategies: cost leadership, differentiation, and focus. The first is based on obtaining lower costs in the industry and thus generating greater profitability for the business by reducing costs and increasing production efficiency. Differentiation, on the other hand, focuses on creating a superior performance to competitors in order to create differentials for the consumer, such as technological innovation, brand image, quality of products/services, or attendance. Finally, the generic strategy of focus is based on serving a specific market, delivering something unique to their customers.

Each type of generic competitive strategy can be broken down into business models, which are discussed in the following section.

10.4 Elaboration of Business Models Based on Circular Economy Strategies

As seen in the previous section, strategic planning for sustainability will guide the organization to define its business models and then generate long-term competitive advantage. Strategies that can allow for cost reduction, differentiation, and focus are derived from the circular economy, whose business models will be detailed here.

It is important to highlight that in order to prevent business models in a circular economy from failing, we aim at avoiding the herd effect, in which a company "copies" the competitor's business model without analyzing the context, available resources, and the key factors for its implementation. In other words, there is no single circular economy solution that fits all situations. The choice and application of the solution in the business model depends on factors such as sector, value chain, available resources, location, and culture (National Zero Waste Council 2016).

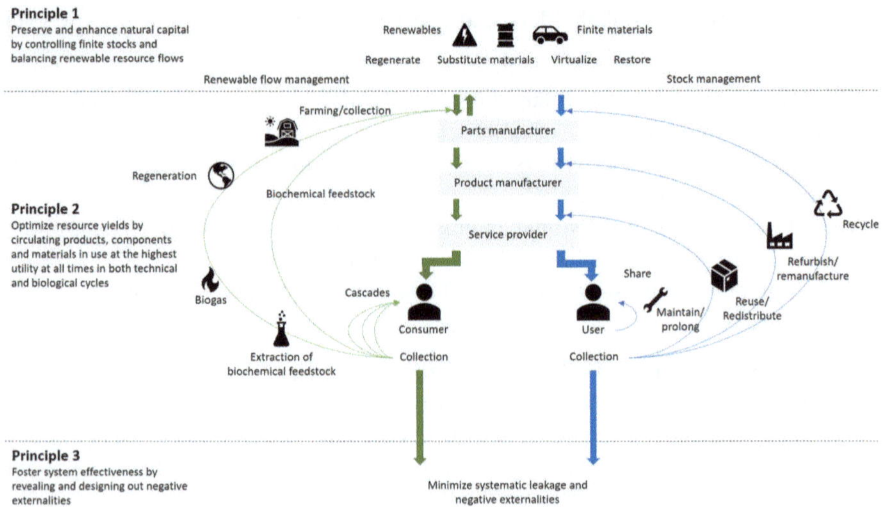

Fig. 10.5 Representation of flows in a circular economy and its principles. *Source* Adapted from Ellen MacArthur Foundation (2013)

As already mentioned, the circular economy has gained attention in recent years especially when it comes to the business model and it can be understood as intentionally created to be regenerative (Ellen MacArthur Foundation 2013). Figure 10.5 shows the diagram representing the circular economy and its principles.

As shown in Fig. 10.5, the circular economy has three principles: (1) intentionally eliminate and generate waste, impacts, and negative externalities in the product life cycle through the preservation of resources and replacement of non-renewable ones; (2) keep materials and products in use, which can occur through the technical (recycling, remanufacturing, reuse, sharing of assets and increasing lifetime) and biological (producing biogas from animal waste and manure from anaerobic digestion processes, for example) cycles; and finally (3) regenerating nature, minimizing negative externalities.

The shorter the reverse cycle, the lower the value lost. In other words, we mean that sharing, maintaining, and extending products are more effective options compared to recycling. However, the most appropriate choice for each case also depends on other factors, such as available technology, technical viability, and market interest.

Next, we present types of circular business models that can be used to put one or more of these principles of circular economy into practice.

10.5 The Five Types of Circular Business Models

Business models are fundamental to the circular economy since they represent the way the company does business, creates and delivers value to its customers and

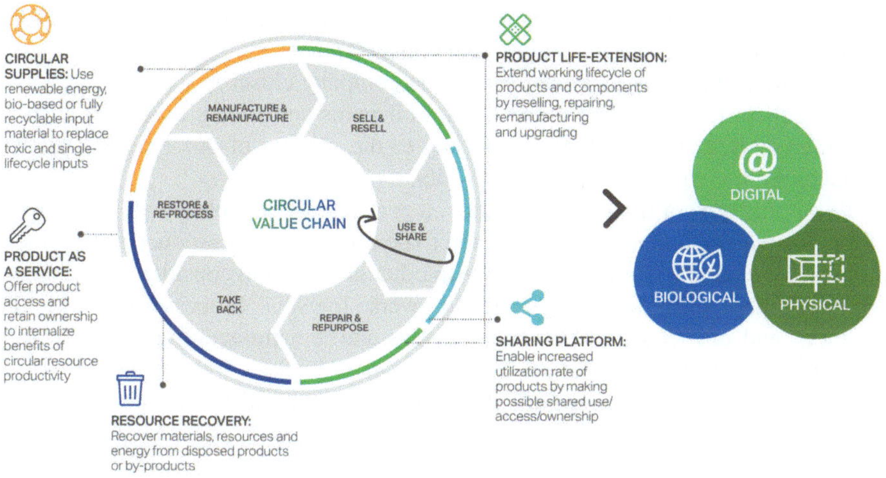

Fig. 10.6 Categories of circular business models. *Source* WBCSD (2017)

stakeholders. When a company has a circular business model, it means that it has integrated into its business some circular strategy that produces more value for a specific audience and for the environment.

According to Accenture (2015), business models related to circular economy strategies can be classified into five categories: product as a service, sharing, product lifetime extension, circular supply chain, and recovery and recycling (Fig. 10.6). It is worth noting that some categories adopt more than one option and in some cases, they are similar to other categories, as we will see below.

10.5.1 Product as a Service

This type of business model is a global trend, enough noting the amount of services available today that were previously offered through the purchase of products (e.g. Netflix, Uber, Spotify). In the product-as-a-service business model—also known as PSS (Product-Service System) model—the focus of the value offer is on function and services, user experience, and quality of attendance [see details on PSS in Chap. 9: Product-Service Systems (PSS)].

Durability, longevity, capacity for reuse and maintenance, product lifetime extension, and sharing become drivers for business success since they are directly associated with profitability and cost reduction. In addition, these characteristics also contribute to sustainable development in an effective way: more people are attended with the same amount of products and resources, besides products and parts that are remanufactured, reused, and reconditioned, known as end-of-life strategies and that can be consulted in the Chap. 12: Green Supply Chain Management in this book. In other words, it becomes interesting for the company that the product has an extended lifetime, leaving no room for programmed obsolescence.

In addition, this type of model brings the company closer to the customer/user, allowing them to understand more about their pains and needs. Processes such as customer service and maintenance are key to the success of this type of business model as well as to greater customer loyalty.

To put this type of model into practice, it is necessary to mobilize different areas of the company and many processes need to be reviewed. The product development process, for example, starts to act closer to other areas and to consider requirements such as durability, lifetime extension, and reverse logistics. Processes related to the product's end of life, such as reverse logistics and remanufacturing, are necessary since the product is an asset of the company.

We ask the reader to review the PSS chapter, which provides interesting examples of this circular business model.

10.5.2 Sharing

Sharing business models, as well as offering a product as a service, focus on product use and user experience rather than ownership. In this type of business model, the efficiency of the use of a product's resources is increased by intensifying the product use, that is, a single product serves more than one user. As in the product-as-a-service business model, there is a tendency for the same amount of resources to serve more people through intensification.

In this case, the value is offered through sharing, but the business model can be monetized or non-monetized. Below, two examples of Brazilian cases are presented: monetized and non-monetized.

10.5.2.1 Case Study: Tem Açúcar and Vamo Fortaleza

The *Tem Açúcar* application, launched in late 2014, registered 30 thousand users in just one month of operation. In 2017, 150 thousand users were sharing, free of charge, objects in more than 10 thousand neighborhoods in all states of Brazil. Some application statistics show that half of the requests for objects are answered, 25% in a period of just half an hour. This loan model helped users to save a total of USD1.45 million (Ometto et al. 2018).

The Vamo Fortaleza project presents a case of a monetized sharing business model. Launched in 2016, "the Vamo project (Portuguese for Alternative Mobility Vehicles) aims at promoting urban and sustainable mobility through an electric car-sharing network in the city of Fortaleza" (Ometto et al. 2018). Altogether, there are 10 stations in which the user can find the electric cars and request the service through an application.

10.5.3 Product lifetime extension

Business models that adopt the strategy of product lifetime extension are models that preserve the added value of part of the product, thus reducing the consumption of natural resources and the generation of waste without considering other impacts such as water footprint and carbon.

Some examples to be mentioned are remanufacturing and reconditioning, in which only a few defective parts are replaced with new ones or repaired. For this business model to be successful, the product needs to be designed for reconditioning and remanufacturing, that is, it needs to be easy to recover.

Although little explored by the companies, there is still a repair focused on improving the functions and specifications of the product such as, for example, upgrade of notebooks by replacing the processor. To work, design needs to project the product with the concept of modularity.

10.5.3.1 Case Study: EStoks and Fairphone

EStoks identified a business opportunity to extend the lifetime of defective electronics products before use in the Northeast of Brazil. In short, the business model is based on the collection of pre-consumer products from the retail chain or from partner producers (i.e. Philips, Britânia, Philco, Magazine Luiza, Cadence, Oster, and Arno) significantly reducing the costs of reverse logistics, since most producers are located in the South and Southeast regions.

After collection, the company applies its own algorithm to assess the status and quality of products and, thus, selects the best alternative to keep them at their highest level of utility and value, returning to the market. From the total of returned products, 55% undergo repair, 25% are repaired to be sold at eStoks' own stores at more affordable prices, and the remaining 20%, which cannot be recovered, are disassembled and their components reused.

Another example that combines product lifetime extension and resource recovery is the Fairphone business model. The value proposal is centered on offering a smartphone that unites technology and care for the planet: made with recycled plastic, the cell phone is designed to be repaired based on the concept of modularity (Ometto et al. 2018).

10.5.4 Circular Supply Chain

As we will see in Chap. 12: Green Supply Chain Management, in this type of business model the circular value is centered on the choice of raw materials that can be or have been restored, which includes recycled, renewable, refurbished, and remanufactured products.

A success factor for this type of business model is the use of pure inputs. This means avoiding mixtures of substances that cannot be easily separated at the product's or component's end of life. In addition, the use of toxic substances must be eliminated to avoid contamination on other parts and the potential impact on the environment. Finally, the use of renewable resources must be prioritized over the use of non-renewable resources.

10.5.4.1 Case Study: CBPak

Founded in 2002, CBPak offers an alternative to single-use packaging made of plastic or Styrofoam, offering 100% compostable packaging made with cassava starch. In addition to innovation in the use of natural and renewable materials, CBPak operates a service model where the company retains ownership of the packaging, taking care of the reverse logistics operation to ensure that it is directed to composting through commercial partners in locations close to the products use.

Besides consuming 62 times less water than a plastic cup, a cup produced by CBPak absorbs 3.74 g of greenhouse gases, while a plastic cup generates 16.69 g. In addition, the CBPak cup does not occupy space in landfills and allows soil regeneration—1 million composted cassava cups regenerate $100m^3$ of soil (Ometto et al. 2018).

10.5.5 Recovering and Recycling

The main goal of this business model is to recover value of resources through strategies such as recycling and cascading use in a closed or open cycle. In this model, reverse logistics processes and recycling partners are key to success. In addition, alternatives such as industrial symbiosis, in which the waste of one company becomes the raw material of another, and the cross-sectorial view can boost its application. In addition, in value chains involving final consumers, customers have a key role in returning products.

10.5.5.1 Case Study: Lojas Renner

Lojas Renner (Portuguese for Renner Stores) is working with Circular Economy concepts in its production chain and in its product development process. One of these concepts refers to the recovery of losses from the textile cutting process, which were previously destined for landfill or sold as a low added value product. For this to be possible, the company structured the reverse logistics cycle with its suppliers. As a result, in the first nine months of 2017 alone the company recovered 220 tons of waste. All this material allowed us to create hundreds of items of clothing collections for children, women, and men.

In addition to recycling textile cutting waste, *Lojas Renner* collaborates to develop process improvement projects and new technologies for its suppliers from the circular economy, thus also reducing losses in the early stages of their products' life cycle.

From this project, the company intends to expand its Circular Economy program increasing the recovery of resources, using design methodologies, and engaging more suppliers and customers to build a more responsible and circular custom (Ometto et al. 2018).

10.6 How to Put Circular Business Models into Practice?

Business models are fundamental to the circular economy since they represent the way the company does business, creates and delivers value to its customers and stakeholders. However, modifying the business model of established companies is a long-term task, so it is necessary to have strategic alignment. Philips Lighting, for example, has been building its circular business model for more than 5 years and this process was initiated from the corporation's strategy.

After defining a sustainable business strategy, the company must analyze its current model and then identify which are the business model options that have the greatest adherence to its strategy. The following are the main elements that must be considered, according to Wiithaa (2019):

- Function: What is the main function offered by the business model? What are the needs that the offer needs to meet? How to offer this function by integrating the principles of circular economy?
- Value Proposition: What problem will be solved? What value will the company offer to customers/users? What are the key characteristics? Is it in line with circularity?
- Users and contexts: For whom does the company create value? In what context can the company solve problems? Which stakeholders and in which contexts are they positively affected?
- Key activities: What are the key activities that create value for the business and stakeholders? What expertise does the company have available? Which ones does the company need to acquire?
- Partners: Who are the key partners and suppliers that help the company create value? Which ones could increase the business circularity?
- Natural resources: What natural resources are needed? Is circularity possible? Are they biodegradable and non-toxic?
- Technical resources: What technical resources are needed? How can the company apply circularity strategies to keep them in use?

- Energy sources: What energy resources are needed? Are the sources of fossil fuels or renewable?
- Distribution: How can the company improve its value offer? How does the sales process work? How is the product or service offered? How does the user perceive the value in business sustainability?
- Upcycling: What happens at the end of the product's lifetime? Can the components be reused, repaired, or recycled? Is it possible to reach zero waste? How? Are there partners that can support the company in this process? Is it possible to redesign the product to increase circularity?

10.7 Final Considerations

The definition of a sustainable strategy and then of the business models are elements that are closely linked and need to be aligned so that companies can actually obtain a competitive advantage and offer sustainable and profitable solutions.

We saw in this chapter that including sustainability in the business strategy directs the actions that a company takes, but the first step is related to the change of mindset, where sustainability must be seen as a driver of success and competitive advantage and no longer a cost or legal obligation. Then, companies must build the desired vision for the future and understand the current situation to then define the path and create solutions.

In this process, business models are essential for the implementation of sustainable business strategies as they are the way the company captures, creates, and delivers value. Business models oriented to circular economy offer interesting solutions to help organizations build businesses that promote more sustainable production and consumption.

10.8 Exercises

A. Practical application

(1) Make an analysis from the business model elements presented by Wiithaa on the change of business models in the case of Philips Lighting, which appears in the PSS chapter, which is migrating from a sales-based model to a service offer. In your opinion, what are the main changes?
Answer: by adopting a product-as-a-service model, Philips changed several elements if compared to the sales business model:

- Value proposition: offer goes beyond the product, including personalized attendance, solutions offers that meet what the customer needs, and more convenience and less energy consumption.

- User and context: users no longer have to worry about the maintenance and disposal of lamp waste, which are expensive and time-consuming. If you prefer, users can install more technology in the lighting systems.
- Upcycling: Philips becomes responsible for the end of life of products and the design of products with high technology, performance and durability are now factors of success for the business.
- Distribution: Philips is closer to consumers as there is a project team to serve the customer. In addition, Philips approaches during the product use phase, offering maintenance. This allows for a closer relationship, allowing the company to better understand the market's pains.
- Partners: reverse logistics, maintenance partners are key for Philips to be successful. For this reason, the company now has greater proximity and must develop better relations with these partners, especially those who come into direct contact with the user.
- Key activities: project management, R&D, product development, and customer service are more important activities than in the old model they are successful drivers.

(2) What is a mindset for sustainability?
Answer: Mindset for sustainability can be understood as an attitude, habit, or particular way of thinking of the corporation's top management, in which it is possible to see opportunities related to the generation of value from business practices related to environmental sustainability.

B. Reflection

(1) The companies that have been achieving success through sustainable business models and strategies are those that have already undergone a change of mindset, being part of this process the transfer of focus only on the stockholder to the stakeholders. How does this new focus relate to the company's life cycle? Which tool allows this association, identifying business opportunities from stakeholders in the life cycle?
Answer: when adopting a life cycle thinking, the company automatically sees that its responsibility goes beyond generating profit for its stockholders and that it must include the needs and expectations of other stakeholders, as the company's impacts are a reflection of what happens from the extraction of raw material to its suppliers, logistics operators, customers, retail, final consumer and end-of-life. One tool that allows such an analysis is Life Cycle Stakeholders.

(2) Imagine yourself and three other colleagues as members of the top management of the university where you study and that you are responsible for preparing the institution's strategic plan. For this, you decided to use The Natural Step's ABCD framework. Apply all the steps of the framework and then discuss the vision of future and actions you listed with the rest of the team.

Answer: as it depends on each situation and the students' imagination, there is no answer.

(3) What are the types of circular business models? How do you identify that they are in fact business models aligned with the circular economy? What factors must companies take into consideration when choosing which one(s) to adopt?
Answer: Altogether there are five types of circular business models: product as a service, sharing, product lifetime extension, circular supply chain, and recovery and recycling. Business models are aligned with the circular economy when they apply the principles of this approach, which are: intentionally eliminating through design the generation of waste, impacts, and negative externalities in the product life cycle, keeping materials and products in use, and regenerating nature. The factors that must determine which circular business model the company must build include business strategy, context in which it is inserted, value chain, location, available resources.

C. Case analysis

(1) Ford announced, on February 19, 2019, the closure of one of its factories in South America, more specifically in the city of São Bernardo do Campo, in São Paulo. Although newspapers and television reported this event as something exclusively related to the company's high operating costs in Brazil and the obsolescence of its factory, one of Ford's main reasons is its new competitive strategy and business models based on sustainability. One of these business models involves the digital offer of services for mobility, which in addition to the environmental benefits associated with the various principles of Circular Economy is expected to bring a financial return in the margin of 20–30%, against 10% of the sale of cars directly to the final consumer. As an action already taken, Ford acquired Chariot, a mobility services startup, and is planning to act soon on business models, which in addition to the aforementioned mobility service, includes transportation communication infrastructure services in cities and even sharing of bicycles (Johnson 2018). Discuss from the concepts developed in the chapter the entire process that could have taken Ford to these business models.
Answer: Ford, using a visionary and strategic planning process, could have identified opportunities related to the emergence of new technologies such as electric and autonomous vehicles combined with market demands (need to have the service, not the product; underuse of the good; difficulty parking, maintenance costs) to develop a differentiation strategy and, from there, get to the definition of circular business models such as PSS and sharing ahead of competitors and with great expectation of profitability.

References

Accenture: Circular advantage—innovative business models and technologies to create value in a World without limits to growth (2015)

Barbieri, J.C.: Gestão ambiental empresarial: conceitos, modelos e instrumentos. Saraiva, São Paulo (2016)

Ellen Macarthur Foundation: Towards the circular economy: economic and business rationale for an accelerated transition. Cowes: [s.n.], vol. 1 (2013)

Grayson, D., Jin, Z., Lemon, M., Rodriguez, M.A., Slaughter, S., Tay, S.: A new mindset for corporate sustainability (2008)

Hoffman, A.J.: Competitive environmental strategy: a guide to the changing business landscape. Island Press (2000)

Holmberg, J., Robèrt, K.-H.: Backcasting from non-overlaping sustainability principles—a framework for strategic planning. Int. J. Sustain. Dev. World **7** (2000)

Johnson, M.W.: How ford is thinking about the future. Harvard Bus. Rev. (2018)

Kessel, K., Rimanoczy, I., Mitchell, S.F.: The sustainability mindset: connecting being, thinking, and doing in management education. Acad. Manag. Proc. **1**, 2017 (2016)

Kiron, D., Unruh, G., Kruschwitz, N., Reeves, M., Rubel, H., Felde, A.M.Z.: Sustainability at a crossroads: progress toward our common future in uncertain times. MIT Sloan Manag. Rev. (2017)

MIT: The innovation bottom line—how companies that see sustainability as both a necessity and an opportunity, and change their business models in response, are finding success. Research Report (2013)

National Zero Waste Council: Circular economy business toolkit—steps to starting your circular journey (2016)

Ometto, A.R., Amaral, W., Iritani, D.R. (org): Economia circular: oportunidades e desafios para a indústria brasileira. Confederação Nacional da Indústria. CNI, Brasília (2018)

Porter, M.: Competitive strategy: techniques for analyzing industries and competitors. Free Press (2016)

Puglieri, F.N.: Proposta de um modelo de planejamento estratégico ambiental com visão de ciclo de vida. 2015. 171f. Tese (Doutorado) – Departamento de Engenharia de Produção, Escola de Engenharia de São Carlos, Universidade de São Paulo, São Carlos (2015)

The Natural Step: Disponível em: https://thenaturalstep.org/approach/ (2014). Acesso em: 15 set. 2019

WBCSD: The CEO guide to the circular economy (2017)

Wiithaa: Circulab board—the tool to ecodesign the business models. Disponível em: https://circulab.eu/tools/ (2019)

United Nations Environment Programme (UNEP): Eco-i manual: eco-innovation implementation process (2017)

Unruh, G., Kiron, D., Kruschwitz, N., Reeves, M., Rubel, H., Felde, A.M.Z.: Investing for a sustainable future—investors care more about sustainability than many executives belive. MIT SMR Home (2016)

Chapter 11
Environmental Management Systems and Performance Measurement

Camila Fabrício Poltronieri, Luciana Rosa Leite, and Sabrina Rodrigues Sousa

11.1 Introduction

Management systems help organizations to manage the various interrelated areas of their business to achieve their goals (ISO, 2015). They emerged as a way to assist in the continuous improvement of organizations, collaborating with the formation of a structure that contributes to the management of a specific area.

One of the largest institutions responsible for publishing technical standards relating to management systems is the International Organization for Standardization—ISO. ISO is a non-governmental organization with a global scope, founded in 1947, with headquarters in Geneva, Switzerland. Consisting of members representing 162 countries, all with equal voting rights, it aims, as explained by Moreira (2006), the standardization of methods, measures, materials, and their uses in all sectors with the exception of the area of electrotechnology, which is standardized by the International Electrotechnical Commission—IEC.

In order for management systems to fulfill their role of promoting continuous improvement, it is necessary that indicators are adopted to assess the evolution of these systems. However, the need to measure performance is very old. During the industrial revolution phase, companies lived in an era of strong industrialization in

C. F. Poltronieri (✉)
Faculty of Science and Technology-FCT, Production Engineering, Federal University of Goiás-UFG, Campus Aparecida-R. Mucuri, s/n, Aparecida de Goiânia, Goiás 74.968-755, Brazil
e-mail: camilafabricio@ufg.br

L. R. Leite
Center for Technological Sciences—CCT, Department of Production and Systems Engineering, State University of Santa Catarina—UDESC, Campus Joinville—Address R. Paulo Malschitzki, 200—Distrito Industrial, Joinville, Santa Catarina 89219-710, Brazil

S. R. Sousa
Federal Institute of Education, Science and Technology of Rio Grande do Sul, Campus Viamão—Av. Senador Salgado Filho, 7000, Viamão, Rio Grande do Sul 94440-000, Brazil

J. A. de Oliveira et al. (eds.), *Life Cycle Engineering and Management of Products*,
https://doi.org/10.1007/978-3-030-78044-9_11

which success was determined by the way in which they took advantage of the benefits of scale and mass production economies. At that time, the focus was on measures for financial performance and productivity. Over time, new needs have emerged in line with changes in the economy. As a result, new dimensions of performance such as quality, time, flexibility, and customer satisfaction have emerged.

Around 1960, it was noticed that it was also necessary to measure social aspects and, from that, social indicators related to health, education, equity, among others, appeared. In 1972, the pressure exerted by organizations on the environment, which had already gained particular notoriety in the last two decades, was discussed at the first United Nations Conference—UN in Stockholm, Sweden, making the need for environmental adequacy emerge. Thus, along with the emergence of new and more restrictive legal requirements and technologies, management and performance measurement instruments have been developed, essential for understanding and assessing the environmental area as they provide relevant information that guides decision-making processes.

Thus, in the following sections, the definitions of environmental management systems are explained in more detail and how ISO 14001 standard guides its structuring as well as performance indicators and the way recommended by ISO 14031 for the environmental performance assessment to be driven.

11.2 Environmental Management Systems (EMS)

The first model of Environmental Management System—SGA applicable to all business segments is told to be proposed, in 1991, by the International Chamber of Commerce—ICC, a non-governmental organization structured to help companies around the world to operate responsibly. According to Barbieri (2011), the EMS proposed by the ICC has arisen in response to concerns about the effect of environmental issues on the competitiveness of companies in the international market and consists of an articulated set of administrative processes (planning, organization, implementation, and control) integrated with global business management through an environmental policy formulated by the company itself and consistent with its global policy.

In 1992, the British Standards Institution—BSI published BS 7750, considered the first technical standard on EMS, whose structure organized the requirements based on the PDCA Cycle (Plan—Do—Check—Act), proposed by William Deming in the context of continuous improvement of organizational processes. Both the ICC and BSI models were considered generic, that is, applicable to all types and sizes of organization.

The European Commission - EC, in 1993, published Regulation EC 1836/93—EMAS, which comprises an environmental policy tool conceived in a phase of realization of the community goal for sustainable development. The Eco-Management and Audit Scheme - EMAS was initially applicable voluntarily to companies in the industrial sector, but with the 2001 revision (EC Regulation N°. 761—"EMAS II") there was an extension to all economic activities, including local authorities. From this review, as well as in the subsequent ones (EC Regulation N°. 1221/2009—"EMAS III" and EC Regulation N°. 1505/2017), they started to have an alignment with the ISO 14001 standard, incorporating their changes.

Using these previous experiences and fearing the indiscriminate proliferation of standards and mechanisms for environmental certification, in 1993, ISO created the Technical Committee 207 - TC, to which the specific responsibility of coordinating the elaboration of technical standards with an environmental nature was delegated. The activities of ISO/TC 207 are divided among seven different subcommittees - SC and work groups -WG with the theme "environmental management systems" being attributed to SC1, with the participation of 56 countries.

In the context of ISO/TC 207, Brazil was represented by the Support Group for Environmental Standardization—GANA, a special division of ABNT - Brazilian Association of Technical Standards, created in 1994, responsible for organizing environmental management standards and translating into Portuguese standards published by ISO/TC 207. ABNT also follows the work structure in ISO technical committees and, in 1999, GANA was renamed ABNT/CB-38 (CB—Brazilian Committee).

The work developed by ISO/TC 207 resulted in the series of ISO 14000 standards, which, according to Alberton and Costa (2007), serve as guidance to companies for the insertion of the environmental variable in the business management system, incorporating it into the policy, strategic formulations, objectives, and targets, technological options, and operational routine.

According to Sousa et al. (2010), the standards of the ISO 14000 series can be organized in two groups, taking into account the objective of each one, one group dedicated to establishing guidelines for the assessment of environmental considerations in organizations and the other aimed at the assessment of products (goods and services).

Examples include the ISO 14001 standard, which provides guidance on the EMS, and the ISO 14030 subseries, which deals with the measurement of environmental performance, both of which are dedicated to the organization's processes; ISO 14040 subseries, in turn, described in Chap. 3: Life Cycle Assessment (LCA)—Definition of Goals and Scope and ISO 14020 that we will see in Chap. 13: Communication and Environmental Labeling, which address, respectively, the life cycle assessment and environmental labeling, establish guidelines for the assessment of products.

11.3 ISO 14001

In 1996, the first version of the ISO 14001 technical standard was published, which provides specific requirements for the implementation of Environmental Management Systems—EMS. It underwent a review in 2004 and, in 2015, its third edition was published, this being the current version on which this chapter addresses.

Because it is considered generic, that is, a standard applicable to all types and sizes of organizations, its requirements are not absolute. Thus, each organization uses its own solutions that are compatible with the available resources, whether financial, human, among others, to meet the requirements of the standard. For this reason, it is sometimes observed the case of companies that produce the same type of product, they are also certified but have different levels of environmental performance.

ISO 14001 is originally published in English and must be referenced using the ISO 14001:2015 nomenclature, with 2015 referring to the year of the version's publication. When using the Portuguese version translated by ABNT, the correct way to refer to it is ABNT NBR ISO 14001:2015, where NBR is the acronym for Brazilian Standard. It is worth noting, however, that the version's publication in both languages is not simultaneous.

The ISO 14001:2015 standard is structured as follows: scope (clause 1), normative references (clause 2), terms and definitions (clause 3), organization's context (clause 4), leadership (clause 5), planning (clause 6), support (clause 7), operation (clause 8), performance assessment (clause 9) and improvement (clause 10). Clauses 1 to 3 refer to the text of the standard, while clauses 4–10 define the requirements to be implemented to constitute the EMS. The EMS's structure and the interrelation between the clauses and the PDCA cycle are illustrated in Fig. 11.1.

Clause 1 defines the *scope* of the standard. It must be noted that the main objective of applying this standard is to improve the environmental performance of the organization, making it clear that, for this, compliance with all legal requirements that apply to it is fundamental. Depending on its objectives regarding the EMS, an organization may partially use this standard, however, certification can only be claimed if it fully meets all the requirements applicable to its processes.

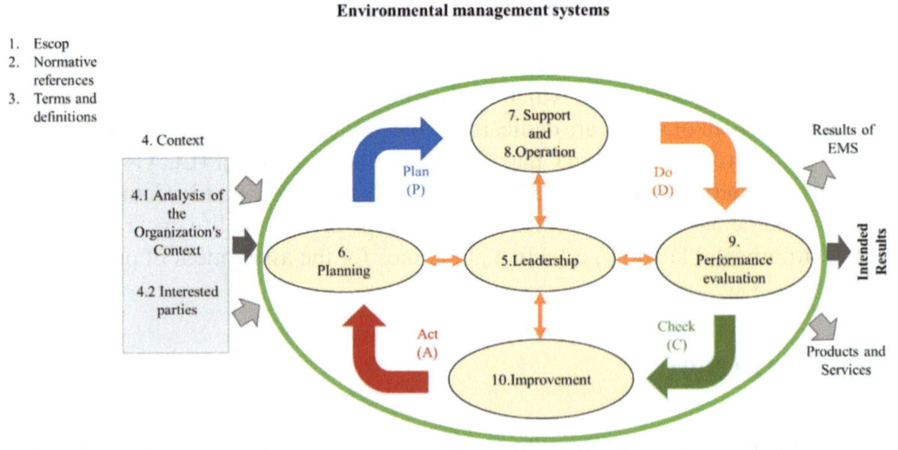

Fig. 11.1 Structure of the management system based on ISO 14001. *Source* Adapted from ISO 14001 (ABNT, 2015)

Clause 2 concerns *normative references*. This field is used by ISO to mention other standards that must be used together in order to achieve the objective of the EMS. In the case of ISO 14001:2015 there is no normative reference or specific content in this clause, it is maintained only for the sake of organization of the standard.

Clause 3 is intended for *terms and definitions*, which are the keywords cited throughout the standard resembling a technical dictionary. This part is of great importance as it allows all organizations to clearly understand all the content present in the standard and this understanding to be the same for all of them.

Clause 4 addresses the *organization's context*. The organization needs to assess internal and external issues that may affect its ability to achieve environmental results as well as to raise and understand the needs of all stakeholders, whether internal or external. How the company will do this, the standard does not specify. One method that can be used is the SWOT analysis (Strengths—Weaknesses—Opportunities—Threats), but this is not the only way to do this survey. Benchmarking can also be used verifying in other companies how this process is done, promoting an adaptation to the company's reality.

Another point that this clause establishes is that the EMS's scope is defined and documented. An organization may decide to implement the EMS in all or part of its processes. For example, a paper and cellulose industry may have the EMS adopted for activities related to paper production, excluding those where cellulose production occurs and vice versa. In the case of a company with more than one brand, business unit, or facilities, not all of them will necessarily be part of the EMS as long as this scope definition is properly documented. In another example, we may use the case of a university that has several campuses. Rectory may decide to implement EMS on itself, on a single campus, on some or all of them.

Clause 5 presents the requirements related to *leadership*. The standard states that the leadership must demonstrate its commitment to the EMS. For this, it must provide the necessary resources for the system's implementation and maintenance, actively participate in the elaboration of the environmental policy, which consists of a declaration from the top management exposing its intentions and principles in relation to its environmental commitment, and take responsibility for the system's success. The company must keep its *environmental policy* updated and documented. The organization must ensure that everyone who works for the company is aware of this policy, even outsourced, temporary, and visiting employees. In addition, the standard establishes that roles and responsibilities are defined inside the organization so that the EMS requirements can be met.

Clause 6 sets out the *planning requirements*. According to this clause, it is necessary to analyze the risks and opportunities related to the organization's environmental aspects.[1] The standard establishes that, in addition to the survey, it is necessary to document the environmental aspects and impacts associated with each organization's activity included in the EMS scope as well as to define the criteria adopted for determining the significance of the environmental aspects and to document them in a

[1] Element of an organization's activities, products, or services that interacts or can interact with the environment (ABNT 2015).

way that you can easily know which ones are more potentially impactful. One of the reasons for determining the significant environmental aspects is for them to be treated as a priority by the organization.

By *environmental aspect* is understood every element of an organization's activities and products that can interact with the environment (ABNT, 2015). In turn, environmental impact represents any change in the environment's physical, chemical, and biological properties, adverse or beneficial, caused by any form of matter or energy resulting from human activities, that is, resulting from an environmental aspect which affects directly or indirectly population's health, safety and well-being; social and economic activities; the biota; the aesthetic and sanitary conditions of the environment; and the quality of environmental resources (ABNT, 2015; Brasil, 1986). For example, the activity of cargo transportation has as an environmental aspect the consumption of fossil fuel and one of the impacts that can originate from it is air pollution by increasing the concentration of polluting gases resulting from the combustion process such as carbon dioxide (CO_2), one of the main greenhouse gases, representing, in this case, a negative environmental impact. In the case of reforestation of an area, this environmental aspect (planting trees) will potentially lead to an improvement in the air quality in the surroundings due to the absorption of carbon dioxide (CO_2) by the photosynthesis process, thus constituting a positive environmental impact. Thus, it is said that the environmental aspects establish a cause and effect relationship with the impacts, which are said to be "potential", given that they will only occur depending on the control conditions established for the aspects.

The identification of environmental aspects must be followed by its analysis, which takes into consideration: (i) the condition and (ii) the temporality in which it occurs, (iii) its nature, and (iv) the organization's influence on it. The condition in which the environmental aspect occurs relates to the activity to which it is associated, which can be a normal activity, which occurs routinely in the organization; abnormal, which includes cleaning, maintenance, etc., which are foreseen but do not occur on a daily frequency; or emergency, when it is about an emergency situation or other occurrences. Temporality refers to the time of occurrence of the environmental aspect which, in general, is present; however, it can eventually be past when the activity originating from the aspect no longer exists, as in the case of environmental liabilities; or even future, if environmental aspects have already been identified although the activity has not started yet—for example, when surveys are used to obtain environmental licenses. In turn, nature can be beneficial when a given environmental aspect results in a positive impact, or adverse when the resulting impact is negative for the environment. Finally, the influence of the organization on the environmental aspect, which can be direct when the original activity is carried out by employees of the organization itself or inside its facilities; or indirect when the activity is carried out by professionals or in outsourced facilities.

After identifying and analyzing environmental aspects, it is necessary for the organization to point out the elements of the environment that can be affected if the environmental aspect results in damage (impact) such as, for example, water, soil, air, natural resources, health and safety of employees, etc. Then, the *associated potential impact significance assessment* is determined, being it a product of the S*everity* (S) of the potential impact, the P*ossibility* (P) of the impact occurrence and the D*etection* (D) forms adopted by the organization in case of occurrence. For each of the items in the potential impact significance assessment, a score (numerical value) is assigned, being considered *significant*, for example, that impact with a significance factor (SF) greater than or equal to 15—see Eq. 11.1.

$$\text{SF} = S \cdot P \cdot D \tag{11.1}$$

where SF $\geq$ 15 means a significant impact. It is recommended that the significance assessment uses the following criteria for grading:

- SEVERITY (S): (1) *Low*—when on soil, water, or air, the impact is easily contained at the place of occurrence; the reaction product with the environment is negligible; there are no legal restrictions or good practices; when it involves only the consumption of a small amount of renewable natural resources; and it does not affect health and safety of the organization's employees; (2) *Moderately low*—when on soil, water, or air, the impact is contained in the place of occurrence; it causes the contamination of these elements of the environment in small proportions, not being noticed in adjacent areas; there are no legal requirements or good practices; and it involves the consumption of a large amount of renewable natural resources; (3) *Moderate*—when on soil, water, or air, the impact caused reaches areas adjacent to the place of occurrence; it causes the contamination of these elements of the environment without being perceived outside the organization; and it fails to comply with the organization's good practices; when it involves the consumption of a small amount of non-renewable natural resources; and it affects health and safety of employees but can be controlled through the use of personal protective equipment - PPE; (4) *Moderately high*—when on soil, water, or air, the impact caused is restricted to the organization's environment; it causes contamination of these elements of the environment; and does not meet legal or organizational requirements, or when there is a possible loss of the organization's reputation with its employees and the external community; and when it involves the consumption of significant non-renewable natural resources; (5) *High*—when on soil, water, or air, the impact caused reaches the organization's external environment; it causes contamination of these elements of the environment; and does not meet legal or organizational requirements; when it involves consuming a large amount of non-renewable natural resources; and it affects health and safety of the external community.
- POSSIBILITY (P): (1) *Low*—when the impact is very unlikely to occur; when it happens in negligible amount; and it is easily controlled at the place of occurrence; (2) *Moderately low*—when the impact can occur during the execution of an activity in negligible amount; when it eventually occurs but in small quantities; and it is easy to control at the place of occurrence; (3) *Moderate*—when the impact is expected to occur or when it occurs during the execution of a small quantity activity; and there are no forms of control in place; (4) *Moderately high*—when the impact is expected to occur or when it occurs during the execution of an

activity in a significant amount; and there are no forms of control in place; (5) *High*—when the impact is expected to occur during the execution of an activity in a significant amount; and there are no forms of control in place.

- DETECTION (D): (1) *Easy*—when there are automatic devices in place; or the occurrence of the impact is easily detected during the execution of the activity; (2) *Moderate*—when detection is limited to inspection in place with recognition for visual aspect, odor, noise, etc.; (3) *Hard*—when the detection is limited to the use of auxiliary equipment not available at the place where the activity is carried out.

Also according to clause 6, the organization needs to establish environmental objectives, define how to achieve them and measure them. One possible way of meeting this requirement is to use the 5W2H (*What—When—Where—Why—Who—How—How much*), through which it defines: **what** the organization aims at in environmental terms (objective); **when** it expects to reach the objective; in which activities, processes, or business units (**where**) the objective is expected to be met; **why**, that is, the justification for the establishment of a certain objective; determining the person(s) responsible (**who**) for monitoring and/or carrying out the necessary actions to achieve the objective; **how** the assessment will be carried out in order to find out if the objective has been fulfilled as planned; and **how much** is expected to be achieved from this objective (goal)—it can also be mentioned how much it will be necessary to invest in order to reach a certain objective. Objectives that involve or have an influence on compliance with legislation must be considered as priorities. Failure to meet an objective can be justified as long as the proofs which demonstrate the reasons why it has not been possible to comply with the requirements are duly documented.

Clause 7 presents the *support* requirements for the EMS's implementation and maintenance. The support provided by the standard involves: providing resources (from human and financial to infrastructure such as facilities, equipment, information technology, among others); ensuring that the team has or develops the skills necessary for the development of its tasks; ensuring that everyone is aware of the environmental policy and objectives as well as its importance within the EMS, in addition to certifying that communication (internal and external) occurs effectively. Another extremely important point addressed by the standard is the guarantee that the documented information is available to all those involved and that the documents are kept up to date.

Clause 8 contains the requirements regarding the *operation*. The standard states that the organization needs to define the relevant processes from an environmental point of view in order to be able to meet the requirements of the EMS. That done, it must establish operational criteria for these processes, control planned changes and critically analyze unplanned ones. The organization also needs to ensure that environmental requirements are considered during the product development process as well as when purchasing external materials and products. This clause establishes that the organization needs to be prepared to respond to potential emergency situations, thus establishing a contingency plan and training those involved through simulations previously scheduled or not.

Clause 9 refers to *performance assessment*. Therefore, the organization needs to define which parameters must be monitored, how and when this monitoring will be done, and which methods and equipment are involved in it. Further details on this clause are provided in Sects. 11.4. Performance measurement and 11.5. ISO 14031.

An important point also addressed in clause 9 is the conduct of *internal audits* to verify the compliance of the EMS, which are called internal (or first part) because they are conducted by qualified auditors directly related to the assessed organization or outsourced auditors who work in its name. Its purpose is to certify that the EMS implemented (or under implementation) in the organization meets all the requirements of the standard. Thus, the organization must prepare an audit program which includes since the selection of auditors, periodicity, method used, and form of disclosure of results.

In addition, the standard requires a *critical analysis* to be made by the organization's top management periodically, for example, annually. The critical analysis must take into account, for example, the results of internal audits, the assessment of compliance with laws and other requirements, the communication received from external parties, the results of meeting environmental objectives and targets, in addition to the status of corrective actions.

Clause 10, finally, deals with the opportunities for improvement to be explored by the EMS. During the audit process, eventually, non-conformities can be observed which are characterized by non-compliance (total or partial) with a requirement of the standard, a requirement established by an interested party or, even, by the organization itself. In this case, for each non-compliance found the organization must establish corrective action to eliminate its cause or prevent its recurrence. The corrective action plan must contain: a detailed description of the verified non-conformity; unit/process/activity where it was observed; the corrective action (or actions) to be taken (which includes immediate actions); schedule for correction; person(s) responsible for monitoring/implementing the plan.

Taking all this information as a basis, top management must propose and discuss opportunities for improvement which involve the adoption of new practices, technologies, and materials that can result in the improvement of environmental management inside the organization.

In addition to the ISO 14001 standard, which provides the requirements, organizations can also use the ISO 14004 standard, which complements the guidelines for the implementation, maintenance, and improvement of the EMS. The requirements of ISO 14001 can also be used to structure any EMS, regardless of certification. This situation is quite common among micro, small and medium-sized companies.

In this case, one option is to follow the instructions defined by the ISO 14005 standard, which guides the implementation of the EMS in phases. In general, it can be said that the implementation in phases works one environmental aspect at a time, especially the one considered as significant for the organization's context. Focusing on just one environmental aspect, for example, water consumption, the organization conducts all the requirements required by ISO 14001, establishing objectives and targets, forms of control, and verification.

At the end of this process (phase), the organization assesses the learning obtained and whether it has the necessary conditions to go with the implementation of ISO 14001 considering all its environmental aspects; otherwise, it can continue working on this same environmental aspect and/or choose another one. If you choose EMS certification according to ISO 14001, first, it must ensure that it fully complies with clauses 4 to 10, which refer to a set of management activities or processes.

Annually, the number of companies that have certified EMS is growing according to the requirements of ISO 14001, as shown in Fig. 11.2. According to ISO (2018), in 2017, this number reached 362,610 organizations with certified EMS, the highest registered so far.

However, the company cannot provide itself with the certificate that it follows the ISO 14001 standard, even though it has undergone an internal audit and it has been found that it follows all the proposed requirements. Therefore, it must look for an accredited certifier. In Brazil, the body that carries out the accreditation of certifying bodies is the National Institute of Metrology, Standardization and Industrial Quality—INMETRO, which is called an accrediting body. In addition to INMETRO, there is another body at the global level that carries out the accreditation of certifying companies, namely the International Accreditation Forum—IAF.

The certifying entity chosen by the organization must carry out external audits (also known as third party). The process is similar to that carried out by the internal audits, however, it is conducted by a team of fully independent auditors. Once the organization's EMS is verified, its certification is recommended and the certificate, which is valid for three years, is issued. During this period, after the external certification audit, and the internal audits that must be carried out regularly, external maintenance audits are carried out annually and, after this period, an external recertification audit.

All management system audits (internal or external) must follow the guidelines provided by the ISO 19011 standard, which is in its third version, launched in 2018

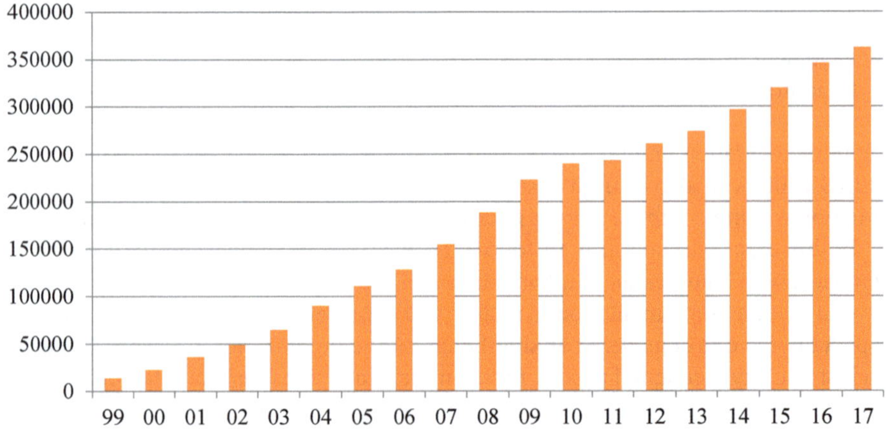

Fig. 11.2 Number of ISO 14001 certificates (world)/year. *Source* Based on ISO (2018)

still without a translation into Portuguese. A curiosity is that until 2002 EMS audits were guided by the ISO 14010 standard; however, it was canceled and its content has been unified with the ISO 10011 standard, which dealt with the quality management system audits (also canceled), originating the ISO 19011 standard. This fact is due to the understanding that the entire verification of compliance of management systems also takes place.

Although the ISO 14001 standard does not include specific requirements for quality, safety, and occupational health, it allows alignment with other management systems. In 2012, ISO proposed a high-level structure known as Annex SL, High Level Structure—HLS, which establishes an identical structure for all management system standards that arise or are revised after this date. This structure contributes to the integration of the different management systems and ISO 14001:2015 already follows this structure. According to Annex SL, the management system standards must be composed of ten previously defined clauses, the content of which is defined according to the standard's object.

Currently, there are several standards for the implementation of management systems, each focusing on a strategic area of the organization; some are generic, being applicable to any type and size of organization as is the case of ISO 9001 (Quality Management System), ISO 14001 (Environmental Management System), ISO 45001/OHSAS 18001 (Safety Management System and Occupational Health) and NBR 16001 (Social Responsibility Management System); others are aimed at specific sectors such as ISO/TS 16949 (Quality Management System for the automotive sector), ISO 27001 (Information Security Management System), ISO 50001 (Energy Management System) and ISO 22000 (Food Security Management System).

When an organization decides to implement two or more management systems, it can opt for integrated management. This is because, regardless of the focus, the management systems present their requirements organized in phases according to the PDCA cycle, that is, for planning the implementation, operationalization, verification and, finally, analysis and proposition of improvements. By integrating the systems, the organization can save resources and time since the integration generates synergy between management activities.

11.4 Performance Measurement

According to Lebas (1995), performance can be defined as the implementation and management of the components of a causal relationship model which leads to the achievement of previously defined objectives for a specific company and situation. Therefore, performance measures only make sense if they are defined from a cause-effect relationship, in which the elements of this relationship are those to be measured.

However, the task of choosing what to measure is not easy and must always be linked to a strategic orientation. We must have a view of the whole, avoiding indicators that reflect only some aspects or areas. This is not an easy task and requires the collaboration of a multidisciplinary team in order to aggregate different views. It is also necessary to be aware of internal and external-to-the-company issues. As an internal dimension, the need for measurement to be aligned with the organizational culture is highlighted, which makes the indicators easier to operate. Regarding the external dimension, two elements must be considered as a priority, customers, and competitors, as these are two important stakeholders for any organization.

The set of different performance indicators used in parallel gives rise to the Performance Measurement system—PM system. Interest in PM system has grown over the years since in order to survive in a constantly changing and fierce competitive world companies need to be prepared and one of the main ways to do this is by setting goals and monitoring indicators to verify if they are being achieved. This all contributes to decision-making as what is not measured is not managed. Therefore, measurement is essential to manage something and it can be created for different users and purposes. Some of the typical situations in which performance measurement can be used are: in monitoring management work, in helping to solve problems, in presenting reports to certain stakeholders, in addition to helping in the establishment of goals.

In Neely's (1998) definition, performance measurement is the process of quantifying an action taken. There are two important concepts when it comes to performance measurement: efficiency and effectiveness. Efficiency refers to the measure of how the resources available in the organization are used to achieve a certain level of customer satisfaction. Meanwhile, effectiveness is the measure of whether the customer's requirements are actually satisfied. Therefore, a performance measurement system allows decisions and actions to be taken based on information about past actions through the collection, examination, classification, analysis, interpretation, and dissemination of the appropriate data.

Franco-Santos et al. (2007) established the necessary elements for the existence of a management system, as shown in Fig. 11.3. The basic structure is defined as: system's characteristics, system's purposes and system's processes. The authors separated into each of these categories: the necessary conditions without which the measurement system does not exist, and the sufficient conditions which condition the measurement system.

Summing up, the use of performance measures or indicators as well as the supporting infrastructure for the collection and analysis of information (which may be the simplest such as a spreadsheet or more sophisticated such as computer programs).

The fundamental purpose of a performance measurement system is to measure performance. However, they can still play a set of other roles that are grouped into the following categories: strategy management, which comprises formulation, execution and focus; communication, which includes both internal and external, benchmarking and compliance with regulations; behavioral influence, which includes monitoring progress and behavioral rewarding/compensation; and learning and improvement inside the organization, which comprises the roles of feedback, double-loop learning, and performance improvement.

Regarding the processes that are essential for the performance measurement system, it can be divided into design and measurement selection, which comprises the process of identifying stakeholder's needs, the planning, definition of strategic objectives, selection and development of measures and targets; data collection and

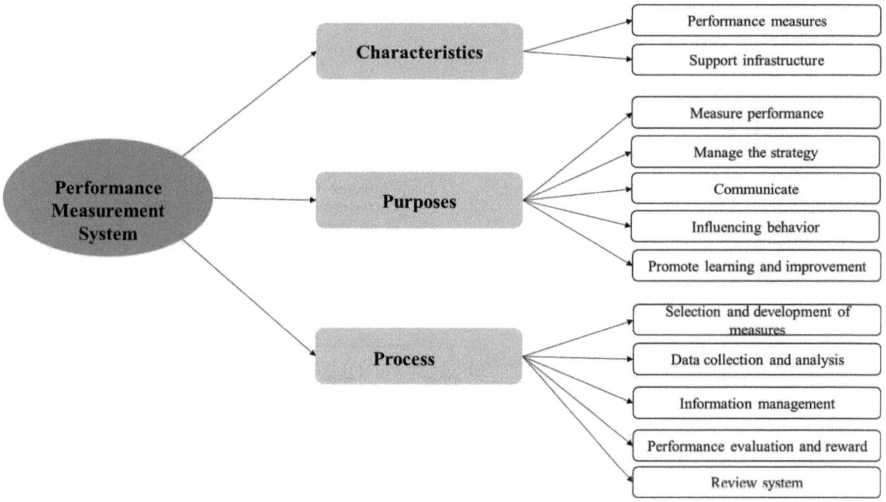

Fig. 11.3 Elements of a performance measurement system. *Source* Adapted from Franco-Santos et al. (2007)

maintenance, which includes data collection and analysis processes; and information management, which includes information management, interpretation, and decision-making processes, that is, it is the use made of the information available in the performance measurement system. Thus, we can define performance measurement system as a set of processes that an organization uses to manage the implementation of its strategy, communicate its position and progress and influence the behavior and actions of its employees.

Performance measurement systems—PM system can be divided into: *traditional* and *non-traditional*. Traditional measurement systems had been widely used until the 1980s. They encouraged a short-term view focusing only on achieving financial results. Even collected with a certain frequency (daily, weekly, monthly, etc.), the indicators provided a misleading view of production, since only the final results of operations were considered. As a result, information was made available late making it impossible to make a decision to improve the production process. For this reason, the purpose of these PM systems was only to monitor and not to promote continuous production improvement. These traditional systems have become inefficient for the reality that appeared in the late 1980s. The need to promote improvements in production processes and new philosophies and management methods (Total Quality Management, Lean Manufacturing, etc.) encouraged changes in models of performance measurement systems that have become more flexible, dynamic, and accessible to all employees of organizations. The production processes started to be monitored more closely, more frequently, making the information more real. Thus, the purpose of non-traditional PM systems is to promote the improvement of production processes and not just to monitor their performance.

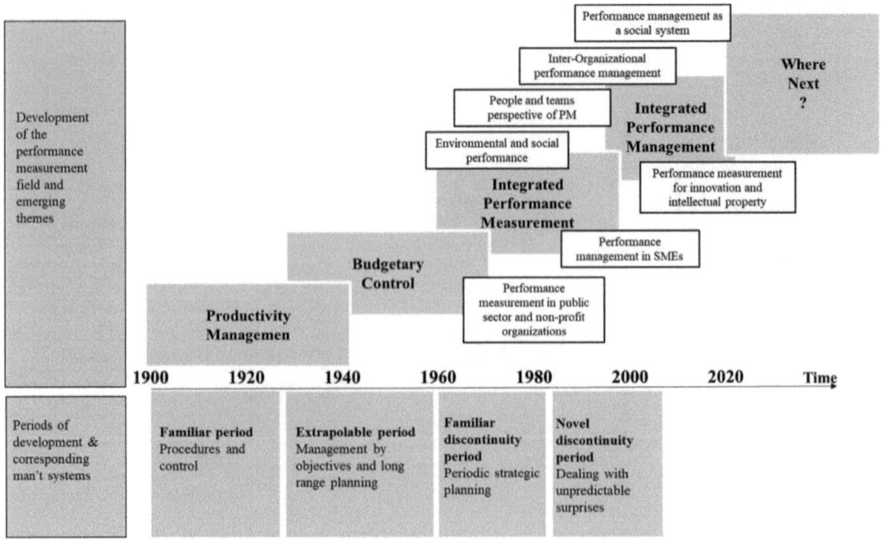

Fig. 11.4 Development of literature on performance measurement and global trends. *Source* Adapted from Bititci et al. (2012)

Through Fig. 11.4 it is possible to visualize the development of PM systems over the years and to verify the change in their intention. PM systems start meeting the demand caused by the increase in industrialization, around 1900, and the need to manage process productivity. With the emergence of more complex organizations, the focus of these systems continued on productivity but there was an increase in concern about budget control. In the following periods, with the fierce global competition and the sophistication of the markets, PM systems started to behave as an integrated performance management system, aggregating several purposes in a single management and measurement system.

Still in Fig. 11.4, it is possible to notice the emergence of important trends for the development of PM systems and especially how the environmental and social aspect has been incorporated into these systems.

Another important point is the need for periodic maintenance and review of measurement systems especially when organizations change their strategies and implement new technologies in such a way that the system always evolves in line with the changes undergone by the organization, fulfilling its role as a major contributor in the continuous improvement process. Without constant updating, it is possible that this system may conflict with the company's objectives and provide measures that do not represent reality. Figure 11.5 shows a model that describes the dynamics of the PM systems evolution process.

The process begins with the existence and use of a measurement system. From the use of this system and driven by external handles (for example, changes in legislation, pressure from stakeholders or the market, or even changes in the organization's

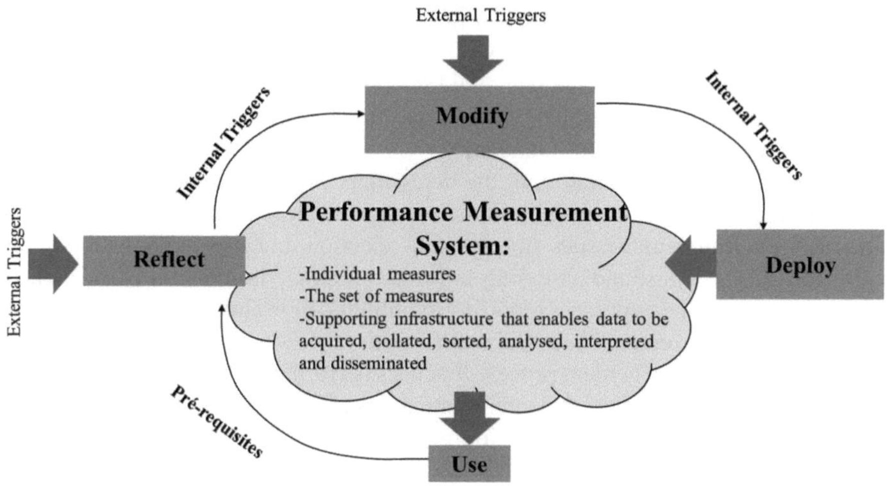

Fig. 11.5 Factors that affect the PM systems evolution. *Source* Adapted from Kennerley and Neely (2002)

ownership), there is a reflection on the existing system to identify failures and possible improvements. Then, the necessary modification is carried out to ensure the alignment of the system with the new contingencies. Finally, a modified performance measurement system is deployed so that it can be used in the organization's performance management. The internal handles (for example, the organization's culture) are responsible for shaping the necessary reflections and modifications.

11.5 ISO 14031

As previously seen, the indicators first emerged as a way of measuring economic growth, and only afterwards other aspects such as the environment were added. Furthermore, you can only manage and improve what you measure. What is not measured, there is no parameter to assess whether the actions taken were effective.

ISO 14031 is a standard that provides guidelines for Environmental Performance Assessment—EPA and is part of the ISO 14000 series of standards. This standard is not eligible for certification as it provides guidelines and not requirements. Although it can support the implementation and certification of ISO 14001, it is not mandatory that it has been implemented for a company to receive ISO 14001 certification.

This standard is in its second revision and its Portuguese version can be quoted as *ABNT NBR ISO 14031:2015*. The English version identical to the Portuguese version was published in 2013 and can be quoted as follows: *ISO 14031:2013*. The first version of this standard in English occurred in 1999 and in Portuguese in 2004.

ISO 14031 is organized into four clauses. The first three deal with Scope, Normative References, and Terms, and Definitions. The fourth clause is where the processes, indicators, and principles of the environmental performance assessment are presented. The standard also presents a series of supplementary guidelines in the Annexes. It is worth mentioning that, like ISO 14001, ISO 14031 is also in line with the PDCA learning cycle, which makes it easier to execute in parallel with other standards that follow this same standard.

It is defined by the standard that the assessment of environmental performance is a continuous process used to facilitate managerial decisions regarding an organization's environmental issues through the selection and use of indicators, data collection and analysis, and assessing information on environmental performance. They are also part of this process reporting, communication and, periodically, critical analysis, and improvement of the indicators used.

Thus, EPA is a management process that uses performance indicators to compare an organization's past and present environmental performance with its environmental objectives and targets. The information generated by the EPA can help organizations to: identify their environmental aspects and determine which ones must be prioritized, establish objectives and targets to improve environmental performance, identify opportunities for improving management of environmental aspects, identify trends in their performance, analyze critically and improve the efficiency and effectiveness of its operations, identify strategic opportunities, verify the regularization of environmental and other requirements that must be met by the organization in the environmental sphere, and report and communicate its environmental performance internally and externally.

According to ISO 14031 (ABNT, 2015), a performance indicator is defined as the measurable representation of the condition or state of an organization's operations. When it comes to the environmental approach, these indicators can be divided into two major groups: the Environmental Condition Indicators—ECI and the Environmental Performance Indicators—EPI.

ECIs are used to check the condition of the environment that can be impacted by an organization's action. This information is important to improve organizations' understanding of the actual or potential environmental impacts of the environmental aspects generated by their operation. It is not always easy to establish this direct connection between the ECI and the operation that generates a certain aspect. In fact, this only occurs when there is a single operation responsible for generating an specific impact. Thus, it is recommended by the standard attention in accounting of any other type of source or factors that may generate similar environmental impacts. For example, a certain indicator is used to monitor the condition of a pollutant emission. However, there may be other processes inside the organization that emit the same pollutant. Thus, it is difficult to make a direct connection with only one operational process inside the organization since the pollutant monitored has several sources of emission.

ECIs are commonly used by regulatory bodies and government agencies to provide an environmental performance baseline, monitor trends, establish permitted limits for some pollutants and create incentives to foster process improvement. As an example, in terms of air pollutants, we can cite the limit indexes of particulate material permitted by law according to Resolution n° 382 of the National Environment Council—CONAMA (Brasil, 2006). In this case, it is necessary to look for the environmental laws and resolutions of each country.

The second group of indicators defined by ISO 14031 are EPIs. These provide information on the management of significant environmental aspects of the organization and demonstrate the results of environmental management programs. This group is further divided into two types, the Management Performance Indicators—MPI and the Operational Performance Indicators—OPI. While the former provides information on management's commitment to influencing the organization's environmental performance, the latter report on the environmental performance of the organization's operations.

Within the context of ISO 14031, these indicators can be used to demonstrate organizations' adherence to the three pillars of sustainability (social, economic, and environmental). In fact, organizational management decisions and actions are directly linked to the performance of its operations. Thus, Fig. 11.6 presents an outline of how EPA can be put into operation by organizations.

In general, ISO 14031 suggests that the first step in implementing an EPA system is to plan its operation. It is noteworthy that EPA can be implemented in companies with or without an implemented and/or certified EMS. However, for the EPA to be effective, the organization must define its environmental policy, targets, objectives and the legal requirements that are applicable. Therefore, the most appropriate indicators must be selected to describe the organization's environmental performance based on its needs.

In selecting indicators, the standard recommends some specific uses for MPIs and OPIs. Within the first group, the information provided must be effective in helping organizations to predict performance changes, identify causes of unplanned performance (both positively and negatively), and identify preventive actions. In this sense, the MPI can be used to monitor the level of commitment to environmental management, the resources for implementing policies and programs linked to the organization's mission, compliance with legal requirements (or other requirements subscribed by the organization). Chart 11.1 shows some examples suggested by ISO 14031 of management performance indicators.

The second group of environmental performance indicators, the Operational Performance Indicators—OPI, are responsible for generating information regarding the environmental performance of the operations themselves. These indicators can be selected based on the inputs, operational processes and equipment, and the outputs generated by the organization's production processes, as shown in Chart 11.2.

Most of the examples provided are expressed in the form of direct measurements, events or numbers, to illustrate how organizations can assess the factors that are useful for monitoring them. However, the standard clarifies that organizations, their policies, objectives, and structures vary widely and a good system for assessing environmental performance must be developed considering these peculiarities.

Once selected, the environmental performance indicators are used and the information generated is processed: this is the second stage of the environmental performance assessment and comprises the processes of collecting data, making the data

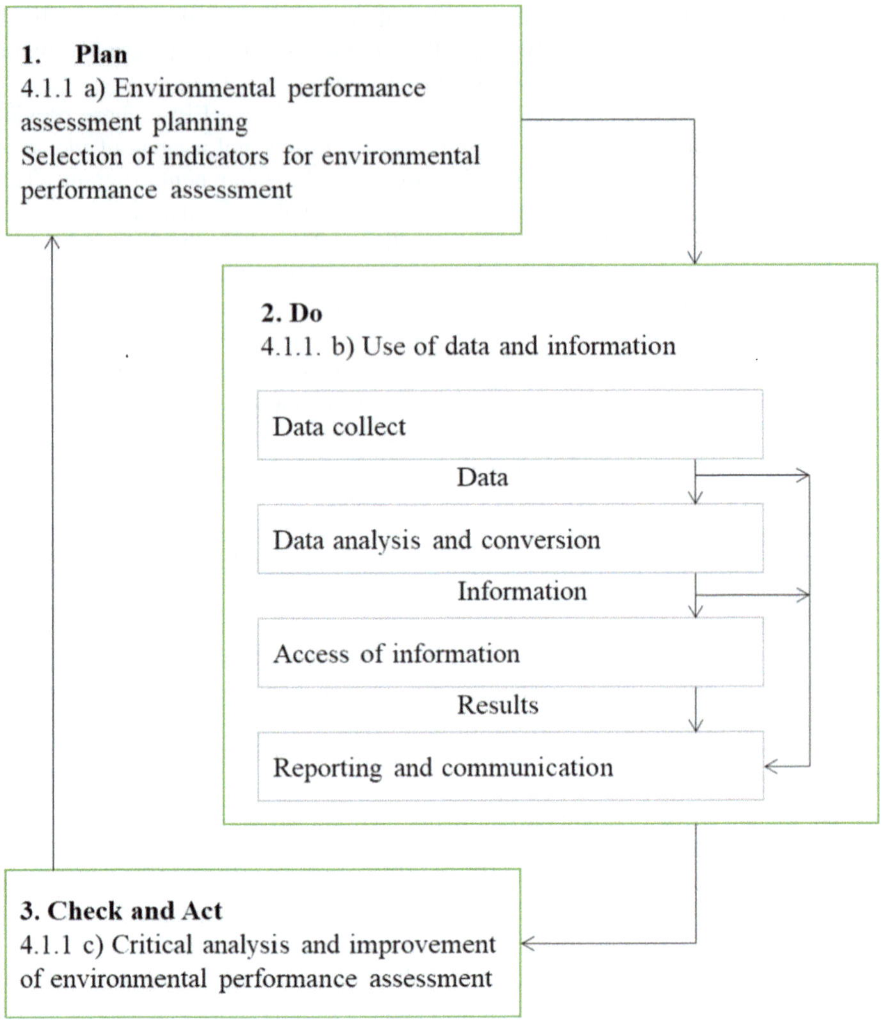

Fig. 11.6 Outline of the environmental performance assessment system following the PDCA standard. *Source* Adapted from ISO (2015)

analysis and conversion, critically assessing the information obtained, and reporting to communicate the results achieved internally and externally.

The source for data collection can be the EMS itself, when it exists, or other sources such as process records, interviews and observations, reports, customers, and other stakeholders, among others. The most important thing is that data collection is carried out in such a way as to guarantee the reliability of the data obtained as well as their validity. Thus, the ISO 14031 standard advises the establishment of a procedure that includes appropriate forms of data identification, archiving, storage, retrieval, and disposition. For the analysis and conversion of the collected data, the standard

Chart 11.1 Examples of management performance indicators - MPIs

Performance area	Indicator
Related to management policies and programs	Resources to implement environmental management policies and programs
	Success of pollution prevention initiatives
	Percentage of collaborators trained versus the percentage of collaborators who need to be trained in environmental management
	Results of research with collaborators on their knowledge of the organization's environmental issues
Related to regulatory compliance	Number and severity of compliance violations
	Time to respond to environmental incidents
	Number of audits
	Frequency of emergency preparedness and response drills
Financial performance related to environmental performance	Costs (operational and capital) associated with the environmental aspects of a product or process
	Return on investment for environmental improvement projects
	Savings obtained by reducing the use of resources, preventing pollution, or recycling waste
	Research and development funds applied to projects with environmental significance
Related to community	Number of external consultations or comments on issues related to the environment
	Resources applied to support environmental programs in the community;
	Approval indexes and surveys in the communities
	Number of press reports on the organization's environmental performance

Source Adapted from ABNT (2015)

recommends statistical tools to be used to increase the reliability of the decisions made and the guarantee of having objectives actually achieved.

In sequence, it is advisable to carry out a critical analysis on the information obtained by the EPA, comparing its results with the objectives and targets defined by the organization. In this process, the EPIs and also the ECIs are compared with the environmental performance objectives proposed by the organization at the beginning of the assessment period. The information from this comparison must be forwarded to the management in order to assist in the process of improving management actions.

Chart 11.2 Examples of operational performance indicators - OPIs - according to the category to be assessed

Category	Indicator
Materials	Material used per product unit
	Raw material reused in the production process
	Water used per product unit
Energy	Energy used by service or customer
	Energy generated by byproduct or process flows
Support services	Quantity of toxic material used by contracted service providers
	Type of waste generated by contracted service providers
Physical facilities and equipment	Percentage of equipment parts with parts designed for easy disassembly, recycling, and reuse
	Carbon dioxide equivalent per unit transported
	Percentage of vehicles in the fleet with technology to reduce pollution
Supply and Distribution	Number of deliveries of goods by means of transportation per unit of time
	Number of business trips for each mean of transportation
Products	Number of products that can be reused or recycled
	Quantity of resources consumed when using the product
	Product durability
Services provided by the organization	Resource consumption per unit of service provided
	Quantity of pollutant per service provided
Waste	Total waste discarded per category
	Quantity of waste converted to reusable material per unit
	Quantity of hazardous waste discarded due to pollution prevention programs
Atmospheric emissions	Specific emissions per year
	Wasted energy released into the atmosphere
Emission of effluents to soil or water	Specific material released per year
	Effluents per service unit or customer
Other emissions	Indicator of noise measured at the specified location
	Quantity of heat, vibration or light emitted per unit

Source Adapted from ABNT (2015)

The last element of the data and information use phase is reporting and dissemination. As a general guideline, ISO 14031 suggests useful information about the organization's environmental performance to be passed on to internal and external stakeholders. This communication must be incorporated into the organization's general communication plan and can bring benefits such as increased awareness and dialogue about environmental policies, environmental performance criteria, and achievements relevant to the organization; and also demonstrate the organization's commitment and effort to improve its environmental performance. More specifically, when it comes to internal reporting, it is important to make it clear that the necessary and appropriate information, which describes the organization's environmental performance, must be available to collaborators in a timely manner so that decisions can be made considering this data. This practice helps employees, suppliers, and other parties related to the organization, to fulfill their responsibilities and goals in the established environmental criteria. Externally, organizations are required to issue environmental reports on their actions, and having the information previously organized in the EPA can speed up this process. In addition, it is common for organizations to wish to voluntarily expose their environmental performance in reports to improve their position in business and their relationships with stakeholders, such as the community in which the organization operates its production processes. Thus, it is essential that this communication is reliable and represents, in fact, the organization's environmental performance in the monitored processes.

Finally, the last stage of the environmental performance assessment is the critical analysis and improvement of the process as a whole. The intention, in this case, is to identify opportunities for improvement and to contribute to the improvement both in management performance and in the organization's operations performance, resulting in the improvement of environmental conditions. This assessment must be periodic in such a way that its result is the improvement and continuous enhancement of the organization's EPA.

Thus, it stands out that EPA is a continuous process of collecting and assessing environmental data and information and aims at providing a current assessment of the organization's performance in this regard. ISO 14031 defines a generic model of how this assessment can be organized, however, it is not the objective of this standard to establish levels for the performance of the indicators used.

11.6 Final Considerations

The standardization of management systems can generate several benefits for the organizations that use them, such as the reduction of costs and waste, improvement of the image with stakeholders, access to new markets, greater guarantee of compliance with the applicable legal requirements, reduction of chances of divergences in production activities performance and, especially, the increase in customer satisfaction.

The incorporation of the environmental variable in organizations, in general, begins with the monitoring of environmental performance indicators, for example, water and/or energy consumption indicators. However, as the company acquires environmental maturity, there is a tendency to expand its management to a systemic process and, for that, technical standards can be used to structure an EMS. To increase its level of environmental maturity, an organization can, for example, adopt cleaner production programs or start an EMS in phases, as suggested by ISO.

Systemic management is important because it ensures that all the organization's environmental aspects are known, controlled and it is working to minimize and/or eliminate them, when possible. However, it must be kept in mind that there is no management without the control of indicators and this premise is valid even for the environmental area.

According to Höjer et al. (2008), the tools available for the study of environmental systems can be classified into: (i) *procedural*, as they are focused on procedures and connections with the social and decision-making contexts; or (ii) *analytical*, as they are directly associated with the technical aspects of the analysis and can be integrated into the structure of a procedural tool. Finnveden and Moberg (2005) add that the combination of tools is justified when there are differences between them related to some criteria such as, for example, degree of quantification, system limits and included impacts, etc., as they add complementary parts to the analysis results and, consequently, decision making. In this sense, EMS can be classified as a procedural instrument, while the PM system is among the analytical tools and can therefore be used in a combined way for better effectiveness of environmental analysis.

A premise of EMS that follows the ISO 14001 standard is the organization to be responsible for the environmental aspects associated with its activities and influence stakeholders to manage their own. This has been even more striking in the standard's 2015 version since it recommends considering the perspective of the life cycle of products delivered by the organization, which goes beyond the organization's physical boundaries. This prevents environmental loads from being transferred from one stage to another, or from one organization to another, as for example with the contracting of outsourced services without due consideration, which could illusively hide the related environmental aspects.

Thus, it is evident that according to ISO recommendations the tendency is for the scope of EMS to be expanded, including the various stages of the life cycle of the organization's products, making the assessment of environmental systems more complete and realistic. In this sense, Life Cycle Assessment—LCA appears as a complementary technique for this new way of managing organizational environmental performance which can be consulted in Chap. 3: Life Cycle Assessment (LCA)—Definition of Goals and Scope of this book.

11.7 Exercises

Question 1. ***Reflect on the following statement: "it is only when an organization has a certified EMS that it can obtain the expected benefits". Considering the above, is this statement correct?***

Feedback:

Answer suggestion: No. By following an EMS model, the organization promotes the environmental adequacy of its processes and activities and, thus, guarantees the benefits of this action, such as: reduction of costs and waste, improvement of the organization's image, access to new markets, greater guarantee of compliance with applicable legal requirements, reduced chance of divergences in the production activities performance, and, especially, increased customer satisfaction. However, when aiming at certification, the organization undergoes periodic external checks (audits) that verify the information provided and consolidate the benefits achieved.

Question 2. ***Do EMSs based on the ISO 14001 standard ensure that certified companies will act in an environmentally correct manner? Justify your answer. Present real examples to complement your answer.***

Feedback:

Answer suggestion: No. The certification guarantees the existence of an EMS but not that the company is acting in an environmentally correct manner. When an organization has an implemented EMS, it certifies that it knows, controls, works to reduce and eliminate its environmental aspects and impacts. This does not mean that it cannot cause negative environmental impacts or that it has environmental liabilities.

*There are cases of companies with EMS certified by the ISO 14001 standard and, even so, they were involved in environmental accidents. An example occurred with the mining company Samarco, operating in the Brazilian state of Minas Gerais. Even having certified EMS, and having adopted environmentally accepted solutions by the competent environmental agency for the management of its waste (namely, dams), the company, on November 5th, 2015, was responsible for one of the biggest environmental accidents in the history of Brazil. (suggestion for a story: https: //*www.istoedinheiro.com.br/blogs-e-colunas/post/20151109/caso-samarco-desmoronamento-responsabilidade-social-corporativa/7737*).*

Question 3. ***In your opinion, is ISO 14001 a product certificate? Which comes first: the implementation of the EMS or an environmental performance measurement system? Justify.***

Feedback:

Answer suggestion: ISO 14001 certification refers to the process and not to the product, that is, a certified company guarantees that its process is structured and ensuring the promotion of continuous improvement. In addition, the adoption of an environmental performance measurement system does not depend on an implemented EMS. When monitoring at least one environmental aspect through an indicator, the company measures its performance in relation to this environmental aspect even if it does not have an EMS. However, on the other hand, for the implementation of an EMS it is necessary that there are performance indicators that periodically monitor the organization's environmental aspects.

Case 1

A company in the paper and cellulose sector has a certified EMS based on the ISO 14001 standard and its environmental performance assessment system is based on the ISO 14031 guidelines. A theme that has been considered strategic for the organization is its role in the face of changes that, following scientific studies, are occurring on the planet. In this case, the main indicator monitored by the company is the emission of greenhouse gases—GHG, calculated based on the emission of kilograms of carbon dioxide equivalent (kg CO_2-eq). The company has signed a sector agreement in which it commits to reduce this indicator by 7% over the next 10 years. For this, an indicator of GHG emissions by the company's production process is measured daily. Another action of the organization has been to calculate the CO_2 emission for all its processes in addition to production and, thus, to make better decisions about the energy matrix used by the company as well as about logistical issues that involve the distance to its suppliers and the vehicles used for cargo transportation. Another point in which the company has invested is to encourage its employees to use means of transportation with low GHG emissions to make the journey from their homes to work.

The company's CEO says: "This is a bold goal that depends on the effort of all areas of the company, not just on production. From the input purchasing sector to the maintenance sector, everyone must think about how to reduce the global CO_2-eq emission indicator, otherwise we will not be able to reach our goal."

Given the efforts of this paper and cellulose company, indicate:

1. **What environmental condition indicator - ECI is being considered in this case?**
   The related ECI is that of greenhouse gas emissions, measured in kilograms of CO_2-equivalent which, in turn, contributes to impact indicators at the global level such as those that indicate climate change. This indicator shows the current situation of the environment at the moment and does not consider the company's performance.
2. **What environmental performance indicators - EPI are described in the presented case?**
   The EPIs described in the case are: CO_2-equivalent emission indicator from production processes, logistic indicators of distance of suppliers and type of cargo transportation, indicator that assesses the employees' means of transportation. All these indicators will later be converted into terms of CO_2-equivalent emissions and thus will contribute to the monitoring of this environmental performance indicator considering the company as a whole.

3. **What management performance indicators do you suggest to be monitored?**
A potential indicator is: percentage of collaborators who use alternative transportation (e.g. bicycle) versus collaborators who use traditional transportation (bus, car, motorcycle). Thus, it would be possible to assess how collaborators are perceiving their contribution to the goal of reducing CO_2-equivalent emissions proposed by the company. Another possibility is to monitor the company's energy matrix considering the contribution of renewable and non-renewable sources.
4. **What operational performance indicators do you suggest the company to monitor?**
Operational performance indicators are responsible for generating information on the environmental performance of the operations themselves. These indicators can be selected based on the inputs, operational processes and equipment, and the outputs generated by the organization's production processes. As an example, we can mention: the company's total energy consumption (MWh), energy consumption by productive/administrative processes and energy consumption per kilogram of product produced. It is still possible to consider other indicators that have an indirect impact on CO_2emissions, with the quantity of products produced, the way in which waste is recycled (for example, external recycling requires transportation and fuel expenditure), among others.

Case 2

In a small plastic packaging factory, Gustavo, the general manager, was visited by Guilherme, manager of a cosmetics company that is one of his main clients. The purpose of the visit was to address environmental issues. According to Guilherme, the company he manages is very concerned with the use of raw materials from non-renewable sources, with the energy consumption in the manufacturing stages for the production of the packaging and with the generated process leftovers (solid waste). He pointed out that, when approaching his customers, it was possible to notice a large increase on the part of them regarding environmental concerns. In addition, one of its main competitors recently implemented an EMS and received ISO 14001 certification. All of this has made Guilherme and his company start to move in a direction where their processes are environmentally better. However, Guilherme states that it is not enough just to be concerned with producing cosmetics with less environmental impact, but it is necessary that the entire supply chain also gets involved and works to minimize its environmental aspects and impacts. After this contact, afraid to lose the client, Gustavo decided to look for Luiza, the production manager to ask for help in face of this request from Guilherme. It is worth noting that the company in which Gustavo works neither has any environmental certification nor is so concerned with any environmental issue inherent in its production processes.

Based on the text, answer the questions:

1. **Considering the structure of the ISO 14001 standard, what would be the first step for the company where Gustavo works to start implementing a**

certified EMS? Would it be towards adopting an EMS or an PM system? Justify your answer.
According to ISO 14031, environmental assessment systems can exist without being associated with a certified EMS. However, for customer demand, certification would be the most relevant step. Thus, it can be suggested Gustavo to initiate this issue that is happening to the company's top management and then proceed with the application of the ISO 14001 standard. After complying with clauses 4 and 5, the company can move on to clause 6, which includes the survey of the environmental aspects and impacts of its production process to then define indicators for them. The ISO 14005 standard can be indicated for this beginning since it structures an EMS based on one or a few environmental aspects and then it expands to a wider EMS. After taking this consolidated first step, the company is safer to expand its EMS scope and, finally, apply for ISO 14001 certification.

2. **What are environmental aspects and impacts? Analyzing the case presented, which environmental aspects are mentioned in it? What are the impacts potentially originating from the environmental aspects analyzed?**
   It is important to understand that by environmental aspect it is understood every element of an organization's activities, products and services that can interact with the environment, while the environmental impact, in turn, represents any modification, adverse or beneficial, resulting from an aspect. In the case presented, the environmental aspects mentioned are: using raw materials from non-renewable sources, energy consumption for the production of packaging, and the leftovers generated during this manufacturing process. For these aspects, the possible environmental impacts are: air contamination, soil contamination, depletion of non-renewable natural resources, etc. *Energy consumption is associated with the source. If it is thermal power, it burns coal and pollutes air; if it is nuclear, it generates highly hazardous waste. Regarding waste leftovers, it involves the correct destination (reuse, recycling, landfill,* etc.*), but if discarded incorrectly they can lead to contamination of water, soil and, if burned incorrectly, air pollution.*

7. **To obtain an ISO 14001 certification, is it necessary that the entire chain and its suppliers are certified or show environmental concerns? Justify your answer.**
   No. ISO 14001 certification refers only to the applicant company and not to its chain. However, the more links in the chain become certified (or show concern for environmental aspects) the better. Although it is not mandatory for suppliers to have certified EMS, it is natural that the company requires more from its suppliers than it previously required in relation to meeting environmental requirements. For example, in the presented case, Gustavo's company is a supplier and does not have a certified EMS yet, while the company where Guilherme works already has an EMS certified by ISO 14001.

References

Alberton, A., Costa, N.C.A.: Meio Ambiente e Desempenho Econômico-Financeiro: Benefícios dos Sistemas de Gestão Ambiental (SGAs) e o Impacto da ISO 14001 nas Empresas Brasileiras. Revista de Administração Contemporânea - RAC-Eletrônica **1**(2), 153–171 (2007)

Associação Brasileira Da Indústria Química – Abiquim. Atuação Responsável – Histórico. [s.d.]. Available at: https://abiquim.org.br/programas/historico. Accessed: 18 September 2018

Associação Brasileira De Normas Técnicas – ABNT. ABNT NBR ISO 14001: sistemas de gestão ambiental – requisitos com orientações para uso, 41 p. ABNT, Rio de Janeiro (2015)

Associação Brasileira De Normas Técnicas – ABNT. ABNT NBR ISO 14031: Gestão ambiental – Avaliação de desempenho ambiental – diretrizes, 44 p. ABNT, Rio de Janeiro (2015)

Barbieri, J.C.: Gestão Ambiental Empresarial. 3ª ed. São Paulo: Saraiva, 376 p (2011)

Bititci, U., Garengo, P., Dorfler, V., Nudurupati, S.: Performance measurement: challenges for tomorrow. Int. J. Manag. Rev. **14** (2012)

Brasil: Ministério do Meio Ambiente. Resolução CONAMA nº 001, de 23 de janeiro de 1986, estabelecendo as definições, as responsabilidades, os critérios básicos e as diretrizes gerais para uso e implementação da Avaliação de Impacto Ambiental como um dos instrumentos da Política Nacional do Meio Ambiente. Diário Oficial da União. Imprensa Oficial, Brasília, DF (1986)

Brasil: Ministério Do Meio Ambiente: Resolução CONAMA nº 382, de 26 de dezembro de 2006, estabelecendo os limites máximos de emissão de poluentes atmosféricos para fontes fixas. Diário Oficial da União. Brasília, DF: Imprensa Oficial (2006)

Finnveden, G., Moberg, Å.: Environmental systems analysis tools: an overview. J. Clean. Prod. **13**, 1165–1173 (2005)

Franco-Santos, M., Kennerley, M., Micheli, P., Martinez, V., Mason, S., Marr, B., Gray, D., et al.: Towards a definition of a business performance measurement system. Int. J. Oper. Prod. Manag. **27**, 784–801 (2007)

Höjer, M., Ahlroth, S., Dreborg, K.H., Ekvall, T., Finnveden, G., Hjelm, O., Hochschorner, E., Nilsson, M., Palm, V.: Scenarios in selected tools for environmental systems analysis. J. Clean. Prod. **16**, 1958–1970 (2008)

ISO: Management system standards. Available at: http://www.iso.org/iso/home/standards/management-standards.htm (2015). Accessed: 19 June 2015

ISO: ISO Survey. [s.d.]. Available at: http://www.iso.org/iso/home/standards/certification/iso-survey.htm?certificate=ISO%209001&countrycode=AF (2018). Accessed: 18 September 2018

Kennerley, M., Neely, A.: A framework of the factors affecting the evolution of performance measurement systems. Int. J. Oper. Prod. Manag. **22**, 1222–1245 (2002)

Lebas, M.J.: Performance measurement and performance management. Int. J. Prod. Econ. **41**, 23–35 (1995)

Moreira, M.S.: Estratégia e Implantação do Sistema de Gestão Ambiental Modelo ISO 14000, 2ª edn., 320 p. INDG Tecnologia e Serviços Ltda., São Paulo (2006)

Neely, A.: Measuring business performance: why, what, how, p. 224. Economist Books, London (1998)

Sousa, S.R., Sanches, R., Ometto, A.R., Pacca, S.A.: A utilização da avaliação do ciclo de vida em sistemas de gestão ambiental: modelos de aplicação. Revista INGEPRO **2**(6), 90–98 (2010)

Chapter 12
Green Supply Chain Management

Fabio Neves Puglieri and Yovana María Barrera Saavedra

12.1 Introduction to Green Supply Chain Management

Before addressing the elements that structure the entire integrated strategy to implement and operationalize Green Supply Chain Management (GSCM), it is important to bring the reader the definition of the main concepts that guide this chapter, that is, what is the Supply Chain Management (SCM) and the green management of this chain.

A Supply Chain (SC) can be defined as any set of processes that link suppliers, companies (factories), retail and customers, that is, from the supply of raw materials to the consumption of the finished product by the customer. Figure 12.1 illustrates all the elements that make up a SC.

As we can see in Fig. 12.1, in every SC there is a focal company, that is, the company that governs the entire supply chain. Upstream of it are your suppliers, which may be a direct supplier (first level) or suppliers of suppliers (second level). Downstream of the focal company is the entire product distribution chain, including retail and the final consumer.

But what is SCM after all? Simply put, SCM deals with the integration, planning and management of all business processes, from final consumers to the first suppliers of inputs and raw materials in order to add value to customers and stakeholders (Pires 2016; Grant 2013).

The Green Supply Chain Management (GSCM) can be seen as an extension of SCM's way of adding value. According to Srivastava (2007), GSCM has its roots in

F. N. Puglieri (✉)
Federal University of Technology - Paraná—UTFPR, Rua Dr. Washington Subtil Chueire, 330, Jardim Carvalho, ZIP Code: 84017-220, Ponta Grossa, PR, Brazil
e-mail: puglieri@utfpr.edu.br

Y. M. B. Saavedra
Federal University of São Carlos, Buri-Lagoa Do Sino Campus, Lauri Simões de Barros Highway, km 12, SP-189, Aracaçú, Buri, São Paulo, Brazil

J. A. de Oliveira et al. (eds.), *Life Cycle Engineering and Management of Products*,
https://doi.org/10.1007/978-3-030-78044-9_12

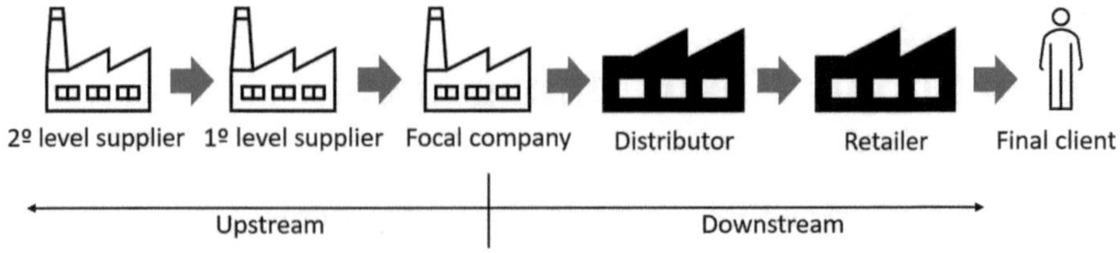

Fig. 12.1 Main components of a supply chain. *Source* Adapted from Pires (2016)

SCM and environmental management, therefore, GSCM includes within SCM the influences and relations between SCM and the environment. Ageron et al. (2012) add that GSCM offers an expanded perspective on environmental management and adopts the application of tools and practices both inside and outside the company.

A very widespread definition for GSCM, and put forward by Srivastava (2007), is that it is the integration of environmental thinking in supply chain management, including in this process activities related to product design, selection of suppliers, material purchases, manufacturing, distribution of products to consumers, including products' end-of-life management.

In other words, we can define GSCM as the incorporation of the environmental issue in a systemic way throughout the supply chain management, which means that in all processes and activities, from purchasing decisions and suppliers, going through design, to manufacturing, inventory management, marketing, transportation and distribution, and reverse logistics, environmental criteria need to be considered together with the other criteria used in SCM such as cost, time and quality.

In this sense, it is clear that GSCM involves the company's entire life cycle, which can bring numerous opportunities for improving the environmental performance of products and services in addition to generating value for the business. It is in GSCM that suppliers of materials and components with better environmental performance can be selected, where better green manufacturing alternatives are considered, where decisions related to the adoption of more efficient transportation of materials and products can be made, in addition to which products' end-of-life strategies can be implemented and operationalized through reverse logistics mechanisms. This is a still recent approach in the area of supply chain management, with Sarkis (2003) and Srivastava (2007) as two of the precursors in the subject in academic literature.

An important factor to highlight is the role of Ecodesign (see Chap. 8) inside GSCM. This is because many opportunities for environmental improvement in the supply chain such as end-of-life strategies, which will be discussed later in this chapter, depend on changes in the Product Development Process (PDP). In addition, the Life Cycle Assessment (LCA), technique presented in Chap. 3, can assist in several decisions related to changes in products to assist in GSCM, since it is one of the main Ecodesign techniques.

Now that the fundamental concepts have been explained so that you, dear reader, can understand what GSCM is and its relation to the company's life cycle, we can

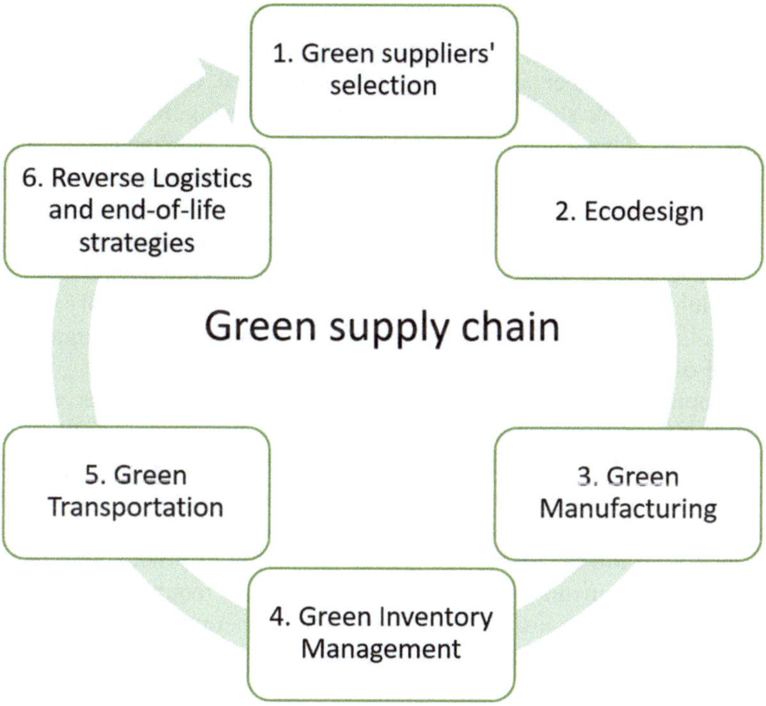

Fig. 12.2 Main component elements of Green Supply Chain Management

present you to Fig. 12.2. Figure 12.2 shows the representation of a model to develop and implement GSCM and will be used as the basis for this chapter.

From Fig. 12.2, this chapter is divided into eight parts. First part dealt with the explanation of the SC, SCM and GSCM concepts. Then, the process of selecting green suppliers will be discussed. In the third moment, the importance of green design (Ecodesign) for GSCM will be quickly highlighted and green manufacturing is discussed in the sequence. Later on, we will discuss opportunities to act at GSCM through green inventory management, which involves material handling and inventory management, and then transportation activities. Finally, we will explore reverse logistics and the main end-of-life strategies that can be adopted, which are fundamental to implement many of the Circular Economy opportunities that were addressed in Chap. 10.

12.2 Selection of Green Suppliers

As already seen in Fig. 12.2, GSCM is structured in six areas of activity and the first of which refers to the selection of green suppliers. Also called sustainable purchasing or Green Purchasing, the objective in the selection of green suppliers refers to the procedure of purchasing materials, components, products or services on the market that cause less environmental impacts. At the same time, there is also the concept of selecting sustainable suppliers and sustainable purchasing. In this particular case, in addition to environmental requirements there are also social and ethical business criteria, however, this concept will not be detailed in this section, being restricted only to the environmental issue.

We know that products and services can cause environmental impacts at all stages of their life cycle. Thus, the importance of considering environmental requirements when selecting potential suppliers is precisely to reduce the impacts that can be generated in the following phases of the life cycle, that is, in manufacturing, transportation, use and end-of-life.

For example, imagine the following situation: you are a manufacturing company and manufacture a certain product. If your company has a business strategy to compete for low cost, with the only criterion for choosing suppliers those who deliver materials and components at the lowest possible value, it would be natural to expect a possible consequence of this to be greater losses in the production process (due to the low quality of the raw material), greater possibility of producing products out of the quality requirements (which may become waste or require rework, leading to impacts related to the disposition or consumption of energy/other inputs), and even a reduction in the product's lifetime at the use phase due to components' premature wear. In other words, there is a strong relation between what the supplier delivers to the company/factory and the possible impacts that this can have on the product's life cycle.

Other examples can also be mentioned such as reducing the possibility of contamination of users/consumers with products whose materials and components are produced without toxic substances. Once again, we emphasize that there is a repercussion throughout the life cycle from decisions related to the selection of suppliers.

We must note that, in addition to the environmental performance itself in the life cycle being affected by such decisions, the quality of product and process can also suffer from such decisions, which may lead to the loss of customers and other types of financial losses for the business.

Thus, it is necessary to adopt procedures to select suppliers that can deliver materials and components to the company with a better environmental performance level. Figure 12.3 presents a sequential scheme structured in stages and activities, which is related to the selection of green suppliers.

Through Fig. 12.3, observing step (1), green purchasing is first made by the selection of green suppliers and this stage is divided into six other activities as shown in (2): identification of needs and specifications; formulation of criteria; call for proposals; qualification; selection; and finally performance assessment. Feedback from performance assessment of current suppliers is constantly used to assist in this decision-making process.

The identification of needs and specifications will depend on the company that is selecting suppliers, that is, their degree of requirement in the face of the expected

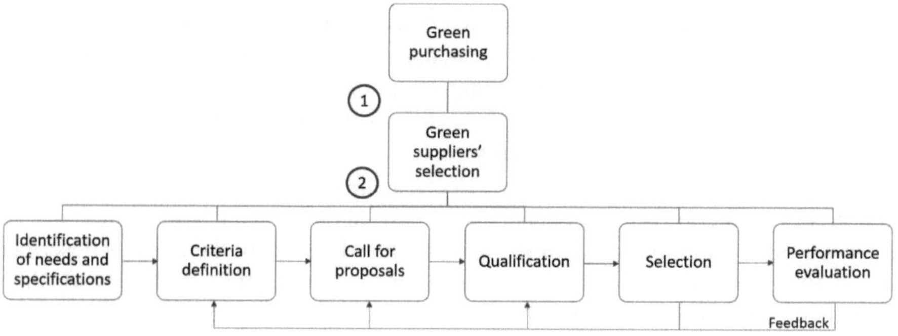

Fig. 12.3 Sequence for selecting green suppliers. *Source* Adapted from Rodrigues (2016)

environmental requirements of the materials/components to be purchased. Many of these needs and specifications can come from an Ecodesign study or even from hotspots of an LCA study. From this, such needs and specifications must be translated into criteria for supplier selection. Figure 12.4 provides a simplified view of how these criteria can be.

The criteria can be divided into categories of criteria and then into sub-criteria, which can be both qualitative and quantitative. Despite the emphasis on environmental issues, other economic criteria, such as cost, quality, degree of innovation, technological degree, flexibility, distance, among others, can be added in this decision-making process.

In addition to those presented in Fig. 12.4, other environmental criteria and sub-criteria can also be cited, such as the adoption of ecodesign and circular economy practices, the presence of ISO 14001 certification, or even others based on a life cycle perspective or LCA.

It is interesting to note that some economic and environmental criteria may relate. For example, a decision involving the choice of a company's more distant supplier

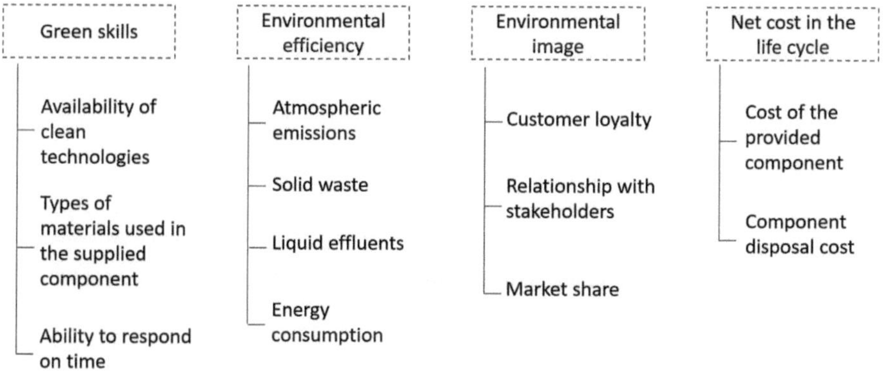

Fig. 12.4 Criteria for selecting green suppliers. *Source* Adapted from Noci (1997)

can lead to greater generation of resource consumption and emissions from supply activity as well as the low quality of the material or component provided for leading to process losses and rework, generating waste or re-consuming natural resources and energy. In addition, there are some trade-offs that need to be dealt with, such as environmental performance and supply cost.

Formulated the criteria for supplier selection, the activities of calling, qualification and supplier selection are started and can be carried out with the help of methods. As there are often a number of suppliers greater than 1 in the market, and as described above, there may be some criteria (economic and environmental, in addition to others such as social, that are not treated here), this kind of problem may be called multi-criteria decision problem (Rodrigues 2016).

Govidan et al. (2015) demonstrate an extensive number of multi-criteria decision-making approaches for the selection and assessment of green suppliers. One of the most cited methods to assist in the green supplier selection is the Analytical Hierarchy Process (AHP), including fuzzy AHP, followed by the Analytic Network Process (ANP). Figure 12.5 shows a typical case involving the decision-making of green suppliers using the AHP method.

Observing Fig. 12.5, a green supplier selection, as already described, will depend on the definition of criteria and sub-criteria, which may be environmental indicators that are addressed in Chap. 11. At the end of the process, the suppliers are valued through the relative importance of each one in relation to the objective of the selection, which in this case would be to find the best green supplier. These green suppliers become important so that the materials used in the product can help meet Ecodesign requirements.

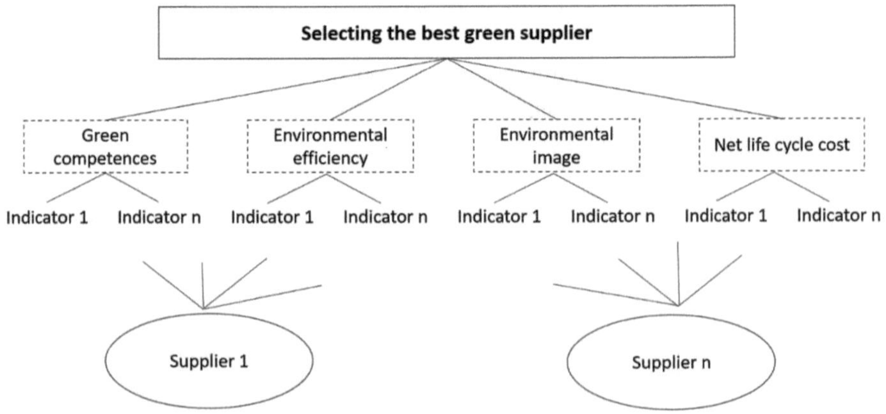

Fig. 12.5 AHP method used for green supplier selection. *Source* Adapted from Noci (1997)

12.3 Ecodesign

Dear reader, as we saw in Chap. 8 and in the introduction of this chapter, Ecodesign, which aims at integrating environmental issues inside the Product Development Process (PDP), is one of the key elements inside GSCM. But why is Ecodesign important to GSCM? The answer is that Ecodesign decisions can support and be supported by other elements in the supply chain.

Let us bring out some examples. If in the company there are already green suppliers, this will enable the use of high quality raw materials and have less environmental impact, helping to meet one of the Ecodesign strategies that deals with the extension of product lifetime and reduction of impacts on the life cycle. In the case of green transportation and green inventory, Ecodesign can help to make decisions aimed at minimizing energy consumption during transportation and storage, for example, by reducing the volume and packaging of products. In addition, by reducing weight and volume in products, Ecodesign also promotes the reduction of atmospheric emissions which can cause global warming.

Now, let us look at a more practical example: imagine a company that manufactures computers and distributes it to stores in individual boxes, containing inside a plastic bag with the computer, a mouse and a cable. In order to be able to distribute the quantities requested by the store, in this condition, the company must pay high costs in transportation generating more emissions and impacts. However, if the customer buys the computer in the store, he ends up discarding the box in it, generating a large amount of waste. Thinking about it, through Ecodesign, the company can develop a more dematerialized computer whose packaging is just a sturdy suitcase with the computer and accessories inside. To transport the computers, boxes with capacity for 5 computers can be used and, at the time of arriving at the store, the computers are unloaded and the store returns the boxes to the company, which takes them back to make new deliveries. To implement this strategy, there is a need to develop the entire reverse logistics plan that is part of GSCM.

For more details about Ecodesign, we ask you to read Chap. 8.

12.4 Green Manufacturing

Green Manufacturing (GM) is defined as a mode of manufacturing in which its production processes are less environmentally impacting and that integrates the reduction and preservation of natural resources and the minimization of waste generation (Dornfeld et al. 2013). The transformation, manufacturing and assembly processes of semi-finished and finished products are also part of GM. Srivastava (2007) adds that in GM tools and techniques are used to help reduce the use of energy, water and raw material consumption, highlighting LCA as one of the most used techniques in recent years.

Defined in Chap. 2, Ecoefficiency is determinant for GM processes and can be calculated according to NBR ISO 14045 by Eq. 12.1:

$$\text{Eco-efficiency} = \frac{\text{Product value}}{\text{Environmental impact}} \tag{12.1}$$

In addition, NBR ISO 14045 recommends that to compare the eco-efficiency between products, Eq. 12.2 must be used:

$$\text{Eco-efficiency} = \frac{\text{Evaluated product Eco-efficiency}}{\text{Base product Eco-efficiency}} \tag{12.2}$$

As an example of its use we can mention the research developed by Santos et al. (2016) that performs an LCA combined with the assessment of Ecoefficiency in a thermoelectric plant to assess the alternatives of waste recovery. The Ecoefficiency assessment has been applied in the process of reusing heavy ash generated in the combustion of mineral coal to generate the plant's electric power and then compared with the production of sand for the ceramic coating industry and clinker production for the Portland cement industry. With this it was possible to demonstrate that the ashes can be used in these types of companies.

To perform the calculation of eco-efficiency, Santos et al. (2016) considered the results of impacts related to human health, ecosystems and resources (Table 12.1) and the product value was the generation of 804.65 t of ash per day by the thermoelectric plant.

According to Eq. 12.1, the results were:

I. Eco-efficiency of the thermoelectric plant in the generation of heavy ash = 804.65/196.4 = 4.10.
II. Eco-efficiency of the thermoelectric in the use of heavy ash for ceramic coatings = 804.65/196.5 = 4.09.
III. Eco-efficiency of the thermoelectric plant in the use of heavy ash for Portland cement production = 804.65/198.3 = 4.06.

Table 12.1 Environmental impacts calculated by the ReCiPe endpoint H single score method

Process	Score			
	Human health	Ecosystem	Resources	Total
Ash generation by thermoelectric plant	163.6	32.6	0.21	196.4
Ash generation for ceramics	163.7	32.6	0.21	196.5
Ash generation for cement	164.9	33.2	0.212	198.3
Clinker production	28.2	16.2	0.084	44.5
Sand production	1.6	0.7	0.001	2.3
Total	522	115.3	0.72	638.1

Source Adapted from Santos et al. (2016)

With these results, the authors made the comparisons using Eq. 12.2. For the Eco-efficiency of heavy ash generation processes used in ceramic processes and Eco-efficiency in the generation of heavy ash of the thermoelectric plant the results were 4.09/4.10 = 0.9975 (99.75%), which means that using heavy ash for ceramic coatings it is not possible to improve the Eco-efficiency of the thermoelectric plant, as it would be lower than the disposal of heavy ash made in ponds (0.25%) (Santos et al. 2016).

In the comparison of the generation of heavy ash used in Portland cement production with Eco-efficiency of the generation of heavy ash of the thermoelectric plant, we have 4.06/4.10 = 0.99 (99%). With this result, it is observed that the use of heavy ash in the Portland cement production industry also would not increase the Eco-efficiency of the thermoelectric plant being 1% less in relation to the disposal of the ashes in ponds (Santos et al. 2016). As observed in the example presented, Eco-efficiency is not always easily or directly achieved, which leads to the need to analyze several alternatives that consider, in addition to the environmental and economic impacts, the legislation related to the assessed processes.

Another frequent indicator in GM is the so-called Overall Equipment Effectiveness (OEE), that is included inside Total Productive Maintenance (TPM) to verify how much the company is using available resources such as: machinery, manpower, and materials. In the OEE, efficiency is deployed into three indicators: availability, performance and quality. According to Nakajima (1989), OEE has a basic role in achieving and maximizing equipment efficiency. Hansen (2001) and Chiaradia (2004) add that losses and inefficiencies represent that in the company a part of the resource is not being used in its entirety. In this sense, we can point out that from these losses and inefficiencies more natural resources are used and more environmental impacts are generated by the equipment.

An Opportunity related to the OEE for the calculation of the Eco-efficiency of manufacturing operations can also be established by means of an equation, such as the following:

$$\text{Eco-efficiency} = \frac{\text{Overall Equipment Effectiveness (OEE)}}{\text{Environmental impact from LCA}} \tag{12.3}$$

To learn more about OEE indicators, see the works by Chiaradia (2004) and Santos (2018). For more details on the potential impacts from LCA, see Chap. 5.

Also in GM, actions can be inserted to control air pollution resulting from the use of equipment in different processes. In this sense, Lona and Leme (2013) mention the importance of considering in the selection of equipment factors such as: emission patterns, energy consumption, efficiency, investment cost, cost of operation and maintenance, physical and chemical nature of particulates and their dangerousness.

The same authors also highlight that all equipment has its efficiency (E) calculated by its mass balance which in this case is the relation between the mass of particles that comes out (mo) and the amount of particles that goes in (mi), measured by the system participation rate (Pt). In other words, it refers to the fraction of mass that is not be ported by the network, represented by Eq. 12.4.

$$\mathrm{Pt} = \frac{\mathrm{mo}}{\mathrm{mi}} \quad (12.4)$$

In which:
Pt = System Participation
mo = mass of particles that comes out of the system
mi = amount of particles that goes in the system
Equation 12.5, in turn, is related to the

$$E = 1 - \mathrm{Pt} \quad (12.5)$$

In which:
E = Efficiency
Pt = System participation (calculated with Eq. 12.4)
When the system displays devices in series, Eq. 12.6 must be used:

$$\mathrm{Pt}_o = \prod_{i=1}^{n} \mathrm{Pt}_i \quad (12.6)$$

In which:
$\mathrm{Pt_o}$ = global penetration rate,
Pt_i = penetration rate of device *i*.
The overall collection efficiency of the system is given by Eq. 12.7:

$$\mathrm{Eo} = 1 + \mathrm{Pt_o} \quad (12.7)$$

Or it can be calculated through efficiency by fractions:
Efi = Separation efficiency of particles with dpi diameter;
En = Penetration rate of particles with dpi diameter.

According to Lona and Leme (2013), another concept related to cutting diameter (dpc) and that refers to particle diameter with 50% efficiency, i.e. Ef = 0.5, must be considered important. In other words, particles with a diameter greater than dpc are collected efficiently at less than 50%.

As an illustrative example, the authors present an exercise of a particle sampling performed in an industrial process X in order to estimate the minimum overall efficiency of the system. Tables 12.2 and 12.3 must be used for calculating.

Solution:

Table 12.2 Data obtained from sampling

Dpi	2	6	10	20	40	50
mass (mg) (mi)	25	125	100	75	30	5
Massa (mg) (mo)	0,10	0,60	3,75	15,00	37,50	22,50

Source Adapted from Lona and Leme (2013)

Table 12.3 Data obtained from sampling

Dpi (μm)	0–2	2–4	4–7	7–10	15–25	10–15	25–40	40–60	> 60
mass (mg) (mi)	10	25	45	70	85	95	97	98	100

Source Adapted from Lona and Leme (2013)

Table 12.4 Consolidated data

dpi	mi (mg)	*E* (%)	mo (mg)
50	5	98	0.10
40	30	98	0.60
20	75	85	3.75
10	100	95	15.00
6	125	45	37.50
2	25	10	22.50
Total	360		79.45

Source Adapted from Lona and Leme (2013)

To be able to estimate the minimum overall efficiency of the system it is necessary to determine the penetration rate (Table 12.4).

$$\text{Pt} = \frac{\text{mo}}{\text{mi}} = \frac{79.45}{360} = 0.220694 = 22.07\%$$
$$\text{E} = 1 - \text{Pt} = 1 - 0.220694 = 0.7793 = 77.93\%$$

With this result we can observe that new alternatives are still needed to increase the efficiency of equipment used in the analyzed industry's processes. Another important point to highlight is that in this type of analysis it is also possible to use the Eco-efficiency indicators presented at the beginning of the chapter. In this case, the results indicated the most eco-efficient operation and that will be the greenest operation inside the process.

In the next subchapter we will present another important element inside GSCM, the green inventory management.

12.5 Green Inventory Management

In green inventory management, issues related to environmental performance such as material loss and consequent waste generation, in addition to emissions, are now considered along with economic performance, that is, operating costs. In this subchapter, we will particularly talk about two important issues in green inventory management: material movement and inventory management.

Regarding material movement, some practices can be adopted thinking about GSCM. One of them is the change of factory layout which can be optimized to reduce distances between operations and consequently reduce the handling of materials and components (which can be damaged or broken, requiring rework or generating waste), also saving energy and fuel consumption and reducing emissions. In addition, there are alternative technologies that can be adopted such as replacing fossil fuels with renewable sources in equipment such as forklifts and even the use of Automatic Guided Vehicles (AGVs).

Poor inventory management can lead to loss of environmental performance in the supply chain. Materials and components can deteriorate especially those with perishability properties, generating, in addition to the cost to the company, a solid waste. Organic products, especially food, when decomposing are also responsible for carbon emissions into atmosphere. A management practice that can be adopted to avoid such a situation is the First In First Out (FIFO), which consists of removing from the materials in stock first those that arrived first in the factory warehouse.

There is also, in the case of the storage of refrigerated and frozen products, the consumption of electricity and atmospheric emissions associated with the life cycle of its generation. In this case, actions related to the replacement of old equipment with new ones that have better energy efficiency. Practices that can be identified using Cleaner Production (see Chap. 2 of this book) are recommended. The next item brings another important element of GSCM, the green transportation.

12.6 Green Transportation

Another very important part of GSCM refers to the transportation between supplier, factory, point of sale and consumer, in addition to the internal material movement. Transportation activities are an intrinsic part of any supply chain and a product's life cycle. This is evident because since the extraction of raw materials, supply of materials and components in industry, distribution and reverse logistics, transportation is involved. In addition, throughout the life cycle, transportation activities can be configured as one of the activities that most harm a product's environmental performance. Several impact categories such as consumption of non-renewable resources due to the use of petroleum derivatives in the combustion of vehicle engines and global warming potential, by the emission of diesel burning, are some examples of environmental impact categories that are associated with transportation.

LCA and Ecodesign itself, presented respectively in Chaps. 3 and 8, present opportunities to improve the environmental performance of the supply chain in relation to transportation and material movement activities. The identification of hotspots throughout the life cycle and the comparison of alternatives (what would be better in the life cycle: using diesel or hybrid/electric vehicles?) can be obtained by LCA studies, while Ecodesign for example presents opportunities for environmental improvement of these products.

In addition, other opportunities can also be obtained to improve the environmental performance of transportation and material movement. According to Bouchery et al. (2017), in choosing the best green transportation strategies must be considered some variables, among them: the distance to be traveled by the vehicle; the transportation modes to be used; type of equipment used in logistics operations; loading rate of the vehicle used in transportation; and finally, the operations, which involve the skills and environmental awareness of the vehicle operator in addition to the optimization degree of the transportation plan. Next, we will discuss, from these variables, some green transportation strategies that can improve the environmental performance of the entire supply chain.

As already mentioned, some impact categories are directly associated with transportation activities such as consumption of non-renewable resources and atmospheric emissions with global warming potential and one of the factors that most influences these impacts is the distance traveled. That is, the longer the distance, the more fuel will be used, the greater the burning of this fuel and, consequently, the more emissions are generated.

One strategy to minimize environmental impacts related to transportation distance is to opt for suppliers that are located closer to the company. Another possible strategy to optimize routes and reduce transportation distances comes from Operational Research. Routing consists of defining routes that decrease costs and distribution times from one or more distribution centers or warehouses. However, when considering that shorter routes can reduce the consumption of resources, especially non-renewable ones such as diesel in addition to the emissions associated with its burning, we can say that routing can bring benefits to GSCM.

Several methods can be used here and they can be divided into exact methods, heuristic methods and metaheuristic methods. The exact methods allow to arrive at an optimal solution, however, they are of great complexity. Heuristic methods, in turn, generate approximate solutions but usually faster and with less computational effort. Finally, metaheuristic methods aggregate more advanced techniques such as expert and interactive systems.

Examples of heuristic methods widely cited in literature that can be used for transportation routing and improvement of GSCM are the Sweep and Clark and Wright methods. Currently, there is a wide variety of software that can also be used to assist in routing-related decisions.

In relation to the type of transportation mode, it can also have a major influence on the transportation environmental impacts. Data presented by the International Chamber of Shipping show that in a comparison between three different modes, for each ton-kilometer, a container ship emits 3 g of CO_2 eq., a truck 80 g of CO_2 eq., while an aircraft emits 435 g of CO_2 eq. As a result, for better green management of the supply chain, it is necessary to consider transportation modes that are more energy efficient, opting, whenever possible and whenever there is infrastructure, waterway and rail transportation. Another opportunity is related to the installation of factories near multimodal terminals in order to combine different modes in transportation, reducing the dependence on road transportation.

Vehicles in poor condition can also be sources of environmental impact. In this sense, the adoption of preventive maintenance programs can bring efficient results to promote a more environmentally friendly supply chain. At the same time, the search for transportation alternatives that make use of cleaner energy sources and replace older fleets with more modern ones are valid strategies to improve environmental performance in the supply chain.

The vehicle loading rate can also substantially influence the transportation environmental performance and the supply chain. Data from the 2006 U.S. Environmental Protection Agency presented by Elhedhli and Merrick (2012) show that when you double the amount of cargo transported (in tons) in a truck, atmospheric emissions do not increase in the same proportion but increase by approximately 35%. Because of this, making maximum use of the load capacity during transportation is an eco-efficient practice.

Finally, it is important to highlight to the reader the operator's driving skills of transportation vehicles and environmental awareness. While often neglected as a strategy to reduce environmental impacts on transportation, driver decisions and behaviors can affect green supply chain performance. Let us take the following example: imagine a truck driver who always takes congested routes, requiring constant stops and low-speed traffic. It is possible to expect this vehicle to consume more fuel and generate more emissions per ton-kilometer than the same vehicle that is transiting on a free way at 100 km/h and at constant speed, right?

Based on this, as a way to minimize environmental impacts on transportation, the driver must recognize that his decisions and behaviors can impact the performance of the entire GSCM. As a strategy to act in this way, we can mention the use of monitoring systems such as vehicular telemetry. Vehicle telemetry allows an operator to track real-time vehicle variables that may be associated with impact categories such as fuel consumption, average and instantaneous speed, and location. Based on these data, decisions can be made, among them, to conduct training on the best way of driving vehicles, aiming at reducing fuel consumption and emissions, creating in the long run an environmental culture in drivers.

Next, we will talk a little about Reverse Logistics and End-of-Life Strategies and its importance to GSCM.

12.7 Reverse Logistics and End-of-Life Strategies

Now dear reader, we will show some concepts and alternatives that mainly help the products' disposal phase, that is, when the product reaches its end-of-life, it has lost its functionality or its validity. And here you may be thinking: But can I recover the products with their materials after they are discarded? And the answer is yes! A good portion of the products that are currently discarded can be reinstated directly or indirectly into the supply chain and this is where Reverse Logistics (RL) and End-of-Life Strategies (EoL) play a key role.

Something very important that you must remember is that by integrating these concepts with their different tools you can transform the linear production flow into a semi-circular or circular (closed) flow. That is, we will have less input of resources and less output, contributing to GSCM.

Next, let us talk separately about RL and EoL.

12.7.1 Reverse Logistics

RL is defined as the efficient process of planning, implementation of economic flows control and raw materials, product stocks in the process, semi-finished and finished products, and related information from the point of view of consumption to the point of origin with the objective of recovering value or performing the proper final disposal.

For Tibben-Lembke (2002), RL takes place from the point of consumption to the point of origin and in this path are considered the processes of planning, implementation and flow control of finished products with all their information and always with the main intention of achieving their recovery and proper final disposal.

Leite (2017) adds that RL is the new area of business logistics that takes care of the return of after-sales and post-consumer products to the production chain and uses reverse channels for this. After-sales products are understood as those items sent in the wrong way by the distributor, damaged products (such as food and medicines that may have their expiration date expired and cannot be used by consumers), excessive stocks in the distribution channel, consignment and quality problems. Post-consumer products, in turn, are those that are no longer used by the consumer and are discarded after a certain time that can vary from days to years. The products discarded with their components and materials are reinstated to the production chain as a secondary raw material.

For a better understanding of the RL concept, imagine the following situation: you decide to discard your old printer at the collection points of the city's supermarkets. The company that carries out the recovery schedules transportation so that on specific dates these discarded products are collected and go through the recovery process and proper final disposal. Okay, now you know what RL is!

In recent years, RL has become a key alternative to assist in the integration of life cycle thinking into GSCM. In addition, with RL it is possible to obtain benefits that can bring added value in aspects related to economic, legal, environmental, social, customer loyalty, new market entry, transparency, among others.

It is important to note that in Brazil there is specific legislation for solid waste and it must always be consulted together with state and local laws when planning to work with RL and EoL. The Brazilian National Solid Waste Policy (PNRS), Law 12,305 from 2010 with its Decree No. 7,404 of December 23, 2010 brings the main concepts, objectives and instruments to achieve integrated management and adequate waste management. In this PNRS, section II on Shared Responsibility in the Management of Solid Waste appears, in which manufacturers, traders, importers and distributors are obliged to structure and implement RL batteries, tires, oils and lubricants, their waste and packaging, fluorescent lamps of sodium vapor and mercury and mixed light and pesticides with their waste and packaging. Even supply chain actors, such as consumers, also become important elements for the operationalization of RL (BRASIL 2010).

It is important to highlight that, with this, the company must not necessarily make the recovery and final disposal, but it is its responsibility to create paths for RL to actually happen and be carried out properly. In addition, although other products are not mentioned in the PNRS, it recommends that companies integrate the following hierarchical strategies when dealing with solid waste: non-generation, reduction, reuse, recycling, treatment and final disposal. It is important that you are aware that without RL no type of recovery of the discarded product is viable, that is, without RL it is not possible to put into practice the EoL that will be described next.

12.7.2 End-of-Life Strategies (EoL)

Figure 12.6 represents the different types of recovery and final disposal that currently exist.

From Fig. 12.6 we can highlight three points: 1. The main or linear cycle, represented by the black line, is basically a product's life cycle. 2. The dotted blue line is the secondary or reverse flow of the discarded product or waste and it is from that moment that RL happens. 3. It is observed that there are several forms of recovery and proper final disposal and this may vary according to the characteristics of the discarded product and its components, by the level of recovery, by the stage of reinsertion inside the supply chain and by the health risk that represents this waste. In

Fig. 12.6 Product end-of-life strategies and the life cycle. *Source* Adapted from Gehin et al. (2008), Saavedra (2010) and Thierry et al. (1995)

addition, EoL do not necessarily occur in reverse flow because in some cases, as in recycling, it can also occur in the main flow as we will see in the next paragraphs.

To know the EoL, we must first understand their concept. EoL are the alternatives or paths that can be used to recover discarded products with their components and somehow be able to reintegrate them into the production chain. Product EoL vary primarily by the level of ease of disassembly to achieve their recovery and reintegration inside the supply chain.

From the EoL, it is possible to obtain some benefits such as: reduction of areas for waste disposal, reduction of environmental impacts, generation of new business, reduction of costs by the acquisition of raw material, generation of jobs, among others. Something very important that you must remember is that before deciding to opt for the use of some EoL, it is important that you perform an economic, environmental and social viability analysis to make the best decision and for this we recommend the reader to explore the other chapters of this book where various techniques and tools that can be applied for this purpose are presented.

Back to Fig. 12.6, we find three groups and the first one is defined as a direct reuse group. The second group is product recovery management and the last one, waste management that deals with tailings management, because theoretically the only one that must go to landfills are those tailings that were not possible to be recovered. Now in the next paragraphs we will discuss a little more about each of the EoL and the forms of final disposal.

Direct Reuse and resale (1) consists of discarded products that pass to another consumer, what allows to extend this product's lifetime. In this EoL, the product retains its original functionality and design and does not go through any kind of repair, except for the wear of the parts. A very common example is the transfer between people or sale on websites of used products, which often occurs when it happens the purchase of newer devices and with more advanced technologies, such as mobile phones, computers, tablets etc.

The second group, named Recovered Product Management presents the following EoL: Repair (2), Refurbishing (3), Remanufacturing (4), Cannibalization (5) and Recycling (6). We start with **Cannibalization (5)** because it is considered a distribution center for the other EoL, where the products may or may not be disassembled and directed to the other EoL. In addition, cannibalization can receive parts from other EoL for new targeting and recovery. As an example, companies that carry out the recovery of Waste Electrical and Electronic Equipment (WEEE) can be cited.

Repair (2), in turn, is an EoL in which the discarded product has a limited disassembly only for the replacement of small broken or damaged parts. In this EoL, the important thing is that the product's functionality is maintained and the parts replaced are usually used parts of other discarded products that may be of the same or similar products. In this case, the warranty applies only to the replaced parts and not to the entire product. Examples of this type of EoL can be cited for the replacement of parts in cars, replacement of parts in mobile phones, computer cables, etc.

In the **Refurbishing (3)** EoL, the process is made from a defective product by means of a mechanical work where a partial disassembly of the product is performed and usually presents partial cleaning of the components and the replacement is only for the defective part, not being mandatory that they are new parts. In the end, we have a refurbished product that retains the previous standard. The warranty will be only for the refurbished part. As an example, we have a computer that goes through a partial disassembly to replace the HD or the broken screen.

In **Remanufacturing (4)**, the main goal is to return a product with the warranty and quality of a new product at a price that is approximately 40% lower than the price of the same new product. For this, the discarded product (core) goes through a sequence of steps where the product's total disassembly is performed. Then the cleaning of the parts, the inspection and storage of the parts, the refurbishing and replacement of the parts, and the reassembly of the product are carried out. Still throughout this process, several quality tests are performed and the product can receive technological update. One of the biggest advantages from this strategy is to be able to extend the products' complete life cycle, that is, if I have a notebook with a 5-year-span life cycle, through remanufacturing it is possible to achieve at least 1.5 to 2 more complete life cycles for this same product (which can represent between 10 and 15 years). Obviously, the product must always be analyzed and see the percentage made from remanufacturing, because if it is necessary to replace more than 70% of parts, it will no longer be appropriate to perform remanufacturing.

In Brazil, the ABNT 16290:2014 standard establishes the general requirements for classification, type of reprocessing, products to be commercialized and endings, and applicable definitions. This is one of the EoL that offers the most possibilities to extend the complete life cycle of products. Countries such as Europe and the United States consider remanufacturing as a profitable industry that generates several jobs annually.

It is important to highlight that for the remanufacturing to be more successful the ideal is the process to be carried out by the product's original company because it has the design and list of materials and components that make up this product, which ensures that the remanufactured product maintains the characteristics and requirements of the original product. If the process is not carried out by the original company, it is recommended that the outsourced one is trained to perform an appropriate remanufacturing process by the original company that manufactures the product. Another important factor is the importance that the concept with its principles is integrated since from the Product Development Process (PDP) and Ecodesign, as this will facilitate all reprocessing and avoid damage to the parts, optimizing the assembly and reassembly times of the remanufactured product. For more information, see the work by Gehin et al. (2008), Gray and Charter (2006), Ijomah et al. (2007), Saavedra (2010), Saavedra et al. (2013) and Sitcharangsie et al. (2019).

Most common examples of remanufacturing are products with high added value, such as heavy vehicle clutches, electronics, medical equipment, agricultural machinery, among others. In Brazil, we can highlight the National Association of Auto Parts Remanufacturing Companies (ANRAP) which has more than 15 companies that carry out remanufacturing.

Finally, **Recycling (6)** is considered the most historically used EoL for the recovery of materials and aims at transforming discarded products into new raw materials for the manufacture of the same product or other products. A negative point of recycling when compared to remanufacturing is that the product's energy, functionality and geometry are lost in the transformation process. According to Manzini and Vezzoli (2002), recycling is classified according to its life cycle. That is, we can have internal recycling in the company's processes that would be named pre-consumer and we have the recycling that happens in the post-consumer, that is, when the product is discarded.

In this EoL, there are also a series of steps where depending on the product it is carried out its total disassembly, separation and destination of the parts. For example, the computer can go through a recycling process where its plastic parts are recycled and the boards are used for recovery of precious materials such as gold and platinum.

Recycling can be divided into mechanical and chemical. In mechanical recycling, products undergo fragmentation basically for size reduction, using mechanisms such as crushing and grinding. Chemical recycling uses processes such as hydrogenation, pyrolysis, gasification, among others, to recover components that will be used as raw material for the manufacture of new products. According to the Brazilian Association of Public Cleaning and Special Waste Companies (ABRELPE), in 2018 approximately 11,308 t of plastic and 434 t of aluminum cans were collected in Brazil and these products were sent for recycling. It is important to remember that Brazil recycles approximately 98% of aluminum cans, that is, it has a practically closed cycle compared to other countries that perform the same practice (ABRELPE 2018). For more information on recycling data see ABRELPE's 2018 Solid Waste Panorama in Brazil.

Finally, it is important to mention the incineration and landfills that are inserted inside the Waste Management, that is, where are those discarded products that have no recovery options such as products discarded in hospitals, diapers, pads, among others. Although composting does not appear as an EoL, it is important to be cited for recovering and transforming organic waste into fertilizer, which can be used in agricultural and livestock activities.

The following is a tool that can assist in the activities of companies and academy interested in this theme called RemPro matrix. The RemPro matrix allows you to relate a set of desired characteristics with the steps of remanufacturing, seven of which are these steps and nine the product's properties, as listed in Table 12.5. The matrix aims at assisting in determining the properties of a product that is required for the product to be remanufactured. For example, in the case of the cleaning step, the product must have easily accessible characteristics (e.g. modularity) and wear resistance (e.g. use of durable materials to resist contact with corrosive chemicals, etc.) (Sundin 2004).

This is an easy-to-apply matrix where results can create opportunities for new information to be entered during the PDP phases and mechanisms can be created to facilitate the remanufacturing of products developed by a company. At the time of applying the matrix it is necessary to perform or simulate all the steps of remanufacturing to obtain more real results. In addition, it is recommended that you write down all the observations identified in each Table in order to have a detailed follow-up of the assessed product.

Table 12.5 Repro2–RemPro matrix

Remanufacturing steps / Product properties	Inspection	Cleaning	Disassembly	Storage	Reprocessing	Reassembly	Test
Ease of Identification	X		X	X			X
Ease of verification	X						
Ease of access	X	X	X		X		X
Ease of handling			X	X	X	X	
Ease of separation			X		X		
Ease of safety						X	
Ease of alignment						X	
Ease of storage				X			
Wear resistance		X	X		X	X	

Source Adapted from SUNDIN (2004)

12.8 Final Considerations

This chapter brought GSCM as an alternative that must be adopted inside our current production model to achieve a circular and more sustainable model. The adoption of life cycle thinking and the strategy and systemic vision become central elements of the supply chain. In addition, Ecodesign and LCA must be the central basis for this transformation, which must be carried out internally and externally to the chain, i.e., all actors must participate. In this sense, Ecodesign, green manufacturing and reverse logistics, including end-of-life strategies, all serve as support to make processes more environmentally sustainable, help extend the products' life cycle and achieve higher levels of recovery of discarded products that can be reinserted into the chain. However, the green supply chain will not be fully transformed if greener purchasing, smarter and more environmentally sustainable transportation and control in the storage and handling of materials are not considered.

In conclusion, for the coming years we have great challenges with GSCM due to the fact that its insertion is still very early. However, it is expected to make greater progress in its transformation with the insertion and expansion of new practices by industry and governments through circular economy.

12.9 Exercises

Question 1 What are the benefits of GSCM? List at least 5.

Answer:

- It integrates life cycle thinking
- It has a strategic and systemic vision that allows to look at the entire chain together with its internal and external actors to create circular production models
- It creates products with better environmental performance that includes their processes and activities to be more sustainable
- It offers the use of fewer natural resources where there is an extension of materials' life cycle and inclusion of closed or circular cycles
- It offers greater economic and social benefits.

Question 2 For this exercise you must use the data from the work by Santos et al. (2016) that appear in Table 1 of this chapter (Environmental impacts calculated by the ReCiPe endpoint H single score method) and replace the sand with the heavy ashes in the ceramic coating industry and verify if the thermoelectric could increase its eco-efficiency. Remember to use Eqs. 12.1 and 12.2 and adopt 804.65 t daily of heavy ash generated by the thermoelectric.

Answer:

By removing the use of sand in the production of ceramic coatings automatically we would have an impact reduction of 2.3 (which would be the sum of the impacts related to human health, ecosystem and resource, see Table 12.1). In this order of ideas, the new result would be:

Eco-efficiency = 804.65/(196.5 − 2.3) = 4.14 (use of Eq. 12.1 and Table 12.1 data).

Comparing (use of Eq. 12.2).

4.14/4.10 (data already obtained in the example presented) = 1.0097 (100.97%) That is, with this change the thermoelectric could increase eco-efficiency to 0.97% or that would mean that it would avoid the disposal of heavy ash in the pond and avoid various environmental impacts.

Question 3 Research and describe two methods of air pollution control that can be used in point sources (companies).

Answer:

Two of the methods that can be used in air pollution control are The cyclonic separators that aim at removing particles and make the gas to rotate in a spiral pattern inside a tube, where centrifugal force moves to the equipment walls and can fall or slip on the bottom of the cyclone and then be removed.

The second method is the sleeve filter that operates just like a vacuum cleaner where particles are carried through the air and pass through a cloth. Each time the gas passes through the cloth the dirt is forming a type of filter cake and then a clean gas is released.

Question 4 Discuss the main points related to life cycle thinking presented inside PNRS.

Answer:

Among the main points related to life cycle thinking that are present in the PNRS, they can be cited.

- The use of the life cycle concept in the integration of waste management,
- Use and implementation of the management hierarchy, having the following structure: non-generation, reduction, reuse, recycling, treatment and final disposal
- Implementation of reverse logistics to perform the recovery of products with their components and materials;
- The principle of shared responsibility where all actors in the chain are responsible for waste management
- Sectoral agreements that must be established for discarded products, etc.

Question 5 Apply the RemPro matrix to a mobile phone to verify whether the product is easy to remanufacture or not. To do this, write down the assembly and reassembly time and present the filled matrix along with the information identified by each assessed reframe and propose if it is necessary to propose some improvement alternatives that should be made to facilitate the remanufacturing process.

Example of answer:

Property of the product	Remanufacturing steps						
	Inspection	Cleaning	Disassembly	Storage	Reprocessing	Reassembly	Test
Ease of identification	X Obs		X Difficulty to identify the parts	X Obs			X Obs
Ease of verification	X Obs						
Ease of Access	X Obs	X Obs	X Obs		X Obs		X Obs
Ease of handling			X Difficult handling and makes disassembly difficult	X Obs	X Obs	X Difficulty to reassemble the parts	
Ease of separation			X Breaking components		X Obs		
Ease of safety						X Obs	
Ease of alignment						X Obs	
Ease of storage				X Obs			
Wear resistance		X Obs	X Obs		X Obs	X Obs	
Disassembly time					Assembly time		
Improvement proposals	Avoid welding on the phone, Use modular parts for easy disassembly Facilitate technology upgrade to increase mobile phone's life cycle						

Question 6 What are the top five strategies for improving transportation efficiency at GSCM?

Answer:

(1) reducing the distance to be traveled by the vehicle; (2) choosing transportation modes that cause less environmental impact; (3) acting in the choice, conservation and maintenance of equipment used in logistics operations; (4) promoting greater occupation of the loading of vehicles used in transportation; (5) Acting in the vehicle operator's environmental awareness and creating an environmental culture in the company.

References

Ageron, B., Gunasekaran, A., Spalanzani, A.: Sustainable supply management: an empirical study. Int. J. Prod. Econ. **140**, 168–182 (2012)

ABNT (Associação Brasileira de Normas Técnicas): NBR 16290. Bens reprocessados. Rio de Janeiro (2014)

ABRELPE: Panorama dos resíduos sólidos no Brasil 2018/2019. Disponível em: http://abrelpe.org.br/panorama/. Acesso: fevereiro de 2020

Brasil Lei 12.305, de 2 de agosto de 2010 Institui a Política Nacional de Resíduos Sólidos; altera a Lei 9.605, de 12 de fevereiro de 1998; e dá outras providências. Brasília, DF. Disponível em: http://www2.mma.gov.br/port/conama/legiabre.cfm?codlegi=636

Bouchery, Y., Corbett, C.J., Fransoo, J.C., Tan, T.: Sustainable Supply Chains. Springer (2017)

Caterpillar: Peças Cat Reman. Disponível em: https://www.sotreq.com.br/pt-br/pecas/pecas-catr-reman. Aceso: fevereiro de 2020

Chiaradia, A.J.P.: Utilização do indicador de eficiência global de equipamentos na gestão de melhoria contínua de equipamentos: um estudo de caso na indústria automobilística. Trabalho de conclusão de mestrado profissionalizante em engenharia. Universidade Federal de Rio Grande do Sul, 133p. (2004)

Cummins. Peças genuínas. Disponível em: https://www.cummins.com.br/produtos/pecas-genuinas. Acesso: fevereiro de 2020

Dornfeld, D.A. et al.: Introduction to green manufacturing. In: Dornfeld, D.A. (ed.) Green Manufacturing: Fundamentals and Applications. Springer, Berkeley (2013)

Elhedhli, S., Merrick, R.: Green supply chain network design to reduce carbon emissions. Transp. Res. Part d: Transp. Environ. **17**, 370–379 (2012)

Gehin, A., Zwolinski, P., Brissaud, D.: A tool to implement sustainable end-of-life strategies in the product development phase. J. Clean. Prod. **16**, 566–576 (2008)

Grant, D.B.: Gestão de logística e cadeia de suprimentos. Saraiva, São Paulo (2013)

Ijomah, W.L., McMahon, C.A., Hammond, G.P., Newman, S.T.: Development of design for remanufacturing guidelines to support sustainable. Manuf. Robot. Comput.-Integr. Manuf. **23**, 712–719. (2007)

Govidan, K., Rajendran, S., Sarkis, J., Murugesan, P.: Multi criteria decision making approaches for green supplier evaluation and selection: a literature review. J. Clean. Prod. **98**, 66–83 (2015)

Gray, C., Charter, M.: Remanufacturing and product design: designing for the 7th generation. The Centre for Sustainable Design. University for the Creative Art, Farnham, Reino Unido, p. 206 (2006).

Hansen, R.C.: Overall Equipment Effectiveness: A Powerful Production/Maintenance Tool for Increased Profits. Industrial Press, New York (2001)

Leite, P.R.: Logística Reversa: Sustentabilidade e Competitividade, teoria, prática e estratégias. 3ª ed. Saraiva, 360p. (2017)
Lona, E.E.S., Leme, M.M.V.: Processos produtivos e a poluição atmosférica. In: Adissi, P.J., Pinheiro, F.A., Cardoso, R.S. (eds.) Gestão Ambiental de unidades produtivas, 1st ed. Elsevier, Rio de Janeiro (2013)
Manzini, E., Vezzoli, C.O.: Desenvolvimento de Produtos Sustentáveis: os requisitos ambientais dos produtos industriais. Tradução Astrid de Carvalho. Editora da Universidade de São Paulo, São Paulo, 367p. (2002)
Nakajima, S.: TPM Development Program: Implementing Total Productive Maintenance. Productivity Press, Portland, OR (1989)
Noci, G.: Designing 'green' vendor rating systems for the assessment of a supplier's environmental performance. Eur. J. Purchas. Supply Manag. **3**(2), 103–114 (1997)
Pires, S.R.I.: Gestão da cadeia de suprimentos: conceitos, estratégias, práticas e casos, 3rd ed. Atlas, São Paulo (2016)
Rodrigues, L.R.: Seleção de fornecedores sustentáveis utilizando *fuzzy* DEMATEL-ANP. *165f*. Dissertação (Mestrado em Engenharia de Produção), Universidade Federal de São Carlos (UFSCar), São Carlos (2016)
Saavedra, Y.M.B.S.: Práticas de Estratégias de Fim de Vida no Processo de Desenvolvimento de Produtos e suas aplicações em empresas que realizam a recuperação de produtos pós-consumo. Dissertação de Mestrado, Engenharia de Produção. Escola de Engenharia de São Carlos. USP-EESC, 235p. (2010)
Saavedra, Y.M.B., Barquet, P.B.A., Rozenfeld, H., Forcellini, A.F.: Remanufacturing in Brazil: case studies on the automotive sector. J. Clean. Prod. **53**, 267–276 (2013)
SACHS. Mas do que um mero reparo: Remanufatura de embreagens da SACHS. Disponível em: https://aftermarket.zf.com/la/pt/sachs/quem-somos/qualidade-como-novo/. Aceso: fevereiro de 2020
Santos, M.R., Teixeira, C.E., Kniess, C.T., Barbieri, J.C.: O uso da avaliação do ciclo de vida e da ecoeficiência para avaliar alternativas de valorização de resíduos: um estudo em empresa termelétrica. Revista de Administração da Universidade Federal de Santa Maria. Especial Engema **9**, 82–99 (2016)
Santos, P.V.S.: Aplicação de um indicador *Overall Equipment Effectiveness* (OEE): um estudo de caso numa retífica e oficina mecânica. Braz. J. Prod. Eng. **4**(3), 01–18 (2018)
Sarkis, J.: A strategic decision framework for green supply chain management. J. Clean. Prod. **11**(4), 397–409 (2003)
Sitcharangsie, S., Ijomah, W., Wong, T.C.: Decision making in key remanufacturing activities to optimize remanufacturing outcomes: a review. J. Clean. Prod. **232**, 1465–1481 (2019)
Sundin, E.: Product and Design for Successful Remanufacturing. Ph.D. Dissertation, Linköping University (LiU) (2004)
Srivastava, S.K.: Green supply-chain management: a state-of-the-art literature review. Int. J. Manag. Rev. **9**(1), 53–80 (2007)
Thierry et al.: Strategies issues in product recovery management. Calif. Manag. Rev. **37**(2), 114–135 (1995)
Tibben-Lembke, R.S.: Life after death—reverse logistics and the product life cycle. Int. J. Phys. Distrib. Logist. Manag. **32**(3), 223–244 (2002)

Chapter 13
Communication and Environmental Labelling

Cassiano Moro Piekarski, Murillo Vetroni Barros, Rodrigo Salvador, Fabio Neves Puglieri, Felipe Queiroz Coelho, and Beatriz Cristina Koszka Kiss

13.1 Introduction

Given the current environmental and socio-economic scenario, it is important to share quality information about the environmental performance of certain products. This statement is confirmed in the form of demands generated by individual consumers and organizations that increasingly aim at "more sustainable" choices in the purchase and consumption of various products. If, on the one hand, consumers seek information about the environmental impacts generated by the products and the consequences of their decisions, on the other hand, manufacturers must be prepared to provide complete and reliable information that assists this process. But how to choose a better product in a scenario in which more than 600 programs, labels and tools (UNEP 2015) put themselves on the market brings different information about product sustainability?

It is necessary, then, a clear communication of the benefits generated by a product as well as the advantages for the environment. The standardization and regulation of this type of communication, especially environmental labelling, have gained strength in recent years, encouraging organizations from various sectors and preventing this information from confusing consumers.

In addition, the emergence of new business models and products has made managers rethink how to buy, produce, sell and relate to suppliers and customers.

C. M. Piekarski (✉) · M. V. Barros · R. Salvador · F. N. Puglieri
Federal University of Technology - Paraná—UTFPR, Rua Dr. Washington Subtil Chueire, 330, Jardim Carvalho, ZIP Code: 84017-220, Ponta Grossa, PR, Brazil
e-mail: piekarski@utfpr.edu.br

F. Q. Coelho
Vanzolini Foundation, Rua Camburiú, 255, Alto da Lapa, CEP: 05058-020, São Paulo, SP, Brazil

B. C. K. Kiss
Center for Sustainability Studies at Fundação Getúlio Vargas—FGVces, Avenida Nove de Julho, 2029, 11th Floor, Bela Vista, CEP: 01313-001, São Paulo, SP, Brazil

J. A. de Oliveira et al. (eds.), *Life Cycle Engineering and Management of Products*,
https://doi.org/10.1007/978-3-030-78044-9_13

Thus, a prominent strategy focused on gaining competitive advantage can be an environmental labelling program, especially in terms of import and export, as many companies have aimed at developing relationships and partnerships with organizations that hold certain environmental certifications relevant to their areas of business.

Environmental labelling can be considered an economic and communication tool in order to disseminate information that positively encourages production and consumption patterns in addition to potentiating the awareness of production organizations for the need to extract natural resources from renewable sources and in a more responsible way (Moura 2013). ISO defines the objective of environmental labels as

> The overall goal of environmental labels and declarations is, through communication and accurate and verifiable information, which are not misleading on environmental aspects of products and services, to promote the demand and supply of products and services that cause less environmental impact, thus stimulating the potential for continuous environmental improvement dictated by the market (ABNT 2002).

Environmental labelling, first, arises from the need for standardization for a certain product (good or service). The labelling system came after a series of pressures from the environmental movement. The first official environmental labelling program was named Blue Angel and emerged in 1978 in Germany with the purpose of analysing the environmental impacts of products. Blue Angel operates in the market until nowadays, establishing high standards of eco-friendly product design in order to sustain a more responsible consumption (BLUE ANGEL 2019). After that, other certifying bodies in several countries such as the United States, Japan and European countries began the process of disseminating environmental labelling. In Brazil, the first indication of labelling occurred in 1993 with the creation of the Brazilian Environmental Labelling Program, prepared by the Brazilian Association of Technical Standards (ABNT). Overall, the Global Ecolabelling Network (GEN 2019) brings together 27 environmental labelling programs based on ISO 14024 (type II labelling, which will be addressed in Sect. 2.2 of this chapter). ABNT is the Brazilian representative in GEN.

The objective of environmental labelling is to promote the product's environmental quality by applying efforts by different parts of the organization in order to clearly convey environmental communication to the consumer of the product. And it is aimed at the dissemination of environmental care practices and can add value to the product through the differentiation of these in the market. More recently, environmental labelling has also been used to provide greater clarity and prove the environmental benefits or advantages of "green" or "sustainable products" (KISS 2018). The European Commission defines green products as "those that use resources more efficiently and cause less environmental impact throughout their life cycle, from raw material extraction, its production, distribution, use, to the end of the life cycle (end-of-life strategies, including reuse, recycling and recovery, which will be addressed in the Green Supply Chain chapter) compared to other similar products from the same category." "'Green products' exist in any product categories, whether they are eco-labelled or marketed as green; it is environmental performance that defines them as 'green'" (European Commission 2013).

Added to this, the adoption of the practice promotes an environmental management system (according to chapter of environmental management system and performance assessment) more efficient and less impacting on the environment. This message is reinforced by the sustainable development goals (SDGs) and the United Nations 2030 Agenda (UN): in its Goal 6 (clean water and sanitation), 7 (affordable and clean energy), 8 (decent work and economic growth), 9 (industry, innovation and infrastructure), 12 (sustainable consumption and production) and 13 (climate action, against global climate change) (UN 2015). With an emphasis on Goal 12, the UN aims at ensuring sustainable production and consumption patterns. This will be done, among others, through

- Achieve sustainable management and efficient use of natural resources;
- Achieve environmentally healthy management of chemicals and waste throughout their life cycle and minimize their negative impacts on human health and environment;
- Substantially reduce waste generation through prevention, reduction, recycling and reuse;
- Encourage companies, especially large and transnational companies, to adopt sustainable practices;
- Promote sustainable public procurement practices in accordance with national policies and priorities;
- Ensure that people everywhere have relevant information and awareness for sustainable development and lifestyles in harmony with nature.

As a result of the dissemination of labels, organizations have environmentally friendly options to acquire recognition and facilitate entry and stay in the international market, especially in more developed countries where the demand for such products is higher. In addition, the positive impacts of labelling can mean an important advance in terms of product commercialization.

One example is the European Union's The Single Market for Green Products initiative, launched in 2013 (European Commission 2013) which aims at directing the market to the choice of less impactful and more resource-use-efficient products, fostering clear communication on what a "green product" is. In this context, information on the products' environmental performance is essential. Its absence may represent the constitution of non-tariff trade barriers for products that will arrive in Europe, strongly impacting the way of doing business, in the way of producing and communicating the attributes of each product (KISS 2018).

Aesthetic appeal in terms of environmental labelling can play an important role in the information and advertising instrument (SCATOLIM 2009). Communication information can be transmitted in terms of symbols, graphics, images, marks, texts, i.e. any type of illustration. However, environmental information transmitted to the consumer must be transmitted concisely, clearly, relevantly and objectively, facilitating the understanding of environmental language. Figure 13.1 presents some examples of environmental labels which will be covered in more detail in subsequent sections, dealing with Type I, Type II and Type III labels, a subject that will also be dealt with later.

Communication is facilitated through the performance of actors such as manufacturer and consumer. The labelling is based on a tripod of three main actors: (i) public environmental agencies that act in the regulation, standardization and protection of the environment; (ii) production organizations introducing technological actions targeted for the environment and (iii) consumers, making conscious choices, beneficial to the environment through purchasing power (Moura 2013). Moreover, the practice can still be used in environmental communication between the organization's stakeholders, such as stockholders, community, suppliers, customers and employees. This communication process is inherent to the product and can occur through transmitter and receiver. Communication is carried out in terms of advertising, marketing, digital printing, digital media or on the product packaging itself.

As an effective tool for measuring the potential environmental impact of products, services and processes, the LCA stands out acting as a collaborator on the environmental labelling process. For more details on LCA, see Chaps. 4–7.

It is also notelike the benefits of applying environmental labelling to organizations, which can be understood as: efficient management of resources from renewable and non-renewable sources, stimulation of the conscious destination of the product in the post-use, options such as reuse, repair, recycling, reuse and final destination. With this, the potential for improvement in the environmental performance of the product and materials used for manufacturing can be analysed, measured and compared. To this end, changes in the production process can be represented by the replacement of raw materials of fossil origin by renewable; reduction in water and energy use; minimization of air pollutants emissions; treatment and reuse of process water; generation of electricity from a clean source (wind, solar, biomass) as well as in the eco-efficient design, among other examples. In addition, some companies have realized that industrial waste can be treated, processed and reused, resulting in financial, strategic value and fostering initiatives for environmental labelling.

Although the use of product's environmental labels focuses on the voluntary context, different standards, regulations and initiatives mention or use environmental labelling as a way to differentiate products. For example, companies and governments can use these seals and labels as a way to prove better environmental performance of products in the process of procurement and bidding. In Brazil, there are already manuals for sustainable public procurement (Alem et al. 2015). One example is the purchase of paper by the Brazilian Federal Senate (SENADO FEDERAL 2019). For the purchase of A4 paper for reprography, it is required to have certified legal origin, which is verified by the registration of the supplier company at the Brazilian Institute of Environment (IBAMA). The paper can be recycled or white, which does not make use of elemental chlorine on bleaching process.

In addition to the environmental perspective described here, the economic perspective can also be considered as a strong ally in environmental labelling programs. Through the application, promotion and justification of initiatives such as waste reduction and cost reduction—especially in the stage of use of products—they can make up the benefits generated to the consumer and communicated on environmental labels. With this, the awareness of society can lead to changes in behaviour and in decision-making processes in favour of less impactful products, in favour of the environment. The newly launched ISO 20400:2017 standard proposes guidelines for sustainable procurement and reinforces the fact that institutional decisions, through the purchase of products, can influence value chains as a whole, fostering initiatives that reduce the impacts generated by products in all phases of their life cycle. The standard also recommends the use of LCA as a way to measure and manage these impacts (ISO 2017).

Therefore, given all the motivation for the creation and/or achievement of environmental labels, as well as their communication and what they mean, attention must be given to the requirements necessary for their practice as there are standards and patterns at the international and national levels that govern them.

Following are presented the three types of environmental labelling and communication standardized by ISO in their Brazilian versions (by NBR ISO) developed by ABNT.

13.2 Types of Environmental Labelling and Communication

In Brazil, the theme on environmental labelling has been studied, developed and applied in recent decades mainly due to negotiations with an export profile. All markets are looking for constant advances in the economy and industrial sectors. The Brazilian market has been making greater investments in efficiency and environmental communication with a view to competitive advantages. The identification of new market niches, innovation opportunities, financial growth and openness to international trade make up the portfolio of benefits for organizations.

According to NBR ISO 14024 (ABNT 2004), an environmental labelling program aims at increasing awareness in the supply chain of products, such as suppliers, producers, distributors and consumers. In this sense, the application of labelling can mean an efficient action in achieving environmental gain. On the one hand, consumers may charge producers for the commitments required by their environmental certification. On the other hand, producers are interested in reaching new markets, achieving strategic advantages and thereby capital growth. In addition, the program has the function of preventing environmental information failures, increasing consumer environmental awareness and educate them in this direction, providing incentive to the manufacturer in order to produce products with low environmental impact (ABNT 2004).

According to NBR ISO 14020:2002, an environmental label or environmental declaration is a "statement that indicates the environmental aspects of a product or service" whose objective is to promote products and/or services that generate less environmental impact, with clear and non-misleading information, that can be verified (ABNT 2002). The principles of environmental labelling also state that labelling procedures must not be designed in such a way as to create obstacles to international trade.

In addition, ISO 14024:2018 points out some advantage in the use of environmental labelling such as promoting the reduction of waste inherent in the product, optimizing processes, demonstrating to the market and stakeholders that the organization is concerned with actions focused on sustainability, promoting the preservation of the environment by reducing negative impacts and allowing the framing of requirements in terms of sustainable environmental bids.

An environmental label may stand the company out from competitors and help increase profitability. The content of the labelling serves both for direct communication with the final consumer and for business-to-business (B2B).

In direct relationship with the consumer, the labels serve to inform the final consumer about the environmental characteristics of the product being purchased. In the B2B relationship, many businesses give preference to (suppliers with) products (raw materials) and services with environmental certifications/labels as well as others working exclusively with partners who have a certain certification.

In Brazil, the guidelines for environmental labelling are established by the NBR ISO 14020:2002 series. There are three types of labelling, type I (which are environmental labels or seals), type II (which are self-declared environmental claims) and type III (which are Environmental Product Declarations). The requirements for each type are established by a specific standard.

13.2.1 Type I Labelling: Environmental Label (ISO 14024:2018 and NBR ISO 14024:2004)

This standard established the principles and procedures for type I environmental labelling programs, which includes selection of product categories as well as environmental criteria, functional characteristics and means to assess and demonstrate their compliance. NBR ISO 14024:2004 also establishes the procedures for label certification and concession (ABNT 2004).

It must also be emphasized that a type I environmental labelling program, which generates environmental labels (*ecolabels*), must aim at highlighting products/services that evidence the reduction of environmental impacts, considering their life cycle, and not only the transfer of responsibility for impacts, in some way, to third parties. Products/services must demonstrate environmental preference in an assessment process based on solid scientific and engineering principles. Furthermore, all information must be transparently treated and made available to all stakeholders for consultation.

An expiration date shall also be determined to the certification, which may vary according to that established by the certifier third party. At the end of this period, the product/service system will require a new audit for recertification, if it conforms to the requirements. In case of updates related to the certification requirements, the system will keep the certification until the end of its expiration date and will only lose the certification if it does not meet the new requirements on the recertification process.

Particularly, ISO 14024:2018 was updated in 2018. Its goals and principles have been maintained; however, the new version aims at strengthening the documentary requirements for obtaining a label in addition to ensuring transparency and credibility of type I environmental labelling programs and defining the skills necessary for process auditors. The certifications granted by NBR ISO 14024:2004 remain valid until a new certification cycle, with the update of the equivalent Brazilian standard.

Then, it is noteworthy that Type I labelling aims at offering the possibility of promoting products proven (through external verification) to be environmentally friendly to their competitors and can be used to promote the practices of a company or even to establish partnerships in production chains, since some companies may recommend all their partners to have certification in a certain seal to maintain economic and strategic relations. The external verification is carried out by an independent body (authorized by the certifying body) without any relation to the product's manufacturer/seller and aims to determine whether it is in accordance with the guidelines and criteria for granting the label in question. Later, in Sect. 3, some examples of Type I labels will be presented.

13.2.2 Type II Labelling: Self-Declared Environmental Claim (ISO 14021:2013 and NBR ISO 14021:2017)

NBR ISO 14021:2017 sets the requirements for environmental self-declarations. Self-declarations cannot be vague, so terms such as "environmentally safe", "environmentally friendly", "friend of the Earth", "non-polluting", "green", "friend of nature" and "friend of the ozone layer" cannot be used, according to NBR ISO 14021 (ABNT 2017).

It is strongly encouraged that self-declarations are accompanied by explanatory texts, since isolated self-declarations can sometimes lead to misunderstandings. When declarations make use of natural symbols, they must only be used if and when there is a direct relationship between the symbol and the benefit provided by the product. Another care to be taken is not to relate an environmental symbol with a brand.

Self-declarations do not undergo external assessment; however, a person responsible for the information and the method of assessment, which must allow reproduction of the results, must be designated. This person responsible must have the prior knowledge and experience. All documentation regarding the procedures and assessment results must preferably be maintained until the end of the product's lifetime (even after cessation of its commercialization). The standard addresses that there is no standard methodology to be carried out, that is, each certifying body creates its own assessment system. However, the instrument must be reliable, describing in detail so that other individuals can follow.

Still, NBR ISO 14024:2004 provides characteristics of use of terms, qualifications, use of symbols and assessment methodology for the terms "recycled content", "compostable", "degradable", "recyclable, recovered energy", "renewable energy", "renewable material", "design for disassembly", "waste reduction", "reduction in water consumption", "reduction in energy consumption", "reduction in resource use", "reusable and rechargeable", "sustainable", "prolonged product lifetime"; as well as covering statements on "carbon footprint" and "carbon neutral".

But after all, what is a self-declared environmental claim? An example of self-declared environmental claim is when a company claims that a particular product is 100% biodegradable or when it is stated that there has been a 10% increase in the use of recyclable material to produce a given product. This second statement may be confusing as it may represent 10% in relation to the percentage previously used or in relation to the mass of the product. In these cases, it stands out the use of explanatory texts to ensure transparency and better communication, avoiding greenwashing.[1]

13.2.3 Type III Labelling: Environmental Product Declaration (ISO 14025:2006 and ABNT NBR ISO 14025:2015)

ABNT NBR ISO 14025:2015 (ABNT 2015) establishes the principles and procedures for the preparation of Environmental Product Declarations (EPD). An EPD is based on an LCA, which takes into account the environmental aspects of a system in order to measure the potential environmental impacts of the system. Principles and structure for conducting an LCA study can be found in ISO 14040 (ISO 2006a) and NBR ISO 14040 (ABNT 2014a) standards, and requirements and guidelines can be found in ISO 14044 (ISO 2006b) and NBR ISO 14044 (ABNT 2014b) standards. More recently, ISO/TR 14049:2014 has been published, which presents illustrative examples of how to apply ABNT NBR ISO 14044:2014 to the definition of goals and scope and inventory analysis of an LCA, and ISO/TR 14047:2016 which has illustrative examples of how to apply ABNT NBR ISO 14044:2014 to life cycle impact assessment situations, as presented in Chaps. 4–7.

An EPD aims, among other objectives, at highlighting the potential environmental impacts of a product, encouraging improvements in its environmental performance and enabling comparisons between products.

For a type III labelling program, it is necessary that there are product categories what, among other benefits, allows the comparison of the performance of similar product systems as they can be assessed by means of the same functional unit. To do

[1] The term greenwashing refers to the organization's unethical practice of disclosing false information or even inducing the consumer about a false environmental performance of its product/service.

so, there are rules under which products in a given category must be assessed, and these constitute the product category rules (PCR).

A PCR, in a succinct manner, must establish the characteristics of the study and the report of life cycle assessment results for a given product category, including goals and scope topics, functional unit, impact categories, defining the purpose of the LCA and the system's boundaries, that is, which life cycle phases must be considered. In addition, the PCR must also indicate how the results should be disclosed in the EPD.

Taking into account the possibilities of an LCA, for the purposes of an EPD, a product category rule may require the inclusion of (i) all four phases of an LCA, i.e. definition of goals and scope, life cycle inventory analysis, life cycle impact assessment and interpretation or (ii) definition of goals and scope, life cycle inventory analysis and interpretation, not including the impact assessment phase. After carrying out the LCA study, it must undergo external verification, i.e. it must be examined and approved by third parties. This whole process can be carried out with program operators (entities that enable type III labelling). In Brazil, we can mention INMETRO, EPD Brasil (International EPD System), IBU (Institut Bauen und Umwelt), and UL do Brasil Certifications.

But after all, what is an EPD for? And what is the difference between an EPD (Type III) and a seal (Type I)?

Well, an EPD produces a report on the environmental performance of the assessed system, showing what its potential impacts are, through an LCA. The data are quantitative. An EPD, by itself, does not determine whether a particular product is better or worse than another, it exposes the quantitative results of the product's environmental performance. A seal, in turn, is used to certify that a particular product is environmentally preferable to similar products that do not have it considering the characteristics assessed by the seal. An EPD can be used, for example, as part of a Type I labelling program, but alone it does not judge performance.

Finally, due to the practical and economic actions occurred in recent years in Brazil, the trend of the environmental labelling sector shows important advances and growth prospects. Three points support this perspective. The first is the commitment to environmentally friendly actions, fostering initiatives of a clean production, reducing emissions and use of renewable materials. The second point refers to the potentiation of the company's and the product's competitive advantage over competitors. The current globalized market is constantly changing, advances in innovation and technology, so adding environmental seals in products may represent strategic advantage. Finally, the third aspect covers market opening abroad. Many companies have the vision to export the products to a North American, European or Asian market. On the other hand, these markets accept and remunerate environmental seals of companies as a way to show commitment to the environment and to the communication between company and consumer.

Below are some examples of environmental labels/seals in their various uses.

13.3 Practical Examples

In recent years, the creation and dissemination of various environmental labels and seals have invaded markets and consumers with the most varied information. On the one hand, many different labels may help in choosing a particular product; on the other hand, they may also confuse the consumer.

Given this, how can the individual consumer find ways to understand labels and rely on information? And how can companies use this information to develop sustainable procurement policies aiming at using products and services that have high environmental performance?

To this end, we bring in this section examples of labels/seals applied by companies in the market in order to understand some types of labels and that the reader can observe. As shown earlier, there are environmental labels/seals (Type I), self-declarations (Type II) and EPDs (Type III). The following are some examples for each type of labelling.

13.3.1 *Environmental Seals/labels (Type I)*

An example of an environmental seal is the Green Seal (logo on Fig. 13.2). Considered an eco-seal, Green Seal aims at highlighting products that meet a number of performance and environmental health requirements for a large number of products and services.

Companies that own the Green Seal include 3M (e.g. for stainless steel cleaner and protector) and ECOLAB (e.g. glass cleaning solution).

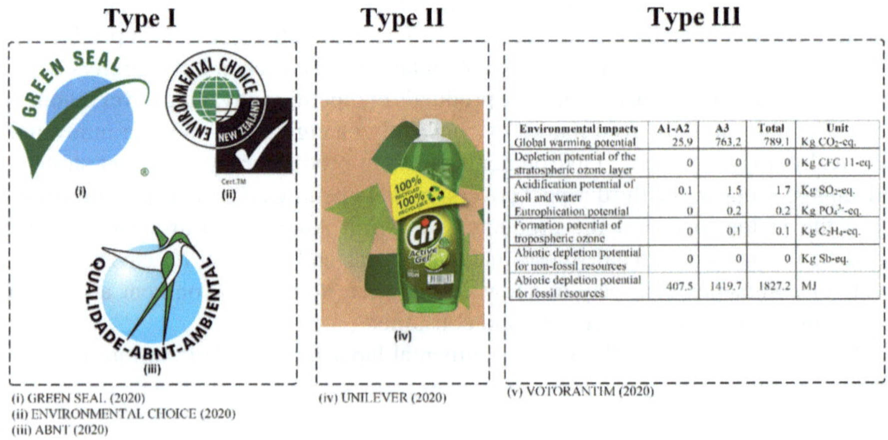

Environmental impacts	A1-A2	A3	Total	Unit
Global warming potential	25.9	763.2	789.1	Kg CO_2-eq.
Depletion potential of the stratospheric ozone layer	0	0	0	Kg CFC 11-eq.
Acidification potential of soil and water	0.1	1.5	1.7	Kg SO_2-eq.
Eutrophication potential	0	0.2	0.2	Kg PO_4^{3-}-eq.
Formation potential of tropospheric ozone	0	0.1	0.1	Kg C_2H_4-eq.
Abiotic depletion potential for non-fossil resources	0	0	0	Kg Sb-eq.
Abiotic depletion potential for fossil resources	407.5	1419.7	1827.2	MJ

Fig. 13.1 Examples of Type I, II and III labels

Fig. 13.2 Eco-seal green seal. *Source* GREEN SEAL (2020)

Fig. 13.3 Environmental choice seal. *Source* Environmental Choice (2019)

There is also the Environmental Choice (in Fig. 13.3). The seal covers a range of products such as buildings and construction products, floors, detergents, recycling products and services, toiletries and office products and services. Companies and products that have earned the label include Canon and Fuji Xerox with, for example, printers and copiers.

It must also be highlighted the existence of a Brazilian environmental quality seal certified by ABNT, the ABNT Environmental Quality, presented in Fig. 13.4.

ABNT environmental quality seal covers requirements such as waste reduction, process optimization, demonstration that the company is concerned about sustainable development by reducing negative environmental impacts and allows participation in sustainable bids. In addition, ABNT Environmental Quality is GEN's only full member in South America.

Fig. 13.4 Environmental quality seal. *Source* ABNT (2019)

13.3.2 Self-declared Environmental Claims (Type II)

A recent example of self-declaration has been presented by Unilever, with the CIF brand (see Fig. 13.5), in Argentina.

The company declares that the packaging used for the commercialization of CIF detergent is 100% recyclable and recycled. That is, all the material used to produce the detergent packaging is recyclable, and all the material used in the production comes from recycling.

Unilever also points out that for this strategy to be possible, it is necessary to alert consumers to the need for a circular economy (UNILEVER 2019).

Fig. 13.5 CIF detergent bottle 100% recycled and 100% recyclable. Source UNILEVER (2019)

13.3.3 Environmental Product Declarations (Type III)

An example of EPD result is presented for Cement CP II E 40 by Votorantim company. The LCA results for the product are illustrated in Table 13.1.

With the functional unit informed and following the PCR specifications, it is possible to compare different products of the same category fairly. It is possible to observe the potential environmental impacts of the assessed product/service, thus defining its environmental performance profile. EPDs are also made available by the program operator for consultation by stakeholders.

As described above, an EPD does not inform preference, but the data can help in obtaining seals and, mainly, reiterate the company's transparency regarding the environmental performance of the products measured quantitatively, and they may include third-party review.

Given the examples presented, consumers may have a notion of how environmental labels work. Several organizations are in search of this practice, and parallel to this, it is important for consumers to understand the seals in order to develop sustainable procurement policies.

13.4 Important Features for an Environmental Label

As presented, environmental labels are a great way to increase credibility and confidence in buying products and services; however, there are many environmental labels on the market, and one label alone is not flaw-free. The following are some tips for organizations or final consumers who make use of environmental labels and statements.

Table 13.1 EPD results for Cement CP II E 40—Votorantim

Environmental impacts	A1–A2	A3	Total	Unit
Global warming potential	25.9	763.2	789.1	kg CO_2-eq
Stratospheric ozone layer depletion potential	0	0	0	kg CFC 11-eq
Soil and water acidification potential	0.1	1.5	1.7	kg SO_2-eq
Eutrophication potential	0	0.2	0.2	kg PO_4^{3-}-eq
Tropospheric ozone formation potential	0	0.1	0.1	kg C_2H_4-eq
Abiotic depletion potential for non-fossil resources	0	0	0	kg Sb-eq
Abiotic depletion potential for fossil resources	407.5	1419.7	1827.2	MJ

Source Adapted from VOTORANTIM (2020)

13.4.1 Know the Types of Labels!

Knowing the types of labelling will help you choose, as each type of labelling may provide more or less information. For example, some labels focus on a single criterion such as type II labelling of recycled content, which is widely used in packaging, this can make it easier to choose if that is what is being aimed, and on the other hand, the lack of knowledge may induce the consumer to think that the packaging or even the product contained in the packaging is "sustainable".

13.4.2 Who is Behind the Label?

With the large number of environmental labels available on the market, it is really important to know who is responsible for the program. A quick search on the organization that operates a particular labelling program can often reveal a lot about its credibility.

13.4.3 Adopted Standards?

Identify whether the methodology is reliable! Verify whether the standards adopted by the seal comply with a well-known standard such as an ISO standard (ISO 14021, 14024, 14025 standards address standards for environmental labels and declarations).

13.4.4 Life Cycle Approach?

The life cycle approach assists in providing relevant and comprehensive environmental information while avoiding possible greenwashing.

13.4.5 Third-Party Verification?

One aspect that creates confusion in choosing an environmental label is the difference between the levels of credibility, reliability and impartiality applied in each labelling program. A program with an independent and third-party verification[2] naturally has a higher accuracy in these three aspects.

---

[2] Third part: Person or body recognized as independent of the parties involved in a given matter. The parties involved generally represent the interests of the supplier (first part) and the buyer (second part).

13.4.6 *Transparency?*

A labelling program must not be reticent about disclosing relevant information, and all issues addressed above must be clearly available to organizations and final consumers.

13.5 Final Considerations

Environmental communication is an important instrument for environmental management and sustainable production and consumption. It allows not only to inform consumers about the environmental performance of the product or service they are acquiring, but also to assist them in the purchasing decision process. This becomes great relevance especially when the consumer market begins to consider the environmental aspects of a product or service as a decision factor for the purchase, in addition to its price, brand and other attributes.

In recent years, the number of published Environmental Declarations has increased. Applications and the various potential uses of declarations are far from running out. Many organizations have innovated not only in the development of EPDs but in how to communicate environmental information.

Sustainability and environmental concerns have changed the way society thinks and people's daily lives, so even though EPDs have first emerged as a communication tool between companies, their use for companies to communicate with the final consumer has expanded.

In this sense, the labelling and environmental declarations of products can also assist in the service and promotion of several UN's SDGs. Example: SDG 6 (clean water and sanitation), SDG 7 (affordable and clean energy), SDG 8 (decent work and economic growth), SDG 9 (industry, innovation and infrastructure), SDG 12 (responsible [sustainable] consumption and production) and SDG 13 (climate action, action against global climate change).

The use of labelling and EPDs today is still limited. Most owners and potential owners of EPDs do not have deep knowledge of them to profitably explore their full sustainable potential yet. However, environmental labelling has benefits, and among the main ones, we may highlight communication, competitive advantage, potential market expansion, as well as the advantages for the environment.

On the other hand, labelling and especially EPDs still present some challenges. In Brazil, these challenges are largely related to the costs of conducting LCA studies and to data reliability due to the lack of national life cycle inventories and regionalized impact assessment methods.

Proposed Exercises

1. What is environmental labelling and for whom is it intended?

2. Fill in the comparative chart between the different types of environmental labelling based on the following criteria:

	ISO 14021—Type II labelling	ISO 14024—Type I labelling	ISO 14025—Type III labelling
Uses external review			
Based on life cycle assessment			
Allows comparability			
Is more prone to greenwashing practices?			

3. According to which Brazilian Standard is environmental labelling regulated? Cite and explain some guidelines from the standard.
4. Fill in with true (T) or false (F).

   a. ( ) it is mandatory to have an EPD to carry out exports
   b. ( ) type I labels are certified with determined expiration date
   c. ( ) environmental labelling can be considered an economic instrument
   d. ( ) greenwashing may induce the consumer to believe in a false environmental performance of a product/service
   e. ( ) a PCR determines how the results of an EPD will be communicated
   f. ( ) there are two types of environmental labels, only seals and self-declarations, because an EPD is not a type of label
   g. ( ) for products without an established PCR, the program operator allows to perform only one LCA study and communicate the results in an EPD
   h. ( ) type II labels do not undergo external assessment
   i. ( ) not all EPDs require third-party review
   j. ( ) the main difference between a type I label and a type III label is that the first one does not undergo a third-party review and the second one does

5. Consider Chart 1 with data from two EPDs from different sectors and answer:

   k. What are the impact categories analysed?
   l. What is the reference flow for each product?
   m. Do the products have the same PCRs and functional units? Justify.
   n. Which of the two products has the lowest environmental impact?
   o. Justify why products cannot be compared.

Feedback

Chart 1 Indicators of potential impacts of two EPDs

Potential environmental impact indicators	Pasta (1 kg of product)[a] Total (cradle-to-grave)	Moisturizing cream (200 ml)[b] Total (cradle-to-grave)
Climate change—fossil contribution (g. CO_2 eq.)	833.80	5.97E−02
Climate change—biogenic contribution (g. CO_2 eq.)	−77.20	–
Acidification (g. SO_2 eq.)	11.29	3.80E−02
Eutrophication (g. PO_4 eq.)	6.12	–
Photochemical ozone (g. C_2H_4 eq.)	0.12	6.10E−02

[a]Information extracted from EPD "dry semolina pasta selezione oro chef", registration number S-P-00492 from the International EPD System

[b]Information extracted from EPD "dei prodotti cosmetici rinse-off", registration number S-P-00867 from the International EPD System

1. What is environmental labelling and for whom is it intended?
   A: Environmental labelling is a communication tool with the aim at promoting the product's environmental quality by applying efforts by different parts of the organization in order to convey environmental communication clearly to the product's consumer. This communication takes place between the product's manufacturer organizations, the public sector and the buyer—both individual and institutional.
2. Fill in with true (T) or false (F).

   a. (F) it is mandatory to have an EPD to carry out exports
   b. (T) type I labels are certified with determined expiration date
   c. (T) environmental labelling can be considered an economic instrument
   d. (T) greenwashing may induce the consumer to believe in a false environmental performance of a product/service
   e. (T) a PCR determines how the results of an EPD will be communicated
   f. (F) there are two types of environmental labels, only seals and self-declarations, because an EPD is not a type of label
   g. (F) for products without an established PCR, the program operator allows to perform only one LCA study and communicate the results in an EPD
   h. (T) type II labels do not undergo external assessment

i. (F) not all EPDs require third-party review
j. (F) the main difference between a type I label and a type III label is that the first one does not undergo a third-party review and the second one does.

1. Fill in the comparative chart between the different types of environmental labelling based on the following criteria (if statement is true, fill the field with an X):

	ISO 14021—Type II labelling	ISO 14024—Type I labelling	ISO 14025—Type III labelling
Uses external review		X	X
Based on life cycle assessment			X
Allows comparability			X
Is more prone to greenwashing practices	X		

2. According to which Brazilian Standard is environmental labelling regulated? Cite and explain some guidelines from the standard.
A: Environmental labelling is regulated in accordance with some Brazilian Standards. Type I labelling is represented in ABNT NBR ISO 14024:2004. A type I environmental labelling program generates the environmental labels (*ecolabels*) with the aim at highlighting products/services which evidence the reduction of environmental impacts, considering their life cycle, and not only the transfer of responsibility for the impacts, in some way, to third parties.
Type II: Self-declared environmental claims is established from ABNT NBR ISO 14021:2017. Type II establishes the requirements for environmental self-declarations.
Type III labelling is called the Environmental Product Declaration (EPD), being regulated the ABNT NBR ISO 14025:2015 standard, establishing the principles and procedures the elaboration of EPDs. A EPD is based on an LCA, which takes into account the environmental aspects of a system in order to measure its potential environmental impacts.
3. Consider Chart 1 with data from two EPDs from different sectors and answer:

4. What are the impact categories analysed?
The impact categories analysed are

i. climate change—fossil contribution;
ii. climate change—biogenic contribution;
iii. acidification;
iv. eutrophication;

Chart 1 Indicators of potential impacts of two EPDs

Potential environmental impact indicators	Pasta (1 kg of product)[a] Total (cradle-to-grave)	Moisturizing cream (200 ml)[b] Total (cradle-to-grave)
Climate change—fossil contribution (g. CO_2 eq.)	833.80	5.97E−02
Climate change—biogenic contribution (g. CO_2 eq.)	−77.20	–
Acidification (g. SO_2 eq.)	11.29	3.80E−02
Eutrophication (g. PO_4 eq.)	6.12	–
Photochemical ozone (g. C_2H_4 eq.)	0.12	6.10E−02

[a]Information extracted from EPD "dry semolina pasta selezione oro chef", registration number S-P-00492 from the International EPD System

[b]Information extracted from EPD "dei prodotti cosmetici rinse-off", registration number S-P-00867 from the International EPD System

v. photochemical ozone.

5. What is the reference flow for each product?
   Reference flow for pasta: 1 kg of product.
   Reference flow for moisturizing cream: 200 ml.
6. Do the products have the same PCRs and functional units? Justify.
   No. Category rules are different, so functional units are different because they are products with different functions.
7. Which of the two products has the lowest environmental impact?
   It is not possible to say because EPDs are not comparable.

1. Justify why products cannot be compared.
   Products cannot be compared because, according to ABNT NBR ISO 14025, it is a requirement for comparability that the definition of the product category and the description (e.g. function, technical performance and use) are identical. Moreover, also according to the same standard, the definition of goals and scope, as well as inventory analysis and impact assessment for the product's LCA, according to the ABNT NBR ISO 14040 series, must be at least equivalent.

References

ABNT (Associação Brasileira de Normas Técnicas): ISO 14044: Environmental management—Life cycle assessment—Requirements and guidelines (2006)

ABNT (Associação Brasileira de Normas Técnicas): NBR ISO 14020: Rótulos e declarações ambientais—Princípios gerais. Brasil (2002)

ABNT (Associação Brasileira de Normas Técnicas): NBR ISO 14021: Rótulos e declarações ambientais—Autodeclarações ambientais (rotulagem do tipo II). Brasil (2017)

ABNT (Associação Brasileira de Normas Técnicas): NBR ISO 14024: Rótulos e declarações ambientais - Rotulagem ambiental do tipo I—Princípios e procedimentos. Brasil (2004)

ABNT (Associação Brasileira de Normas Técnicas): NBR ISO 14025: Rótulos e declarações ambientais—Declarações ambientais de Tipo III—Princípios e procedimentos. Brasil (2015)

ABNT (Associação Brasileira de Normas Técnicas): NBR ISO 14040: Gestão ambiental—Avaliação do ciclo de vida—Princípios e estrutura. Brasil (2014)

ABNT (Associação Brasileira de Normas Técnicas): NBR ISO 14044: Gestão ambiental—Avaliação do ciclo de vida—Requisitos e orientações. Brasil (2014)

ABNT (Associação Brasileira de Normas Técnicas): O que é Rótulo Ecológico? (2019). Disponível em: https://www.abntonline.com.br/sustentabilidade/Rotulo/Default. Acesso 28 March 2019

Alem, G., et al.: Compras sustentáveis & grandes eventos: a avaliação do ciclo de vida como ferramenta para decisões de consumo. Sérgio Ade ed. Programa Gestão Pública e Cidadania, São Paulo (2015)

BLUE ANGEL: The German Ecolabel (2019). Disponível em: https://www.blauer-engel.de/en. Acesso 27 March 2019

Coca Cola: PlantBottle (2019). Disponível em: https://www.coca-colacompany.com/our-company/plantbottle. Acesso 28 March 2019

Coca-Cola Brasil (2019). Disponível em: https://www.cocacolabrasil.com.br/. Acesso 27 March 2019

ENVIRONMENTAL CHOICE: Environmental Choice New Zealand. 2019. Disponível em: https://environmentalchoice.org.nz/. Acesso 28 March 2019

European Commission: 196 final: Building the Single Market for Green Products (2013). Disponível em: http://eur-lex.europa.eu/legal-content/EN/TXT/PDF/?uri=CELEX:52013DC0196&from=EN

European Commission: Single Market for Green Products Initiative. Disponível em: http://ec.europa.eu/environment/eussd/smgp/index.htm. Acesso 30 Apr 2018

GEN (Global Ecolabelling Network): Global Ecolabelling Network (2019). Disponível em: https://www.globalecolabelling.net/. Acesso 02 May 2019

GREEN SEAL: Make Healthier, Greener Choices with Confidence. Disponível em: https://www.greenseal.org/. Acesso 28 March 2019

ISO (International Organization for Standardization): ISO 14040: Environmental management—Life cycle assessment—Principles and framework (2006)

ISO (International Organization for Standardization): ISO 20400:2017—Sustainable Procurement, Guidance (2017)

KISS: Beatriz Cristina Koszka (2018). Análise da aplicação do pensamento de ciclo de vida na gestão empresarial: estudo de casos brasileiros. 158 f. Escola de Administração de Empresas de São Paulo—FGV EAESP. Disponível em: http://bibliotecadigital.fgv.br/dspace/handle/10438/24510. Acesso 15 Apr 2019

Moura, A.M.M.d.: O mecanismo de rotulagem ambiental: perspectivas de aplicação no Brasil. 2013. Disponível em: http://repositorio.ipea.gov.br/handle/11058/5655. Acesso 26 March 2019

ONU (ORGANIZAÇÃO DAS NAÇÕES UNIDAS). Objetivo 12. Assegurar padrões de produção e de consumo sustentáveis. 2015. Disponível em: https://nacoesunidas.org/pos2015/ods12/. Acesso em 02/05/2019.

SCATOLIM: Roberta Lucas. A Importância do Rótulo na Comunicação Visual da Embalagem: Uma Análise Sinestésica do Produto. Unesp, FAAC, Bauru, SP (2009). Disponível em: http://www.bocc.ubi.pt/pag/scatolim-roberta-importancia-rotulo-comunicacao.pdf. Acesso 26 March 2019

SENADO FEDERAL: Compras e Contratações Sustentáveis (2019). Disponível em: https://www12.senado.leg.br/institucional/programas/senado-verde/eixos-tematicos/compras-sustentaveis-1/home. Acesso 11 Sept 2019

UNEP: Product Sustainability Information: State of Play and Way Forward. [s.l.] United Nations Environmental Programme (UNEP) (2015)

VOTORANTIM: EPD—Environmental Product Declaration: CP II E 40, CP III-40 RS and CP V-ARI by Votorantim Cimentos. Disponível em: https://gryphon4.environdec.com/system/data/files/6/11919/epd895%20Votorantim%20Cement.pdf. Acesso 28 March 2019

MIX
Papier aus verantwortungsvollen Quellen
Paper from responsible sources
FSC® C105338

If you have any concerns about our products,
you can contact us on
ProductSafety@springernature.com

In case Publisher is established outside the EU,
the EU authorized representative is:
Springer Nature Customer Service Center GmbH
Europaplatz 3, 69115 Heidelberg, Germany

Printed by Libri Plureos GmbH
in Hamburg, Germany